Birds in Wales
1992 - 2000

by Jonathan Green

Illustrated by
Stephen Culley, Chris Grayell,
Bob Mitchell, Colin Richards and Stephen Roberts

Ringing section by
Peter Howlett and Ian Spence

Published by
**WELSH ORNITHOLOGICAL
SOCIETY**

**CYMDEITHAS ADARYDDOL
CYMRU**
Charity Number: 1037823

Birds in Wales 1999-2000

Illustrations © Stephen Culley, Chris Grayell, Bob Mitchell, Colin Richards and Stephen Roberts, 2002

Text © Jonathan Green, 2002

Ringing section © Peter Howlett and Ian Spence, 2002

Bird Ringing Recovery data © British Trust for Ornithology, 2002

First published in 2002 by The Welsh Ornithological Society
Crud yr Awel, Bowls Road, Blaenporth, Cardigan. SA43 2AR

Distributed by Gardners Books
1 Whittle Drive, Eastbourne. East Sussex. BN23 6QH

Printed & Bound by Antony Rowe Ltd, Eastbourne

All rights reserved. No part of this book may be reproduced, stored in a retrieval system, or transmitted in any form or by any means, electronic, mechanical, photocopying or otherwise, without the permission of the publisher

Text set in 9 pt Times New Roman

A CIP record for this book is available from the British Library

ISBN: 09542145-0-1

CONTENTS

Forward	4
Bird Recording in Wales	5
Environmental Developments in Wales	12
Ringing in Wales	19
Species Accounts	45
References	244
Species Index	247

Birds in Wales 1999-2000

FORWARD

In 1994 Poyser published the most recent avifauna for Wales, *Birds in Wales*. Written by Roger Lovegrove, Graham Williams and Iolo Williams, it summarised the occurrence of birds in Wales up to and including 1991. The status of many species has changed since then, not only with the addition of 14 new species to the Welsh list (3 removed, 2 sub-species upgrades, one species upgrade and 14 new species recorded in Wales), making an overall Welsh list of 419. There has also been an increase in the number of birdwatchers and the number of observations has increased dramatically. With all this in mind the 1994 Welsh avifauna needed to be updated.

This book brings Welsh ornithological knowledge up to date, to the end of 2000 and includes additional records to those that were published in Birds *in Wales*. It is not designed to replace the original work but complement it. The basic reference points for this period were the Welsh Ornithological Society's publications – The Welsh Bird Reports. Additional information was requested from County Recorders, County Report Editors and regional experts. I would like to take the opportunity of thanking them all for their assistance. Thanks also to the RSPB, WWT, CCW and BTO for the use of their data.

To provide a more detailed account on a few key species and concepts, various experts were asked to write a brief summary. Many thanks for their valued contribution.

Mike Shrubb	Farmland
Adam Rowlands	GLWR Goldcliff
Nigel Williams	WWT Llanelli
Ian Higginson	Conwy RSPB
Graham Williams	Wetlands for Wales, Ramsey Island, Inner Marsh Farm, Dolydd Hafren and Lake Vyrnwy
Peter Davis	Red Kite
Patrick Lindley	Black Grouse
Tony Cross	Raven

For certain species various individuals were asked for information and comments. I would like to thank Bob Howells for all his help in checking the accuracy of the counts on the Burry Inlet and particularly for the additional details of the unusual gulls at Blackpill. I would also like to thank David Astins, Steve Sutcliffe, Bob Haycock and Anne Poole for their comments on the seabird sections, Peter Hope Jones for checking the Anglesey data and the Welsh Bird Names, Peter Lansdown for checking the Glamorgan data, Rhion Pritchard for checking the Anglesey and Caernarfon data and David Lamacraft for checking the Chough data. Thanks also to Reg Thorpe for answering the unending list of queries on scarce breeding birds in Wales.

Particularly pleasing is the section on Ringing in Wales. The text was written by Peter Howlett and is complemented by the ringing recovery maps, produced by Ian Spence using data supplied by the BTO from their forthcoming book Atlas of Bird Migration. This section took a great deal of time and computing power to develop and the efforts of both authors are gratefully acknowledged.

I would also like to take the opportunity of thanking the various artists whom kindly their illustrations to be used in this book. The artists are: Stephen Culley, Chris Grayell, Bob Mitchell, Colin Richards and Stephen Roberts. The landscapes by Chris are particularly impressive considering that they were drawn by candle light in Kosovo.

Throughout the species accounts estimates of populations are quoted. These have been produced by Mike Shrubb, using Common Bird Census data and by Tony Prater, using the BTO's data from the New Atlas of Breeding Birds and results from the Breeding Bird Survey. The figures are the first of its kind to be published for Wales using credited data.

I would also like to acknowledge the invaluable support of Mike Shrubb, Graham Williams and Iolo Williams for their positive and helpful comments.

A note of thanks must also be paid to colleagues from Sir Thomas Picton School in Haverfordwest, particularly to Richard Locket & Jamie Buchan from the IT Department (for answering a whole host of "how do you do .." and HELP!!) and Trish George & Lisa Jenkins from the Reprographic Department for help in the production and layout.

Finally I would like to thank my family for all their support and encouragement. Diolch yn fawr.

Jonathan Green

BIRDING IN WALES 1992 – 2000

The map shows the arrangement of bird recording areas for Wales. Following the creation of Unitary Authorities as the basis of local Government, the Welsh Ornithological Society decided to switch to the Watsonian Vice-county system (see the Ray Society, publication No. 146) as the basis of bird recording in Wales. This system largely matches the pre-1974 administrative Counties used by the Society until 1996. Using the Watsonian vice-counties therefore retains historical continuity in recording. This would have been lost in any attempt to follow the Unitary Authorities, the only one of which involves no changes from the county system created in 1974, let alone its historic predecessor, being Powys, which comprises the vice-counties of Brecon and Radnor (accepting the R. Wye as the county boundary between them, as it was throughout the 20th century) and Montgomeryshire. Note, however, that the name of Gwent has been retained for the vice-county of Monmouth to prevent confusion with the new Unitary Authority of Monmouth, which covers only part of the vice-county. The recording areas and Recorders names and addresses along with a brief resume of recent publications / avifaunas are listed below.

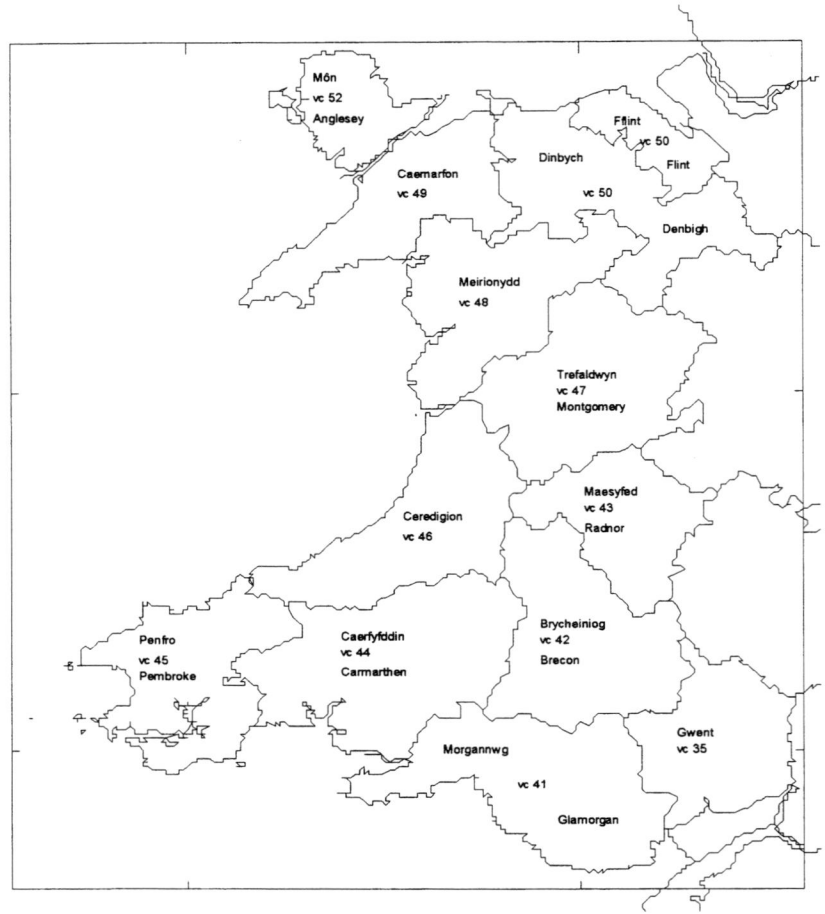

ANGLESEY
Vice-county (vc) 52.
Recorder: S. Culley, Mill House, Penmynydd Road, Menai Bridge, Anglesey LL59 5RT (01248 713091), e-mail SteCul10@aol.com

Current avifauna: The Birds of Anglesey by P. Hope Jones & P. Whalley, is due for publication in 2002.

An annual summary of records is published in the Cambrian Bird Report by the Cambrian Ornithological Society, see below.

BRECON
Vice-county 42.
This area includes a small area around Brynmawr, formerly the NW corner of Gwent. See also under Glamorgan (E).
Recorder: M.F. Peers, Cyffylog, 2 Aberyscir Road, Cradoc, Brecon Powys LD3 9PB.

Current avifauna: The Birds of Breconshire (1990) by M. Peers & M. Shrubb..
Bird Group: Brecon Bird Club. A few meetings a year.
Contact c/o Brecknock Wildlife Trust, Lion House, Bethel Square, Brecon. Powys.LD3 7AY.
The county bird report has been published annually. Annually organise surveys on Kestrel, Lapwing, mixed farmland, commons, Ring Ouzel, Peregrine, MOD bird counts as well as the national WWT & BTO surveys.

CAERNARFON
Vice-county 49.
This vice-county corresponds to the pre-1974 administrative county of Caernarfonshire. Notice that the new county borough of Conwy straddles the old county boundary between Caernarfon and Denbighshire the R. Conwy. Here records from west of the R. Conwy should be sent to the Caernarfon Recorder and from east of the river to the Denbigh Recorder, except for records for the Ormes and Llandudno which remain as Caernarfon. An annual summary of records for Caernarfon is published in the Cambrian Bird Report, see below.

Recorder: J. Barnes, Fach Goch, Waunfawr, Caernarfon. Gwynedd. LL55 4YS.

Current avifauna: The Birds of Caernarfonshire (1998), by J. Barnes.

Bird Groups: Cambrian Ornithological Society and Bangor Bird Group.
In the period there has been considerable progress on several fronts. The Cambrian Ornithological Society (contact via Recorder or web site) has published annual reports and the number of records & contributers has increased considerably over recent years. In 1998 The Birds of Caernarfonshire was published and has stimulated a great deal of discussion and submission of further records. Indeed since its publication there has been no less than 10 species added to the county list. Of note has been the discovery of a small breeding population of Mandarin and the discovery of unprecedented numbers of wintering Hawfinches. The Cambrian Bird Report is available from the Editor - Rhion Pritchard, Pant Afonig, Hafod Lane, Bangor, Gwynedd LL57 4BU.
Bangor Bird Group has monthly indoor and outdoor meetings. Contact: Bangor Bird Group, Treborth Botanic Gardens, Bangor. LL57 2RQ.

An important report is published annually by the Bardsey Observatory. It includes comprehensive reports on other groups of animals and of plants as well as the Systematic List of birds.

CARMARTHEN
Vice-county 44.
Recorder: R.O. Hunt, 9 Waun Road, Llanelli, Carmarthenshire SA15 3RS.

Current avifauna: A Hand List of the Birds of Carmarthenshire (1954) by G.C.S. Ingram & H. Morrey Salmon.

Rob Hunt replaced Dilwyn Roberts as county recorder in 1993. Since that time the county bird report has been published independently by CORC (Carmarthenshire Ornithological Recording Committee) commencing with the 1992 Carmarthenshire Birds. Previously the county bird report was published by the Dyfed Wildlife Trust. CORC is now independent of the WTWW (re-named Dyfed Wildlife Trust).

CEREDIGION
Vice-county 46.
Recorder: H. W. Roderick, 32 Prospect Street, Aberystwyth, Ceredigion SY23 1JJ.

Current avifauna: Ingram, G.C.S., Morrey Salmon, H. & Condry W.M. (1966) *The Birds of Cardiganshire*. The West Wales Naturalists' Trust, Haverfordwest.

An annual report is published by the Wildlife Trust West Wales and is available from the county recorder or from the WTWW headquarters, Welsh Wildlife Centre, Cilgerran, Cardigan. There is currently no organised bird group/society in the county.

DENBIGH
Vice-county 50.
A small part of this vice-county is now in the recording area of Montgomery. See also under Flint below.

Recorder: N. Hallas, 63 Park Avenue, Wrexham LL12 7AW.

FLINT
Vice-county 51.
Denbigh and Flint comprise the area covered by the Clwyd Bird Recording Group, which continues under that title and produces the Clwyd Bird Report covering both vice-counties.
Recorder: N. Hallas, 63 Park Avenue, Wrexham LL12 7AW.
Bird Groups: Clwyd Ornithological Society, Deeside Naturalists Society and Wrexham Birdwatchers.
The Clwyd Bird Recording Group published the 1991-2, 1993-95 & 1999 bird reports and is about to publish a report for the years 1996-98.
The Clwyd Ornithological Society is an informal bird-watching group. In the winter months there are regular indoor meetings and throughout the year there are outdoor visits. Contact: Louise Jones, Sandiway, Llanasa, Holywell. CH8 9NE (01745 852984).
The Deeside Naturalists Society was founded in 1973 with over 350 members. It organises monthly meetings and slide shows. Contact: Mrs. Hazel Jones, 51 Upper Aston Lane, Hawarden. Flintshire. CH5 3EN (01244 533406).
Wrexham Birders styles itself as the small friendly group with members aged between 8 and 80. Meetings are on the first Friday and the (alternating) third Saturday / Sunday of every month. The winter Friday indoor programme runs from September to April, currently at the Memorial Hall, Gresford. The outdoor programme runs throughout the year. A Bulletin is produced regularly along with an annual report. Contact: Hon. Secretary, Marian Williams, 10 Lake View, Gresford, nr. Wrexham. LL12 8PU. (01978 854633).

GLAMORGAN
Vice-county 41.
This area is divided under 2 ornithological Recorders, covering GOWER, comprising the Unitary Authorities of Swansea and Neath/Port Talbot (formerly the county of West Glamorgan), and E.GLAMORGAN, comprising the Unitary Authorities of Bridgend, Rhondda/Cynon/Taff, Vale of Glamorgan, Cardiff and Caerphilly (part) (formerly the counties of Mid and South Glamorgan). Records are normally published under Glamorgan. The East Glamorgan recording area also includes parts of vc 42 (Brecon) and vc 35 (Gwent), which were part of the counties of Mid and South Glamorgan.

Recorders; Gower: R.H.A.Taylor,
285 Llangwfelach Road, Brynhyfryd, Swansea SA5 9LB.
E. Glamorgan: S.J. Moon,
36 Rest Bay Close, Porthcawl CF36 3UN.

Current avifaunas: Birds of Glamorgan (1995), obtainable from Dr. S. Howe, National Museum of Wales, Cathays Park, Cardiff or leading booksellers.
An Atlas of Breeding Birds in West Glamorgan (1992) is obtainable from Dr. D.K. Thomas, Laburnum Cottage, 12 Manselfield Road, Murton, Swansea SA3 3AR.

Bird Groups:
E. Glamorgan: Glamorgan Bird Club. Contact: Hon. Secretary, Richard Smith, 35 Manor Close, Gwaun Miskin, Rhondda Cynon Taff. (01443 205816).
Formed in 1990 from the Cardiff Naturalists Society, it now has around 250 members. The club organises monthly indoor meetings from October to March and runs outdoor meetings during the summer.

Gower: Gower Ornithological Society. Contact: Hon. Secretary, Audrey Jones, 24 Hazel Road, Uplands, Swansea. SA2 0LX (01792 298859).
Throughout the period the membership of GOS has remained steady at around 125 members of which just over two thirds submit records. Although a good number of records are received each year the overall quality of them has dropped and the majority of contributers are reluctant to send in descriptions of rarer species. Many of those received are very brief and of poor quality, a problem being experienced by a number of recorders throughout Wales at the moment.
GOS has monthly indoor meetings between September and April with occasional outdoor meetings throughout the year. The group annually publishes the bird report.

A Guide to Gower Birds was published in 1982 and a Checklist of the Birds of Gower was published in 1999.

GWENT
Vice-county 35.
Vice-county 35 comprises the old county of Monmouth, and now embraces the Unitary Authorities of Monmouth, Newport, Torfaen, Blaenau Gwent and Caerphilly (part). Although the county of Gwent no longer officially exists, the Gwent Ornithological Society is retaining the name to avoid confusion between the new County Borough of Monmouth and the old county of Monmouth, of which the new Borough only comprises part.

Recorder: C. Jones, 22 Walnut Drive, Caerleon, Newport, NP6 1SP.

Current avifaunas: The Birds of Gwent (1976) and The Breeding Atlas of the Birds of Gwent (1987), obtainable from J.M.S. Lewis, Y Bwthyn Gwyn, Coldbrook, Abergavenny, Monmouthshire;
Bird Groups: Gwent Ornithological Society. Contact: Membership Secretary, Gill Jones, Potwallopers, 15 Parva Springs, Tintern, Chepstow. NP16 6TY.

First formed in 1961 as the Pontypool Ornithological Society. Membership quickly expanded forcing the society to broaden its name to the Monmouthshire Ornithological Society in 1963, changing again in 1974, to the present day name upon the local government re-organisation at that time. An active society of approximately 350 members holding 12 indoor meetings at the Goytre Village Hall, Penperlleni near Pontypool during mid September – March and 26 outdoor meetings at a wide range of localities within the County. The Society is currently entering its final year (2002) of fieldwork for the follow up breeding atlas, which will provide an extremely useful comparison with its earlier atlas published in 1987. The Society publishes a quarterly newsletter and an Annual report. Back copies of the reports can be obtained from JMS Lewis, Y Bwythyn Gwyn, Coldbrook, Abergavenny. NP7 9TD.

The recent creation of the Gwent Levels Wetland Reserve, alongside the estuary between Uskmouth and Goldcliff has increased the ornithological awareness of the area with several rarities being recorded within two years of its creation.

MEIRIONNYDD
Vice-county 48.
An annual summary of records is published in the Cambrian Bird Report, see above.
Recorder: D. Smith, 3 Smithfield Lane, Dolgellau, Gwynedd LL40 1BU.

Current avifauna: The Birds of Meirionnydd (1974) by P. Hope Jones.
Bird Groups: Cambrian Ornithological Society.

Dave Smith replaced Reg Thorpe as county recorder in 1999.

MONTGOMERY
Vice-county 47.
The recording area also includes part of vc 50 Denbigh.
Recorder: Brayton Holt, Scops Cottage, Pentre Bierdd, Welshpool Powys SY21 9DL.

Bird Groups: Montgomeryshire Wildlife Trust Bird Group.
Contact: Tony Puzey, Four Seasons, Arddleen, Welshpool.

The bird group of the Montgomery Wildlife Trust was formed in 1997 and currently has more than 80 members. Surveys both at local garden level and BBS for the BTO have enjoyed good coverage, contributing valuable data for the county reports. Relatively "new" reserves have developed and matured with Reed Warblers now established in 4/5 year old reed beds. Corn & Sunflowers, grown for winter feed has attracted 100+ Tree Sparrows to Dolydd Hafren together with c20 Yellowhammers and similar numbers of Reed Buntings. Sadly Lapwing have continues to decline but a scheme is now in place with advice for farmers in an effort to restore this species.

PEMBROKE
Vice-county 45.
Recorders: J.W. Donovan, The Burren, Dingle Lane, Crundale, Haverfordwest, Pembs. SA62 4DJ.
G.H. Rees, 22 Priory Avenue, Haverfordwest, Pembs SA61 1SQ.

Current avifauna: The Birds of Pembrokeshire (1994), by J.W. Donovan, G.H. Rees.

Bird Groups: Pembrokeshire Bird Group (a section of the Wildlife Trust West Wales).
Contact: Mr. T.J. Price, 54 Portfield Avenue, Haverfordwest. SA61 1EG (01437 767190).

The Pembs. Bird Group organises monthly indoor meetings, October – April, and outdoor field meetings in all months except July. The group is responsible for the publication of the annual bird report, organises the longest running annual bird conference in Wales and has an input into many of the WTWW's committees where birds are involved e.g. Chough Study Group, Ornithological Research Committee, County Conservation Committee.
In 1994 the most recent county avifauna was published (available from WTWW) including results of the 1984-88 tetrad atlas.
The Wildlife Trust West Wales also publishes The Island Naturalist, which is the Journal of the Friends of Skokholm and Skomer and includes a good variety of articles on the work at these important Reserves twice annually. Obtainable from their office at WTWW, Welsh Wildlife Centre, Cilgerran, Cardigan.

RADNOR
Vice-county 43.
Recorder: P.P. Jennings, Penbont House, Elan Valley, Rhadader, Powys LD6 5HS.

Current avifauna: The Birds of Radnorshire, by P. Jennings is due for publication in 2002.

Bird reports are published 3 times a year in the Radnorshire Wildlife Trust Newsletter. The Trust also organises regular field meetings.

Within Wales there are three active RSPB Members Groups, each organising a wide range of indoor and field meetings. Contact phone numbers for these groups are:
Cardiff: Margaret Read, 02920 709527
West Glamorgan: Maureen Douglas, 01492 547768
North Wales: Mark Johnson, 01792 882146

CYMDEITHAS ADARYDDOL CYMRU

WELSH ORNITHOLOGICAL SOCIETY

Charity No. 1037823

In March 1988 the Welsh Ornithological Society was launched at a conference in Aberystwyth. Since then its membership has slowly risen to around 200 – 250. Its AGM and conference has taken place annually in Aberystwyth.

With the demise of "Nature in Wales" some years ago there has been no vehicle for the publication of the results of such work. The gap was partially filled, with great success, by the Society's publication of a Welsh Bird Report, initially covering a ten-year period to catch up on a massive backlog and subsequently on an annual basis. As well as a systematic species list and ringing results the Report has included a small number of short papers.

By 1995 the Editor of the Welsh Bird Report has received more papers than he could possibly include in the annual Report and in June of that year launched a twice yearly "Welsh Birds" in which the December issue incorporated the annual bird report and the June issue devoted to the ringing report and papers & articles concerning birds in Wales.

The Society also produces two newsletters each year. The October Newsletter comprises of the Welsh Records Panel Report – summarising the occurrence of Scarce and Rare birds in Wales, while the February Newsletter has articles on the Welsh birding scene.

The cost of subscription is £12 a year, due on the 1st January. Anyone interested in joining should contact the WOS Secretary, Paul Kenyon, 196 Chester Road, Hartford, Northwich, Cheshire CW8 1LG.

WELSH RECORDS PANEL

The Welsh Records Panel is a sub-group of the Welsh Ornithological Society. Its role is to adjudicate all the records of scarce birds in Wales, report its findings to the Welsh Ornithological Society so that they can be included in the Welsh Bird Report and keep the Welsh list up to date. The Panel also publishes its results along with those of BBRC in the Rare and Scarce Birds in Wales Report, in the WOS October Newsletter (copies available at a cost of £1 each from WRP Secretary.

Scarce species are defined as those occurring on average 5 times or fewer annually in Wales and the full list of species for which the Panel considers records is as follows:

Cory's Shearwater, Great Shearwater, Purple Heron, White Stork, Bean Goose, Green-winged Teal, Red-crested Pochard (until 1999), Ring-necked Duck, Surf Scoter, Montagu's Harrier, Rough-legged Buzzard, White-tailed Eagle, Golden Eagle, Spotted Crake, Corncrake, Crane, Stone Curlew, Kentish Plover, Temminck's Stint, Pectoral Sandpiper, Buff-breasted Sandpiper, Red-necked Phalarope, Iceland Gull of the race *kumlieni*, Short-toed Lark, Wood Lark, Shore Lark, Bee-eater, races of Yellow Wagtail (not flava or flavissima), Richard's Pipit, Tawny Pipit, Nightingale, Bluethroat, Aquatic Warbler, Savi's Warbler (until 1998), Marsh Warbler, Icterine Warbler, Melodious Warbler, Dartford Warbler, Barred Warbler, Pallas's Warbler, Red-breasted Flycatcher, Red-backed Shrike Woodchat Shrike, Serin, Common Rosefinch, Cirl Bunting, Ortolan Bunting and Little Bunting.

The Panel consists of five voting members with Reg Thorpe it's Chairman and a non-voting Secretary, Clive Hurford 1992-1995 then Jon Green.

During the 9 years the following have served on the Panel: David Astins – Pembroke, Phil Bristow – Glamorgan, Marc Hughes - Caernarfon, Pete Jennings – Radnor, Steve Moon - Glamorgan, Ian Higginson - Caernarfon and Owen Roberts – Pembroke.

The longest serving member of the Panel retires and a new member is elected annually by the Welsh County Recorders and the Council of the Welsh Ornithological Society. Anyone wishing to be nominated should contact their county recorder or the WRP Secretary.

Descriptions of WRP species should be submitted as soon as possible after the sighting :

Jon Green, Crud yr Awel, Bowls Road, Blaenporth, Cardigan. Ceredigion. SA43 2AR (01239 811561) or e-mail : JGREEN@sirthomaspicton.haverfordwest.sch.uk

Birdline Wales

Birdline Wales is the telephone information service for birdwatchers across Wales. It was established on the 27th March 1991, and has provided daily news without interruption ever since. Run by Alan Davies and based at Penrhyn Bay in Conwy, the news service is available 24 hours a day via a frequently updated taped message. This can be accessed by phoning 09068 700248.

Birdline Wales is always pleased to receive news of bird sightings throughout the Principality. To phone in sightings, not only rarities but any bird news that would be of interest to callers in Wales, simply ring 01492 544588.

Welsh Birding is a quarterly magazine for birdwatchers in Wales. Introduced at the beginning of year 2000, it aims to covers topics of general interest to birders, and was described by one of the national birding monthlies as being "people oriented". It's editors, Birdline Wales's Alan Davies and current Welsh Rarities Panel member Owen Roberts, live at opposite ends of the country, Conwy and Pembrokeshire respectively, giving them contacts throughout the Welsh birding community.

Regular features are guides to Welsh sites, how to find regular but scarce birds in Wales, birder's diaries, and birding personalities.

The editors are always interested in birding articles that have a regional interest. These, and any other correspondence, should be sent to Welsh Birding, 9 Maes y Mynach, St. Davids, Haverfordwest, Pembrokeshire SA62 6QG or by email to: Owen.Roberts@tesco.net

The annual subscription is £10 including P&P and is available from the same address. Cheques should be made payable to Welsh Birding. Some back numbers are still available, please enquire.

ENVIRONMENTAL DEVELOPMENTS

It is impossible to catalogue all the changes that have occurred in Wales during the period 1992-2000 but the following are probably those which had the most impact. On the destructive side three events stand out:

- The Sea Empress disaster in February 1996
- The Cardiff Bay Barrage
- The continued agricultural intensification

Sea Empress

On Feb. 15th 1996 a super-tanker full of oil collided with rocks at the mouth of the Milford Haven Waterway, Pembrokeshire. During the following 5 days, an estimated 70,000 tonnes of oil leaked from the ship and into the Irish Sea. Early fears that the oil would track north and have disastrous consequences to the bird rich islands of Skomer, Skokholm and Ramsey were dispelled as the oil moved north then south east into Carmarthen Bay. A large proportion ended up on the beaches of Freshwater West, Angle Bay, Tenby and the rest of Carmarthen Bay. Most however was never recovered, either mechanically or from the beaches and is thought to have sunk to the bottom of the bay. What effects this will have in the long-term to this region is unclear. It was estimated that a total of 5,000 – 15,000 tonnes of oil and emulsion contaminated about 200km of coastline.

A total of over 7,000 seabirds of 29 different species were found dead or dying with the true number perhaps nearer 70,000. The table below shows the number of birds affected (counted either oiled or dead) in each county.

	Pemb.	Carm.	Glam.	Cere.	Lundy	Ireland	Total Dead	Total Alive	Total
Total	1697	836	518	25	57	362	3495	3440	6935

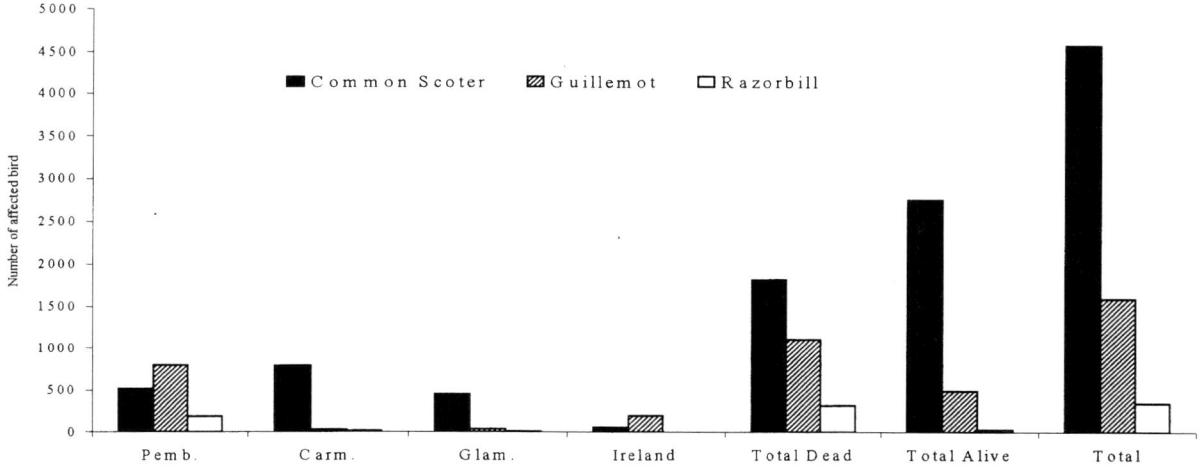

Significant numbers of Guillemots (1,500) and Razorbills (300) were affected but it was Carmarthen Bay's wintering flock of Common Scoters which bore the brunt. The chart below shows the number of deaths in each county of the three most affected species, Common Scoter, Guillemot and Razorbill.

SEEC recommended the establishment of an Impact Assessment Group (IAG) to continue the measurement and monitoring of the long-term effects of the disaster. This would report on

- The fate of oil over time
- The long term and short term effects of the spill
- To determine the concentration of oil at different places
- Better management of the operation in case of a future incident.

Cardiff Bay

The rivers Taff and Ely flow through the heart of Cardiff having commenced their journey in the uplands of the south Wales valleys. Where they enter the Bristol Channel the rivers merge to form a large bay, which prior to the barrage comprised extensive inter-tidal mud-flats, fringed by salt-marsh and surrounded by the famous docklands of the Welsh capital. With the subsequent decline of coal and heavy industry in the region the rivers and estuary gradually became cleaner and healthier but the urban environment of the docklands spiralled into a seemingly unstoppable state of decay and dereliction.

In an effort to revive the fortunes of the area, the Cardiff Bay Development Corporation was formed in 1986 and announced its key project was the construction of a tidal barrage across the mouth of the Taff / Ely Estuary. Parliamentary consent was given in the early 1990's, construction started soon afterwards and was completed in 1999, when on Nov. 4th the sluice gates of the barrage were closed, impounding the rivers as a permanent lagoon of 250 acres.

Prior to its construction, the bay had been designated a SSSI and attracted wintering wildfowl and waders. Within the context of the Severn estuary as a whole, the site held locally significant numbers of Shelduck, Redshank and Dunlin and to a lesser extent Teal, Oystercatcher, Knot, Ringed Plover, Grey Plover and Curlew.

As widely predicted, the completion of the barrage has had a dramatic impact upon the birds of the Taff Ely estuary. Gone are the large flocks of wintering wildfowl and waders. Results from colour-ringing and birds fitted with radio-transmitters, has shown that the principle species affected by the barrage have dispersed to other sites within the Severn estuary, with the nearby Rhymney estuary and Gwent Levels fore-shore the recipient of the majority of the refugees.

On the plus side, small numbers of diving ducks have been seen in recent winters but the most interesting development has been the establishment of a large gull roost, comprising several thousand birds, mainly Herring, Lesser Black-backed and Black-headed Gulls along with the occasional Mediterranean, Yellow-legged and Ring-billed Gull.

Farmland

Lovegrove *et al* stressed the importance to birds of three modern changes in Welsh agriculture, which had largely taken effect since 1970, the decline in tillage, the decline in the area of open rough grazings and the vast increase in sheep numbers. All combined to bring about a major decline in the diversity of Welsh farmlands. The loss of tillage, for example, is probably the most important factor in the virtual extinction of the Grey Partridge in Wales, for this is a bird particularly of arable ecosystems. The importance of arable pockets in pastoral landscapes to a range of farmland birds, particularly seed-eaters such as Tree Sparrow and Yellowhammer, has recently been demonstrated by Robinson *et al* (2001). The decade of the 1990s has seen very little change in the features of Welsh agriculture that Lovegrove *et al* highlighted, as the accompanying figures show (the charts extend the figures shown in Lovegrove *et al* 1994, source WOAD June Census Statistics). The rise in sheep numbers and the decline in tillage have slowed or stopped, although the area of rough grazing and moorland has continued to shrink. There is no doubt that the continued emergence in Wales of what amounts to a sheep-based monoculture and the steep increase in the intensity of grassland management that the increase in stock numbers has required, has seen a marked loss of many ground-nesting species, perhaps particularly waders but also species such as Skylarks, Yellow Wagtails and Whinchats. This has been the most marked effect of modern changes in farmland management in Wales and it is epitomised by the virtual disappearance of the Lapwing as a common farmland breeding species since 1987. It is now one of the RSPB's three most endangered bird species in Wales and subject of a major recovery programme, the core of which is the recreation of extensive areas of habitat, for example in the Dyfi estuary.

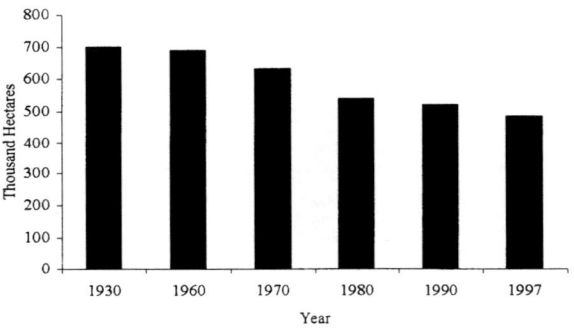

Changes in the area of rough grazing and moorland in Wales during 1930 - 1997

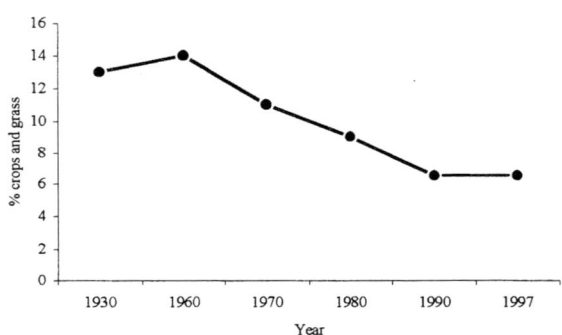

Changes in the area of tillage in Wales as a % of crops and grass during 1930 - 1997

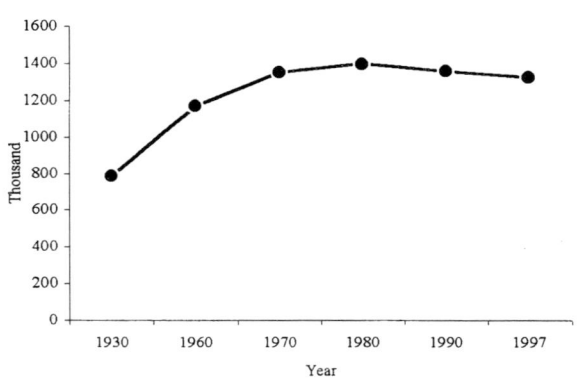

Changes in the total number of cattle in Wales during 1930 - 1997

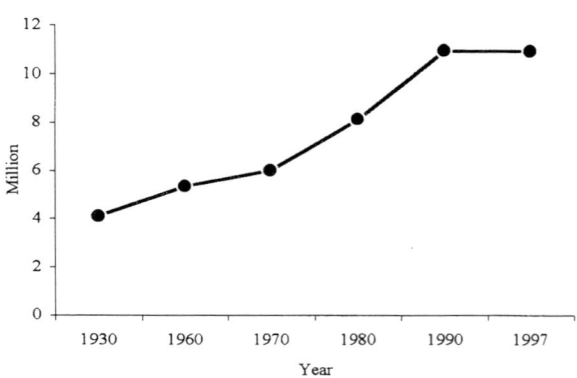

Changes in the total number of sheep in Wales during 1930 - 1997

Birds have tended to dominate the discussion about the impact of declining diversity in farmland, largely because they are conspicuous and relatively easily studied. But the loss of familiar birds should not blind us to the fact that this is only part of the overall loss of diversity which has occurred in Welsh grasslands. For example Kestrels have declined because modern grassland management reduces the extent of good habitat (damp rough grassland) for the voles upon which Kestrels feed. There has also been a marked decline in conspicuous invertebrates such as *lepidoptera* and *orthoptera*, which is strikingly obvious to the ear and eye as one walks across modern pastures (see Cowley *et al* 2000 for example). Increasingly modern rye pastures tend towards botanical and entomological deserts and simply can no longer support the bird populations that once they did.

The main development in Welsh agriculture since 1990 has been the emergence of agri-environment schemes, designed to reduce the impact of intensive farming on the countryside. While it is probably too early to be too dismissive, one has to say that these schemes so far have only had the most marginal impact in restoring farmland bird populations. Nor have they had much obvious impact on the pattern of farming. Sheep numbers, for example, have stabilized at their present very high levels but some significant decline is needed, if birds such as breeding waders are to benefit. Indeed it is not clear whether the stabilization at present visible is the result of agri-environment schemes or simply that Welsh farming is nearing a limit beyond which further intensification will be difficult to achieve. If they are to produce real benefits for wildlife such schemes need to bring about a real reduction in stocking rates and some return to the mixed pastoral systems of the past.

Habitat Management

One of the major changes to ornithology in Wales during the past 9 years, and which will have a major effect in the next millennium is the development of sites for birds – farming for birds and habitat creation. There have been many small projects but the major ones were:

- Gwent Levels Wetland Reserve, Goldcliff in Gwent by CCW
- Penclacwydd Wildfowl & Wetlands Reserve in Carmarthen by WWT
- Ramsey Island in Pembroke by the RSPB
- Ynyshir RSPB extensions in Ceredigion
- Llyn Coed y Dinas and Dolydd Hafren in Montgomery by the Montgomeryshire Wildlife Trust
- Lake Vyrnwy – management of 10,000 ha of moor & farmland by the RSPB / Severn Trent Water
- Conwy RSPB reserve in Caernarfon
- Malltraeth RSPB on Anglesey
- Valley Lakes RSPB on Anglesey
- Inner Marsh Farm RSPB in Flint / Cheshire

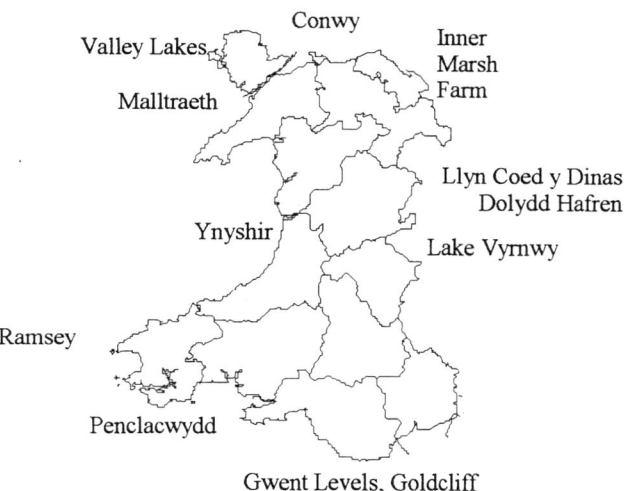

Gwent Levels Wetland Reserve - Goldcliff

The Gwent Levels Wetlands Reserve is a new 440ha nature reserve south east of Newport located on the coastal floodplain from the mouth of the River Usk to Goldcliff. The Reserve is a wetland creation project, specifically designed to provide habitat for wetland bird species, as part of the compensation for the loss of the Cardiff Bay mudflats to the Cardiff Bay Barrage scheme. Full planning consent for the development of saline lagoons, reedbeds and lowland wet grassland was granted in November 1997. The Reserve was created between 1998 and 1999. It is owned and managed by the Countryside Council for Wales (CCW), in partnership with local farmers, Newport County Borough Council (NCBC) and the Royal Society for the Protection of Birds (RSPB). The Reserve consists of four main habitat types: reedbeds (58 ha); lowland wet grassland (150 ha) saline lagoons (11 ha); and saltmarsh (80 ha). Car parking has been provided off West Nash Road at Uskmouth (O.S grid reference ST334834). The car park provides easy access to a series of **tracks around the reedbed lagoons, including way-marked trails. It is proposed to further develop visitor facilities, education initiatives** and interpretation on the site during 2002 - 2006.

The Reserve has an obligation to provide habitat to support at least two species of wintering waterfowl in nationally important numbers. Livestock grazing and hydrological management on the lowland wet grassland will provide habitat for over-wintering and passage wildfowl such as Wigeon, Shoveler, Teal and Gadwall. It should also provide habitat for breeding Shoveler, Garganey, Lapwing and Redshank.

Hydrological management on the saline lagoons will provide feeding opportunities for breeding and passage waders, and the lagoons and islands should provide a site for waders and wildfowl from the Severn Estuary to roost during high-tide periods. The reedbeds, once established, should provide habitat for Bittern, Marsh Harrier, Water Rail, Cetti's Warbler and Bearded Tit.

WWT Llanelli - Penclacwydd

WWT Llanelli is situated on the northern shore of the Burry Inlet, which is regarded as one of the most important estuaries for wildfowl and waders in Wales. Bird numbers on the whole of the Burry Inlet are at their highest in the winter when up to 60,000 birds can be present. Theses include Internationally important numbers of Oystercatchers (21,000), Pintail (3,000), Shoveler (600) and Knot (6,000) and Nationally important numbers of Brent Geese (1,300), Shelduck (2,000), Wigeon (3,000), Teal (2,000), Dunlin (8,700), Lapwing (10,000), Black-tailed Godwit (300), Curlew (2,400), Whimbrel (120 on passage) and Redshank (1,200). As a result of its bird numbers the Burry Inlet is designated a Ramsar Site, SSSI, Special Protection Area (SPA) and a candidate Special Area for Conservation (SAC). The whole area of marsh at WWT (100) hectares lies within these boundaries and many of these birds can be seen from the centre (in smaller numbers).

The Centre opened in 1991 as part of Carmarthenshire County Councils tourism strategy. This strategy was intended to partly compensate for the loss of much of the heavy industry, especially steel and tinplate, for which the area was famous. The creation of a variety of habitats, including brackish & freshwater scrapes, a six acre lagoon, landscaped grounds & reed beds, provides habitats for a wide variety of birds.

In 1999 the reserve expanded to over 600 acres, with the creation of The Millennium Wetland, which is part of The Millennium Coastal Park, a £30 million development stretching 12 miles along the north shore of the Burry Inlet from Loughor Bridge in the east to Pembrey Country Park in the west. The project is funded by The Millennium Commission, Carmarthenshire County Council, The Welsh Development Agency and Welsh Water.

The Millennium Wetland covers an area of 200 acres and has been transformed from low grade agricultural land into deep water lakes, shallow scrapes, reedbeds, wet meadows, ditches and an interconnected system of canal sized channels to provide suitable habitat for a whole range of wetland birds as well as Water Voles, Otters, 19 species of Dragonfly / Damselfly and 23 species of Butterfly. The whole site has been carefully designed to be concealed behind grassed embankments with much effort made to retain and protect the mature hedge banks and ditches which are a feature of the area. A number of viewing screens which are to be replaced by eight proposed bird hides are in the process of being constructed, which will complement the award winning Heron Wings Observatory, all overlook wild bird areas.

The site has been designed for 14 target bird species – Bittern, Bearded Tit, Reed Warbler, Marsh Harrier, Little Egret, Spoonbill, Cetti's Warbler, Black-tailed Godwit, Redshank, Lapwing, Oystercatcher, Skylark, Sand Martin and Kingfisher – some of which no longer breed in Carmarthen and for others it will provide an opportunity to extend their range. To date all but Bearded Tit have been recorded and numbers of breeding Lapwing on site have doubled to around 28 pairs. Other breeding birds so far have included 6 pairs of Redshank, 5 pairs of Little Grebe, up to 3 pairs of Shoveler (first breeding record in Carmarthen since 1950's), 7 pairs of Gadwall, 11 pairs of Pochard, 12 pairs of Tufted Duck, a pair of Barn Owls and while under construction, a pair each of Little Ringed Plover and Common Sandpiper. Other highlights have included all five British Owl species in winter 2000, probably unique for a 200 acre wetland, Blue-winged Teal, Aquatic Warbler, Little Gull, Black Tern, Little Tern and Black-necked Grebe.

The site is entirely fresh water, fed from the neighbouring Welsh Water sewage treatment works and enters the site as high grade disinfected final effluent allowing complete water level management to encourage breeding birds and wintering wildfowl.

Ramsey Island

The purchase of Ramsey Island (253 hectares) by the RSPB in 1992 was greatly welcomed in completing the protection of all the major Pembrokeshire islands by nature conservation bodies. Ramsey is notable for its maritime heath and grassland and for its long established breeding population of Choughs (7 to 8 pairs). 8 pairs of Lapwings maintain a tenuous foothold at the only site left in Pembrokeshire and it is hoped that the Manx Shearwater population of about 1000 pairs will increase in response to the elimination of rats in 2000. For migration enthusiasts Britain's only Indigo Bunting graced the island in 1996 and Wales' only Yellow-rumped Warbler in 1994.

Wetlands for Wales – Ynyshir, Malltraeth and Valley Lakes

One of the most positive and encouraging developments in the period under review was the setting up of the Wetlands for Wales project, a partnership in which the Environment Agency played a leading role and with the RSPB, Countryside Council for Wales and North Wales Wildlife Trust as the other partners. The project, which covers Wales north of the Dyfi Estuary, has benefited from substantial contributions from the Heritage Lottery Fund and from CCW.

Sites which are in course of development as part of the project are Malltraeth Marsh RSPB on Anglesey where land was first purchased in 1994 and recent extensions have built up a holding of 206 hectares of lowland wet grassland, reedswamp and pools. Substantial shallow excavations and reed planting have taken place here, in part as a contribution to the RSPB's Action Plan for Bitterns. Likewise at Valley Lakes recent purchases by RSPB have increased the reserve to 174 hectares of lake and reedswamp with an intensive programme of scrub removal and reed-bed creation. There are important wildfowl breeding numbers at both sites and recent colonisation of Cetti's Warbler at Valley Lakes.

A further significant step forward for the Wetlands for Wales project took place in 2000 with the purchase by RSPB of two areas of lowland wet grassland on the edge of the Dyfi Estuary amounting to 159 hectares, adjoining and close to Ynys-hir. These are being managed principally for the important Lapwing breeding population, these areas forming part of the largest remaining concentration in Wales outside the Dee floodplain.

Conwy RSPB

Over the last 10 years, the RSPB has developed a new reserve on the east bank of the Conwy estuary between Glan Conwy and Llandudno Junction through a lease from the Crown Estate. Formerly saltmarsh and intertidal mudflats, 32 ha was embanked and used as a spoil dump site for the Conwy tunnel section of the A55 expressway during the late 1980s/early 1990s. Although primarily developed as a visitor/educational site, the site has quickly become established a superb birdwatching site with nature trails, visitor centre and 4 birdwatching hides.

Remnants of the former saltmarsh still exist and attract up to 1000 wildfowl during the early winter. A feature of recent years has been a regularly returning drake American Wigeon. Within the bunded body of the reserve, 2 fresh water lagoons (12ha) have been created incorporating a series of grass and shingle islands. These islands provide a safe haven for roosting waders from the adjacent estuary (primarily Redshank, Curlew and Oystercatcher).

Breeding waders on the lagoons include Lapwing, Redshank, Oystercatcher and Ringed Plover, while the reserve attracted Gwynedd's first breeding Little Ringed Plover in 1993.

As the lagoons have established during the 1990s they have become increasingly popular with passage waders with Wood and Green Sandpipers, Ruff, Black-tailed Godwits and Spotted Redshank being annual.

Varying numbers of Little Stints and Curlew Sandpipers occur each autumn while a scattering a vagrants have occurred including 5 Pectoral Sandpipers and the trio of Marsh, Broad-billed and Terek Sandpipers.

4 ha of reed-swamp has been established within the reserve which attracts increasing numbers of breeding Reed and Sedge Warblers. Bare spoil areas have slowly colonised with a mixture of neutral and calcareous grassland. These open areas have attracted an impressive list of scarce passerines for a mainland site including 2 Bluethroats, 2 Richard's Pipits, Black-headed Wagtail, Wryneck, Short-toed Lark and Hoopoe.

Inner Marsh Farm

An RSPB reserve of 30 hectares which straddles the Cheshire/Flintshire border, converted in the late 1980's/early 1990's from arable land to a mosaic of freshwater pools. Strategically located at the head of the Dee Estuary it attracts nationally or internationally important numbers of Teal, Pintail, Shoveler and Black-tailed Godwit and an impressive variety of passage and wintering waders and wildfowl including a wintering flock of Bewick's Swans.

Llyn Coed y Dinas

Out of the development of a by-pass for Welshpool in 1992, a wetland habitat emerged. A mature oak tree on a hillock became the central feature, an island of about half an acre and by stopping excavation at the west side at an appropriate point, a high bank was formed, at once attracting Sand Martins. Further scraping and fine tuning produced lots of shallow muddy areas and gravel islands and planting of a variety of trees around the perimeter and phragmites in some sheltered shallows set the shape of things for the future. The 13 acre lake has attracted various wader species including Little Stints, Wood Sandpiper, Turnstone, Avocet and Little Ringed Plovers. Great Crested Grebes nest and the Sand Martin colony had grown to 80 pairs in 3 years.

The site is managed by the Montgomeryshire Wildlife Trust who helped design the reserve and erect an observation hide.

Dolydd Hafren

A recent reserve of the Montgomeryshire Wildlife Trust this covers 42 hectares of riverside flood meadow and remnant oxbow lakes at the junction of the rivers Severn and Camlad. Management work is being undertaken to enhance the reserve for breeding and wintering waders and wintering passerines; other important species include breeding Yellow Wagtail and there is a wintering flock of Tree Sparrows.

Lake Vyrnwy

A major step forward has taken place recently in the management of the RSPB reserve, with the Society taking responsibility for Severn-Trent Water's farming operation on their Vyrnwy Estate. STW's welcome co-operation in this has enabled full integration to be achieved between farming activities and habitat restoration on the Vyrnwy moorland, with resultant benefits for breeding species such as Black Grouse, Red Grouse and birds of prey. With the support of CCW the Vyrnwy sheep flock has been reduced by about half, whilst the entire enterprise is now run on an organic basis.

BIRD RINGING IN WALES 1992 - 2000

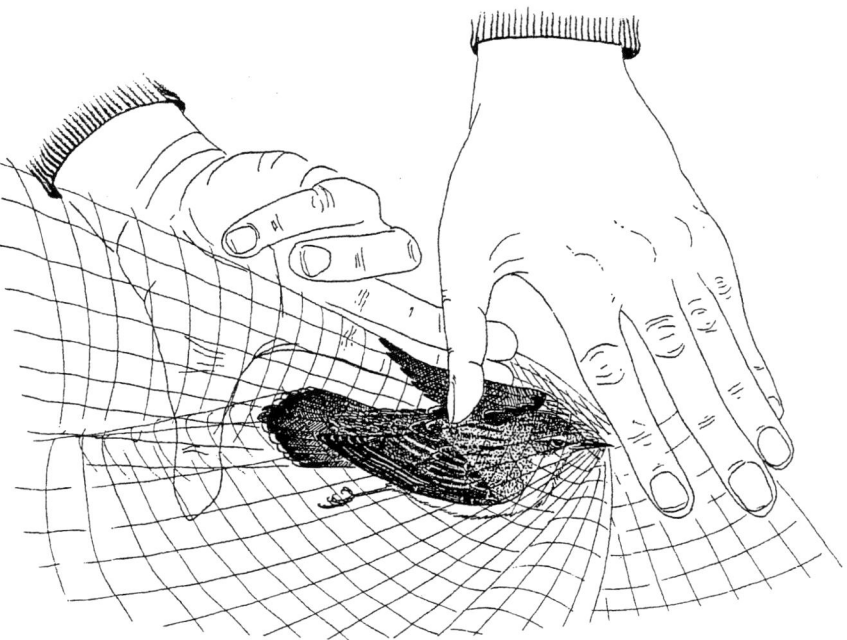

Introduction

Ringing has been carried out for the best part of 90 years in Britain and in that time about 30 million birds have been ringed. Despite this seemingly large number we are still finding out new information on migration routes and wintering grounds all the time. It must also be remembered that the majority of birds have a very short life-span so to be able to successfully monitor population trends it is important that fairly substantial numbers are ringed each year.

Ringing in Britain is carried out under the auspices of the British Trust for Ornithology (BTO) who grant licences on behalf of the Joint Nature Conservation Committee (JNCC). The primary aim of the ringing scheme is contribute to the understanding of population changes by monitoring survival rates, productivity and dispersal. This is increasingly being achieved through highly structured projects such as the Constant Effort Sites (CES) scheme and the Retrapping Adults for Survival (RAS) scheme, although general ringing is still required to maintain sufficient numbers of ringed birds to monitor general trends.

However, the ringing of the bird is only one side of the equation, for most of the population work done by the BTO a ringed bird needs to be recovered. The 2,000 or so ringers operating in Britain provide much information through retrapping live birds (the CES and RAS schemes were designed mainly for that reason). However, the public has a very large role to play as they are responsible for reporting about 40% of the 12,000 recoveries each year. Without this contribution the effectiveness of the ringing scheme would be seriously diminished because recoveries tell us where birds move, how long they live and often some information as to how they died. This information is particularly important in highlighting potential threats to vulnerable species.

The number of ringing recoveries is not just dependent on the number of birds ringed. The size of the bird, the habitat it lives in, whether a particular group is targeted by ringers, and whether plastic 'darvic' rings are used also play a part. These plastic colour-rings with large 3 letter codes are mainly used on larger species such as gulls and swans (although un-numbered colour-rings have been used for years on a wide variety of species). They can be read in the field through a telescope, or with the naked eye in the case of swans if they come close when being fed and have resulted in a massive increase in the reported sightings.

Although the emphasis is on population monitoring the ringing scheme is also still concerned with movement and migration for evidence of this look no further than the *Atlas of Bird Migration for the Birds of Britain and Ireland* (due October 2002), this shows maps of all the ringing recoveries received by the BTO for migratory species. This is a landmark publication and should prove fascinating to everyone who has the slightest interest in birds.

The maps included in this chapter have been derived from data supplied by the BTO, from the new Atlas of Bird Migration, using DMAP by Dr. Alan Morton.

Birds in Wales 1992-2000

Bird Ringing in Wales

While there are over 2,000 ringers throughout Britain there are less than 100 operating in Wales many of whom ring as part of one of the six groups that are either based or have ringing sites in Wales. Since 1988 an annual report has been published by the Welsh Ornithological Society (WOS) detailing the activities of many of them. They have ringed 32-52,000 birds annually between 1988 and 1999 this compares to the British totals of 777-868,000 annually over the same period. The Welsh totals are not absolute as the totals are calculated from returns sent in voluntarily, whereas all ringers must submit their totals to the BTO each year. There is also no definitive list of people who ring in Wales, there may well be others who come into Wales to ring on their holidays who cannot be traced, however, it is likely that all the really active ringers are included.

Throughout much of the 1990s general ringing effort in Wales has remained remarkably constant with the same sites being used for similar amounts of time each year. This means that, for some species at least, changes in the annual totals correlate directly to changes in population.

The totals for Red Kite and Goshawk nestlings reflect steadily increasing Welsh populations (graphs 1 & 2) - although the growth in Goshawk population is almost certainly not quite as dramatic as the graph suggests as the last few years have seen an increase in effort but that effort has been quite stable in the last few years. Graphs 3-5, however, show the worrying declines in Dipper nestlings and full-grown Linnet and Reed Bunting. A decrease shared with the rest of the UK and highlighted by the BTO over the last few years.

Populations can also be affected by severe weather, particularly hard winters or excessively wet summers. The effects of a hard winter are clearly demonstrated in graph 6 which shows how the numbers of Goldcrest plummeted after a cold snap in the winter of 1990-91. Graph 7 shows how the totals for Blue and Great Tit nestlings have dropped, almost certainly due to prolonged heavy rain when chicks are in the nest in the 1998 and 1999 summers. Populations generally recover after one or two breeding seasons but the Goldcrest recovery seems to be very slow.

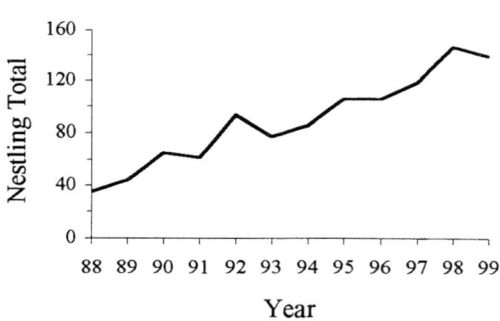

Graph 1: Red Kite

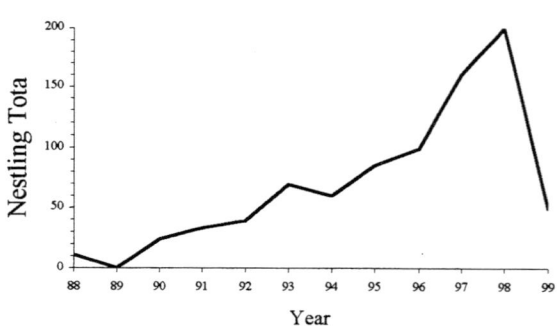

Graph 2: Goshawk

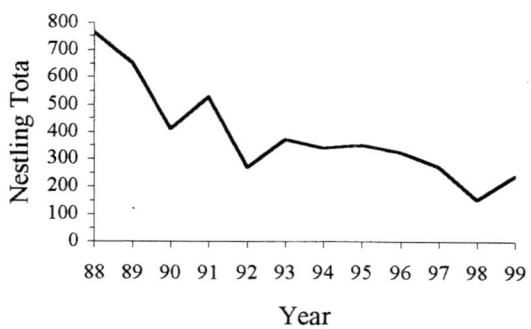

Graph 3: Dipper

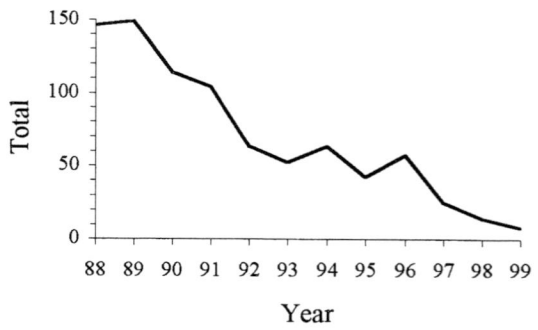

Graph 4: Linnet

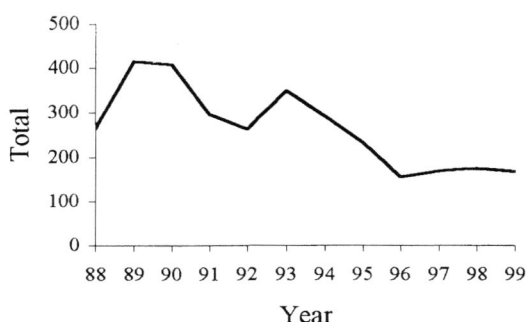

Graph 5: Reed Bunting

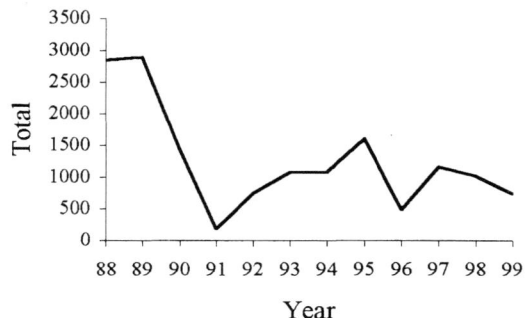

Graph 6: Goldcrest

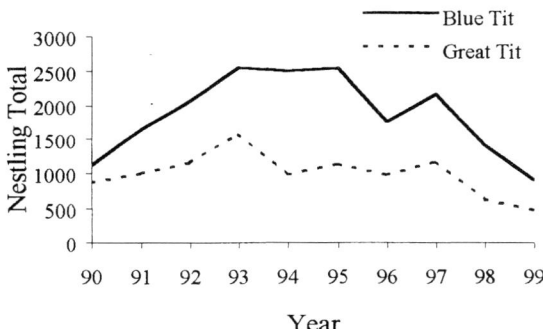

Graph 7: Blue Tit and Great Tit

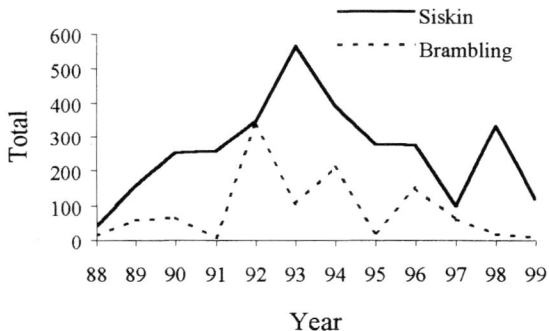

Graph 8: Siskin and Brambling

Some species are highly eruptive, their numbers dependent on abundance of seed crops on trees, population explosions, or severe weather elsewhere in their range. The totals for two species, Brambling and Siskin, show these changes very clearly, graph 8 demonstrates how dramatically their numbers have fluctuated over the past 11 years. Only data from a select few species can be presented in such simple graphs, for most species meaningful analysis is considerably more complex due to the number of variables which need to be taken into account.

The totals for full-grown birds and nestlings are differentiated in the annual report, the BTO has always split them and the WOS reports have followed the same format. In many respects it is more useful to ring birds as nestlings as their age and breeding locality is known precisely, something rarely obtained from full-grown birds. Obviously the locality of breeding adults is known precisely but rarely their age and for the vast majority of birds ringed neither are known with any great accuracy.

When the Welsh full-grown and nestling totals are compared to the British totals some interesting differences emerge. On average nestlings make up 23% of the British total and 30% of the Welsh total. Nestlings ringed in Wales also make a larger contribution to the British total than full-grown birds too, 10% and 5% respectively. Whether these differences are significant is debatable but it shows that Welsh-ringed nestlings probably make a more important contribution to the Ringing Scheme than the full-grown birds.

This higher proportion of nestlings is due to the number of nestboxes that have been put up in Wales. Many ringers check several hundred boxes a year, no mean feat given the terrain most Welsh woodlands occupy. The attraction of running nestbox projects is the relatively high breeding densities of species such as the Pied Flycatcher, indeed it is likely that the breeding numbers are actually helped by the provision of artificial nest sites. At least 30% of the British total Pied Flycatchers nestlings were ringed in Wales (this could of course be higher if the Welsh total is missing the results of a large nestbox project).

Birds in Wales 1992-2000

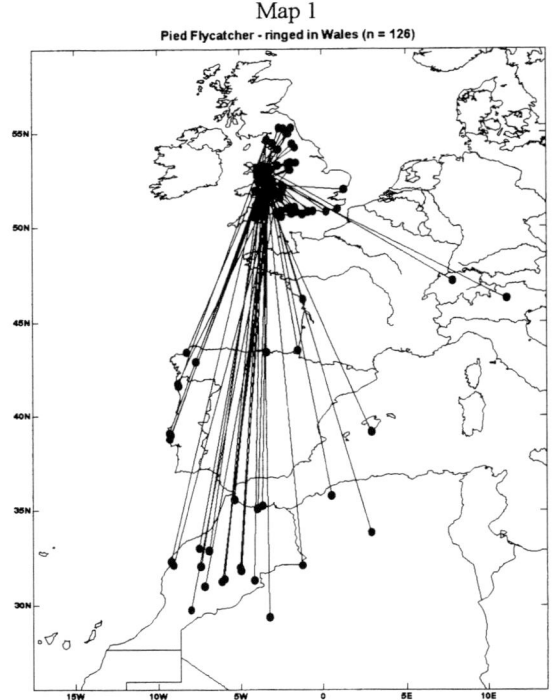

Map 1
Pied Flycatcher - ringed in Wales (n = 126)

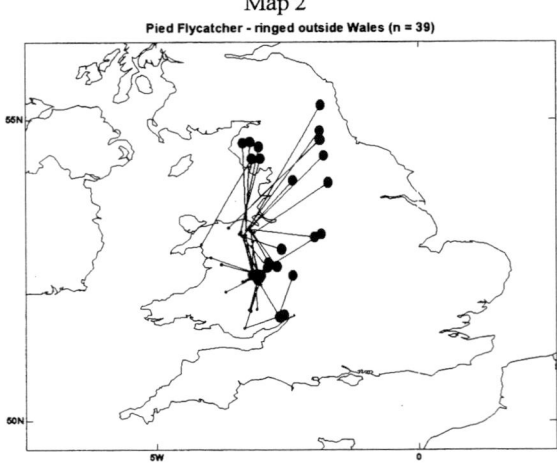

Map 2
Pied Flycatcher - ringed outside Wales (n = 39)

Such a large number of schemes and consequent high totals of birds ringed should provide for many ringing recoveries. For one species, Pied Flycatcher, it does, maps 1 and 2 show the recoveries for birds ringed within Wales and those ringed outside Wales but recovered within respectively. Map 1 clearly shows the North African wintering area and the coastal routes birds take to get there. The dots within Britain to the north and east of Wales show how far this species can move from its natal area to find a breeding territory with considerable interchange between Cumbrian, Northumbrian and Welsh populations.

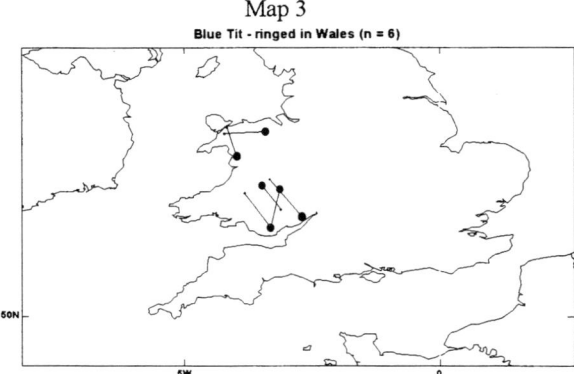

Map 3
Blue Tit - ringed in Wales (n = 6)

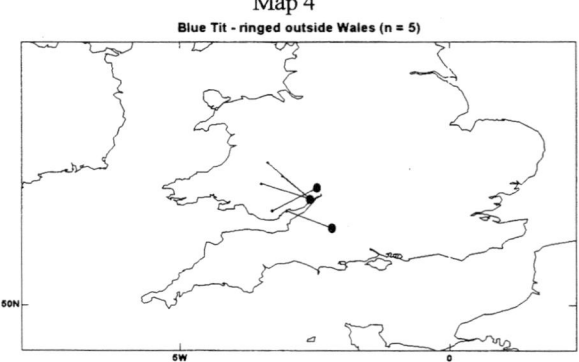

Map 4
Blue Tit - ringed outside Wales (n = 5)

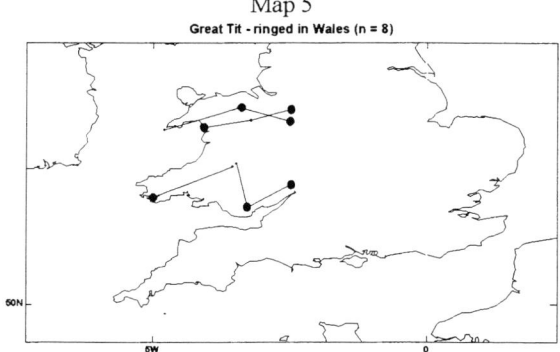

Map 5
Great Tit - ringed in Wales (n = 8)

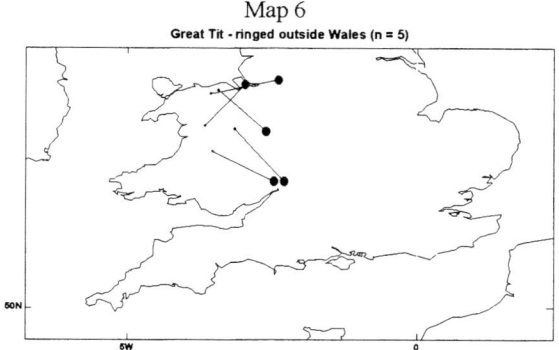

Map 6
Great Tit - ringed outside Wales (n = 5)

For other species it does not. Blue Tit and Great Tit are amongst the most commonly ringed birds (not just in Wales but for the ringing scheme as a whole) but maps 3 - 6 show how few recoveries there are. The nature of the movements gives an indication as to why this should be - they do not move a great distance and are more likely to be retrapped by the people who ringed them which means they do not show up on the maps. The other frequent occupier of nest boxes is the Redstart, the handful of recoveries for this species show a similar pattern to Pied Flycatcher (map 7).

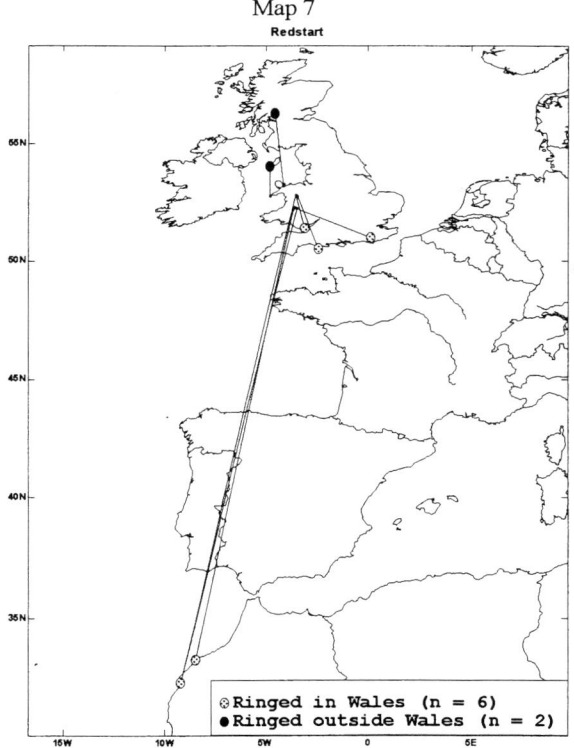

Map 7
Redstart

Away from nestboxes the Welsh moors are home to large numbers of Whinchats, which have been the target of three different ringers for several years. In 1998 they ringed 814 nestlings, virtually the entire British total for that year. As the projects have been running for some years light is starting to be shed on the lifespan of the adults and site fidelity of both adults and returning first year birds. Unfortunately, despite the large number of birds ringed, there have been disappointingly few recoveries. These birds not only breed in open habitats they also favour them on migration too resulting in very few being caught at the various ringing stations elsewhere in Britain and Europe. Map 8 shows the three reported birds which really do not give us much indication as to where Welsh Whinchats winter or the routes taken to get there. Another common moorland species is Wheatear for which there have also been three recoveries, despite many fewer birds being ringed each year - all have been in their North African wintering grounds (map 9).

Birds in Wales 1992-2000

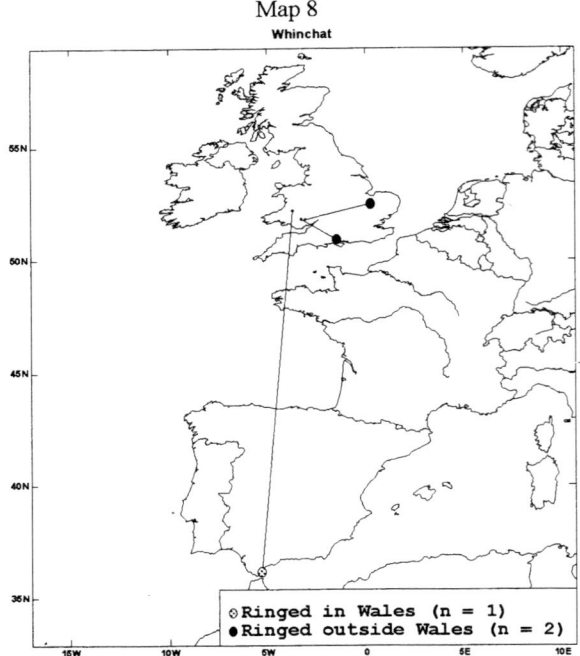

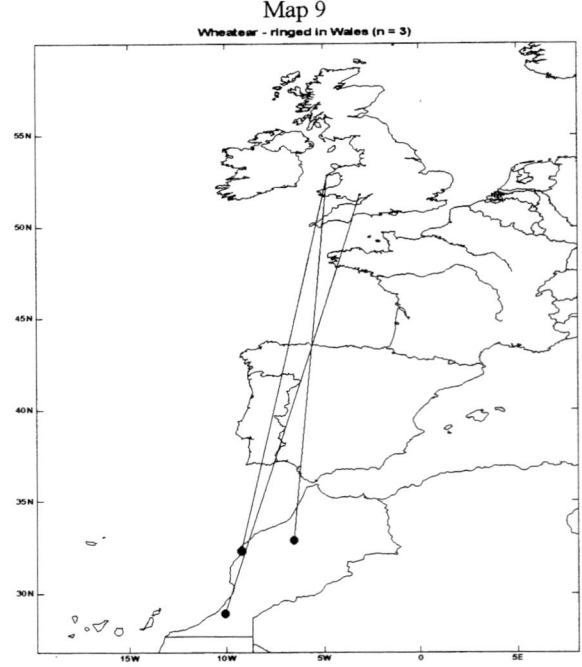

The number of nestlings ringed is also boosted by the presence of a few species which are of conservation concern such as; Red Kite, Goshawk, and Chough and the presence of two islands where seabirds are ringed, namely Skomer and Bardsey. The proportions of the British total ringed in Wales for the first three species are 71%, 50% and 80% respectively. Although the seabird nestlings boost the Welsh total they make very little impact on the British total because the seabird colonies (with the exception of Manx Shearwater) are small in comparison with those in Scotland.

Map 10 shows the sedentary nature of Welsh-bred kites and it is only in the last few years as numbers have increased that the longer-distance movements have appeared.

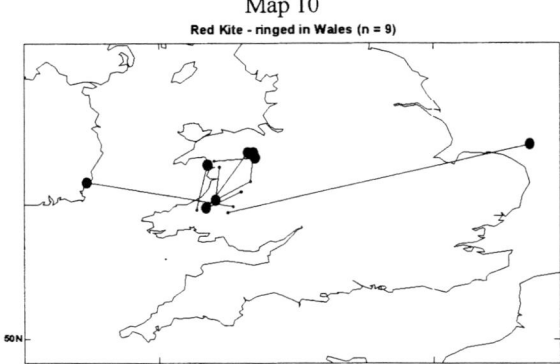

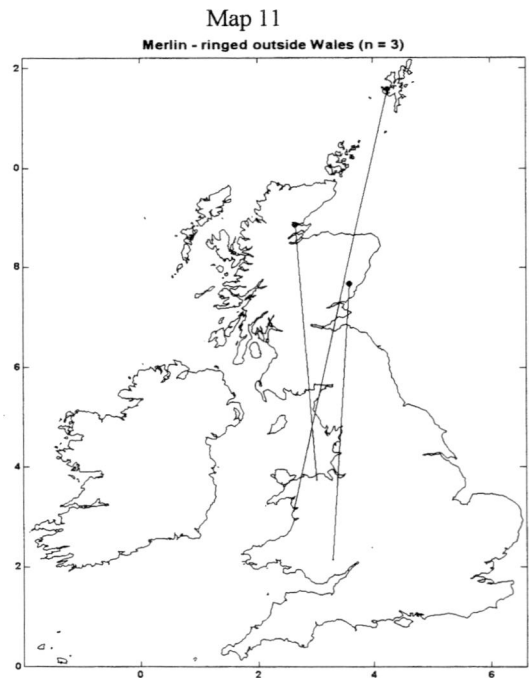

The bird seen on a gas drilling rig 40km off the Norfolk coast less than 30 days after fledging is the record for a Welsh-bred bird. There has also been one record of a bird from the re-introduction scheme recovered in Wales, from the Scottish release site of all places.

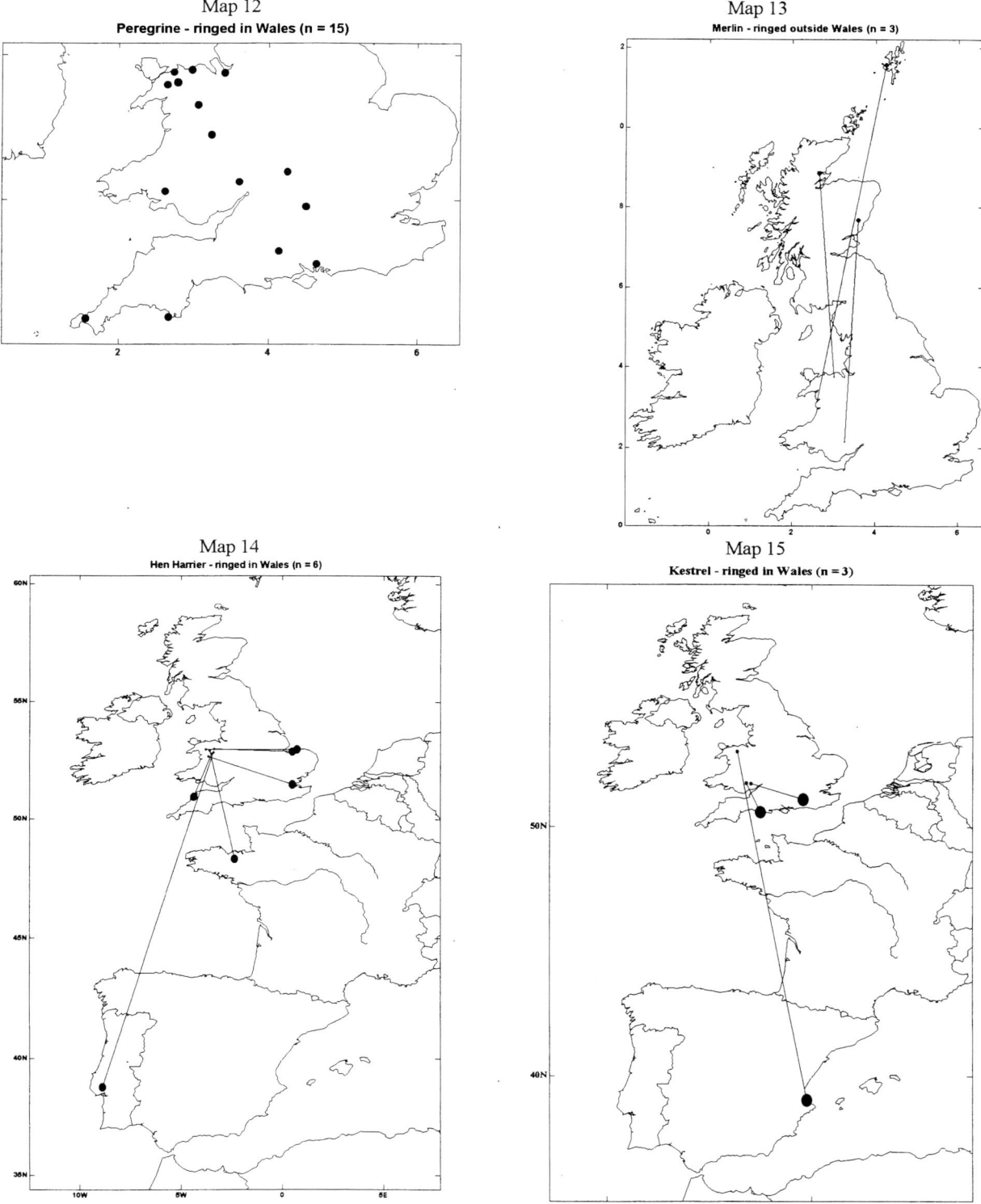

Map 12 Peregrine - ringed in Wales (n = 15)

Map 13 Merlin - ringed outside Wales (n = 3)

Map 14 Hen Harrier - ringed in Wales (n = 6)

Map 15 Kestrel - ringed in Wales (n = 3)

Birds in Wales 1992-2000

Apart from the Common Buzzard (which is even more sedentary than the Red Kite) most of the other birds of prey breeding in Wales show some degree of wanderlust, particularly in their first year. This ranges from a generalised juvenile dispersion as shown by Merlin (map 11) and Peregrine (map 12) - it is interesting to see that winter Merlins in Wales are perhaps just as likely to be from more northerly breeding areas (map 13) - to rather longer movements as seen in Hen Harrier (map 14) and Kestrel (maps 15 &16), the recovery of a Finnish-ringed Kestrel being particularly impressive.

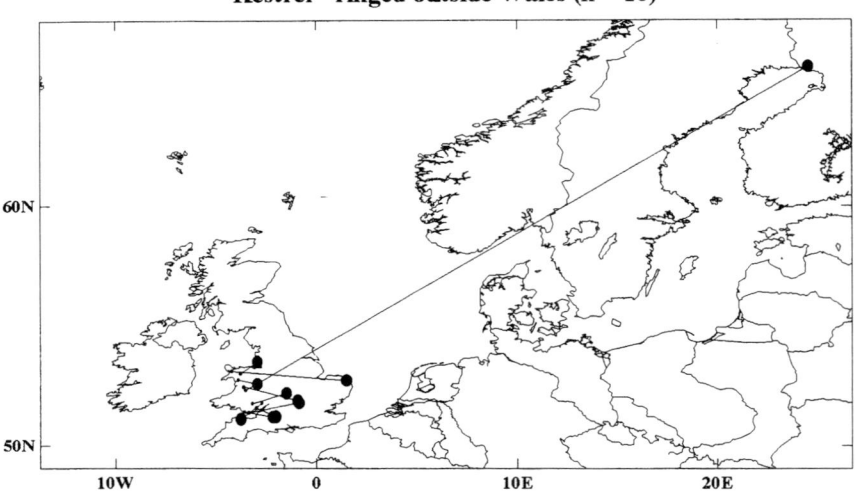

Map 16
Kestrel - ringed outside Wales (n = 10)

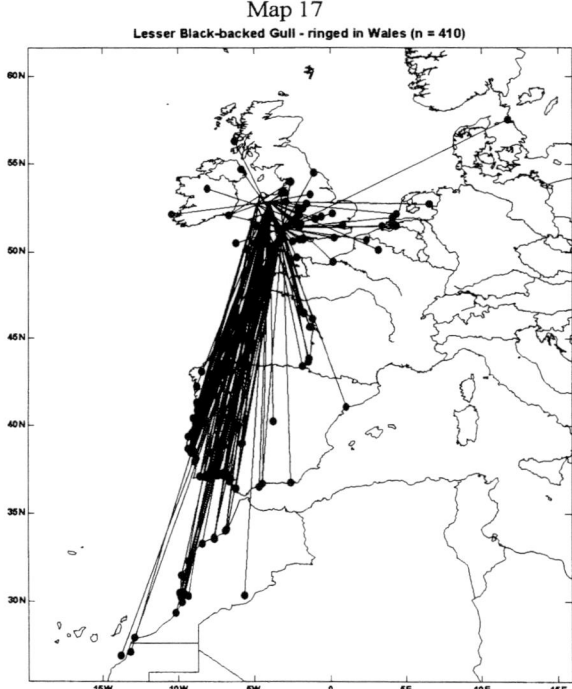

Map 17
Lesser Black-backed Gull - ringed in Wales (n = 410)

Although seabirds are not ringed in huge numbers in Wales the ones that are generate large numbers of recoveries. Two Welsh islands in particular are targeted for ringing nestling seabirds, Bardsey (including the small nearby Gwylans) off the tip of the Lleyn peninsula and Puffin Island off Anglesey. Unfortunately the large colonies on Skomer see only a limited amount of project-based ringing and those on Skokholm are not ringed at all. Several other gull colonies and isolated colonies of birds like Common Terns are also visited and in recent years many Lesser Black-backed Gulls have been fitted with plastic 'Darvic' rings as well as the metal BTO rings. When this is combined with people travelling down through Portugal and Morocco searching for birds the number of recoveries rockets.

Map 17 shows the recoveries of Welsh-ringed Lesser Black-backed Gulls, leaving little doubt as to the main wintering areas. It is quite interesting that the behaviour of Lesser Black-backed Gulls appears to be changing. In the 1960s and 1970s it appeared that nearly all young Lessers left Britain and wintered off Portugal or Morocco and returned in their second year. Nowadays more seem to be staying in Britain all winter and those that do go south do not go quite as far as previously, although as the map shows a great many still do go as far south as Morocco. The increase in sightings because of the Darvic rings gives us much more information and it will be interesting to see if this trend continues.

Birds in Wales 1992-2000

Map 18
Herring Gull - ringed in Wales (n = 25)

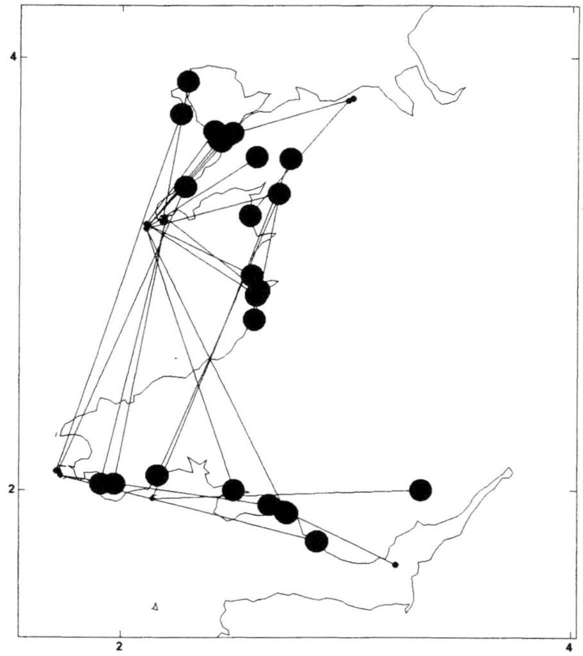

Map 19

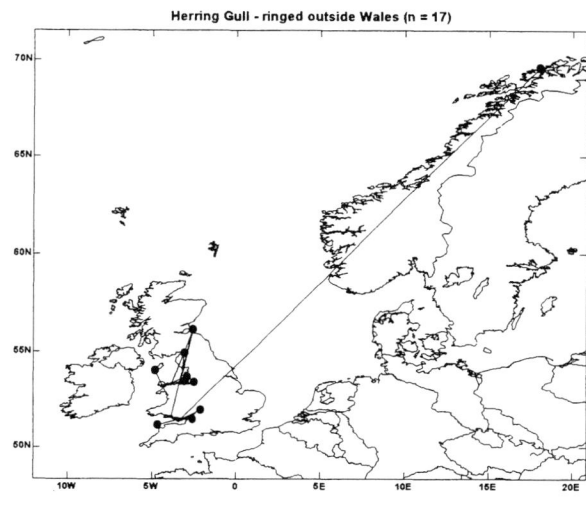
Herring Gull - ringed outside Wales (n = 17)

In stark contrast, for such closely related species, Welsh-bred Herring Gulls rarely move outside the Irish Sea and Bristol Channel area, map 18 shows the recoveries of Welsh ringed birds (nearly all will have been ringed as nestlings). The movement of a bird fledging from Ynys Gwylan Fawr in 1985 to be recovered in Iceland in 1997 is therefore quite exceptional. Given the number of gulls which come to winter in Britain it is perhaps not surprising that a Norwegian-ringed Herring Gull turned up (map 19) but more so that there have been no foreign-ringed Lessers found in Wales (map 20).

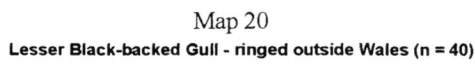
Map 20
Lesser Black-backed Gull - ringed outside Wales (n = 40)

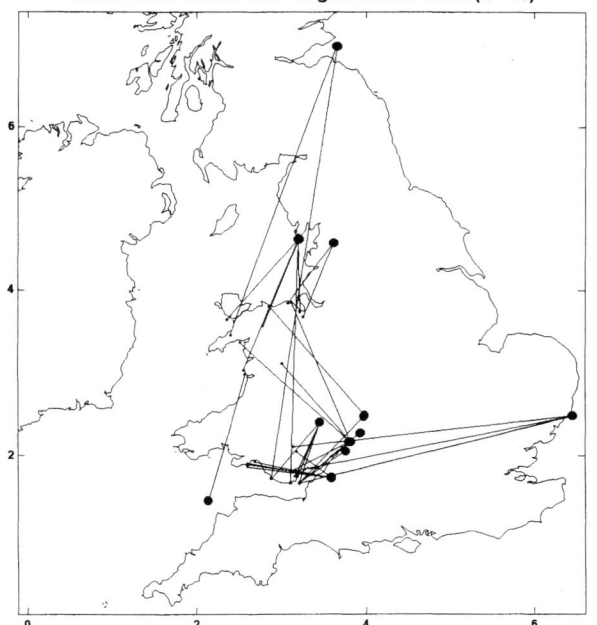

Map 21

Black-headed Gull - ringed outside Wales (n = 43)

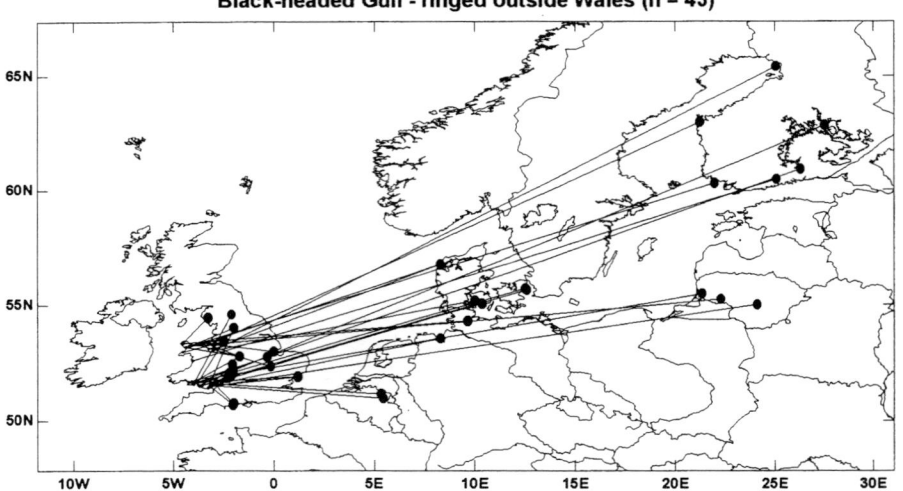

A graphic illustration of the origin of wintering Black-headed Gulls is given by the recoveries of birds ringed outside Wales shown in map 21. Juveniles from the Baltic region can be in Wales by early July, sometimes within a week or two of fledging. The map of birds ringed in Wales (22) is more ambiguous because it includes recoveries of many adult Black-headed Gulls ringed in the winter months.

Map 22

Black-headed Gull - ringed in Wales (n = 16)

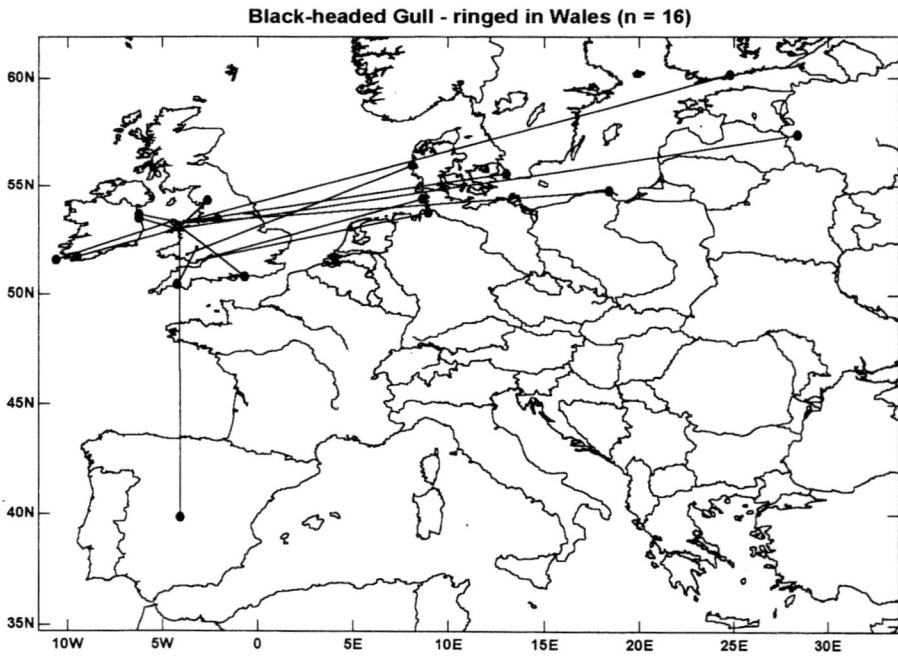

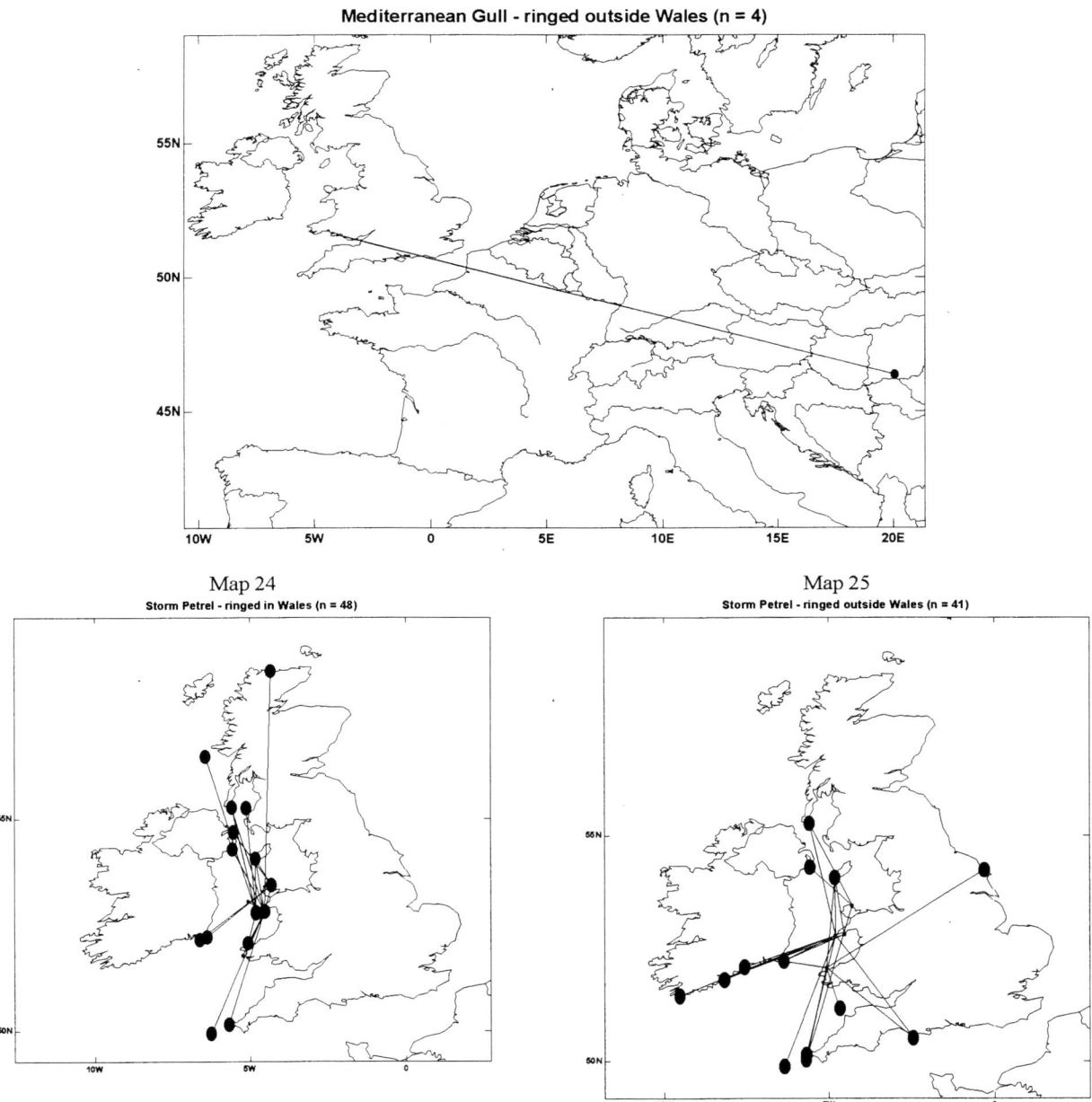

Map 23
Mediterranean Gull - ringed outside Wales (n = 4)

Map 24
Storm Petrel - ringed in Wales (n = 48)

Map 25
Storm Petrel - ringed outside Wales (n = 41)

A close relative, the Mediterranean Gull, has become an increasingly common sight in Wales as it has expanded its range. As with other gulls many are now colour-ringed as chicks and there have been four sightings of colour-ringed birds, all from the same colony (map 23). Most colour-ringed sightings in Britain relate to birds ringed in Holland, Belgium and Germany so these from one of the colonies in south-east Europe are unusual.

Three of the most migratory seabirds to occur in the British Isles are ringed in moderate numbers in Wales. One, the Storm Petrel, generates huge numbers of recoveries because ringers operate at various headlands around the coastline of Britain and Ireland playing tapes of their calls which attracts them to a catching area. Maps 24 & 25 show the recoveries of birds ringed in Wales and those ringed elsewhere and captured in Wales. Amazingly for such a small bird it is capable of travelling well over 200km in a 24 hour period.

Birds in Wales 1992-2000

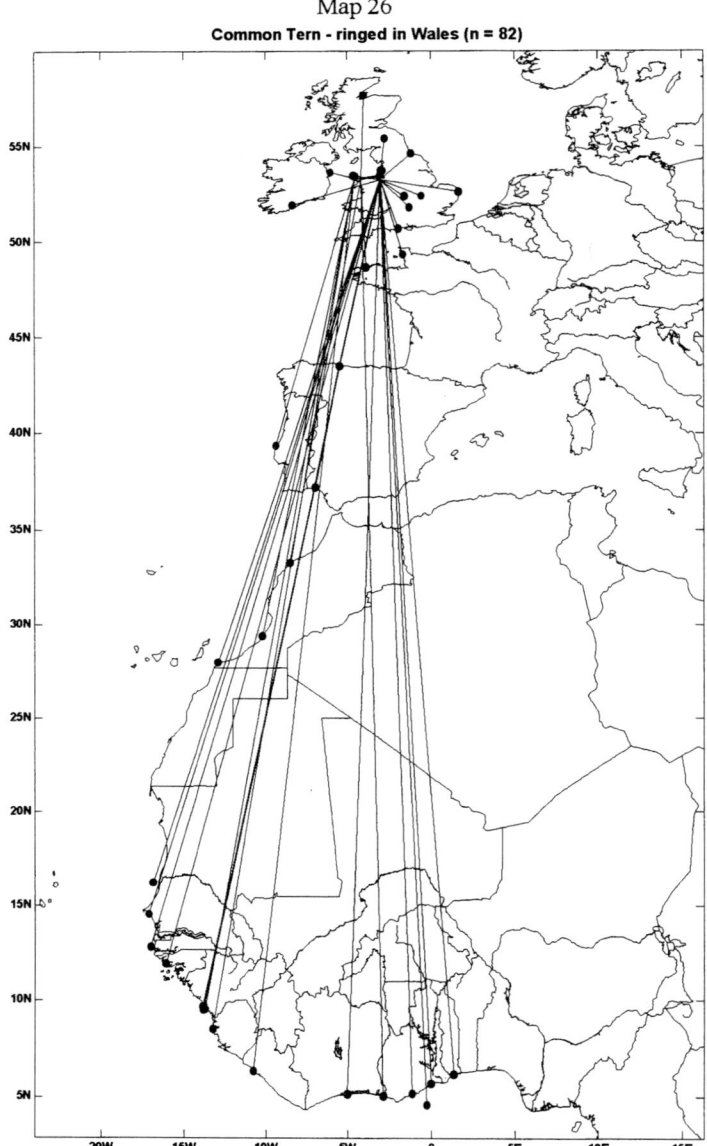

Map 26
Common Tern - ringed in Wales (n = 82)

The wintering grounds of another long-distance traveller are clearly indicated in map 26 which shows recoveries for Common Terns ringed in Wales.

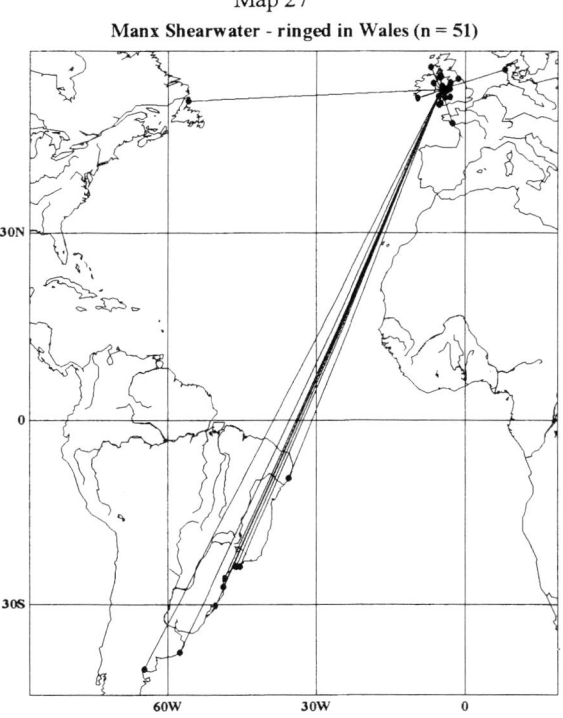

Map 27
Manx Shearwater - ringed in Wales (n = 51)

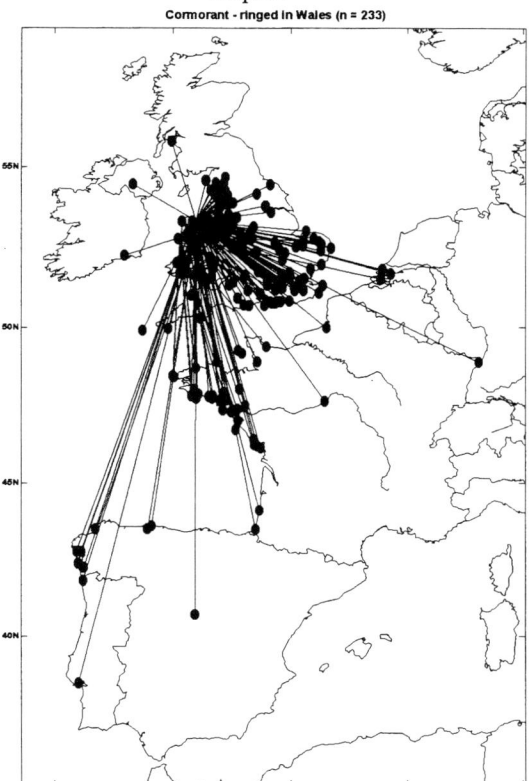

Map 28
Cormorant - ringed in Wales (n = 233)

Despite being home to over 50% of the world's population of Manx Shearwater only around 1-2,000 are ringed each year and map 27 shows the spectacular distances they travel to get to their winter quarters off the coast of south America. There have also been a couple of recoveries in North America where they were recently proved breeding for the first time. Closer to home there are frequent recoveries of storm-blown 1st year birds at inland sites dotted around the British Isles.

Like Lesser Black-backed Gulls, Cormorants also appear to have altered their dispersal. In the early 1990s a great many of the chicks ringed on Puffin Island moved down to eastern England, particularly the reservoirs in Essex. In the last couple of years this has not been as clear cut and it seems that birds are dispersing in a far more random manner. Surprisingly, given this, the number of recoveries to the Biscay coast have not really changed. Map 28 shows all recoveries of Welsh ringed birds showing the cluster in Essex and another up into Cumbria with far flung birds into Portugal, a lake near Madrid and central Europe.

Birds in Wales 1992-2000

As a general rule the smaller and more numerous a bird is the lower the chances of a ringed bird being recovered, as was demonstrated with the tits. There are, fortunately, one or two exceptions to this rule and this happens where a species uses a habitat that is either uncommon or of conservation concern. This can be seen with the Sedge and Reed Warbler (maps 29-32), both of which occur in reedbeds, a habitat of conservation concern. Reedbeds also act as magnets for birds and keep them in a relatively confined area and hence many ringers target them as ringing sites. The National Museums & Galleries of Wales ran a project at Kenfig National Nature Reserve for five years during the 1990s and because of the interest in reedbeds nation-wide and internationally (ringers were operating a site in Senegal over the same period) a relatively large number of recoveries were generated. The recovery rate was about 1 in 200 or so for the two species, much higher than for many other passerines.

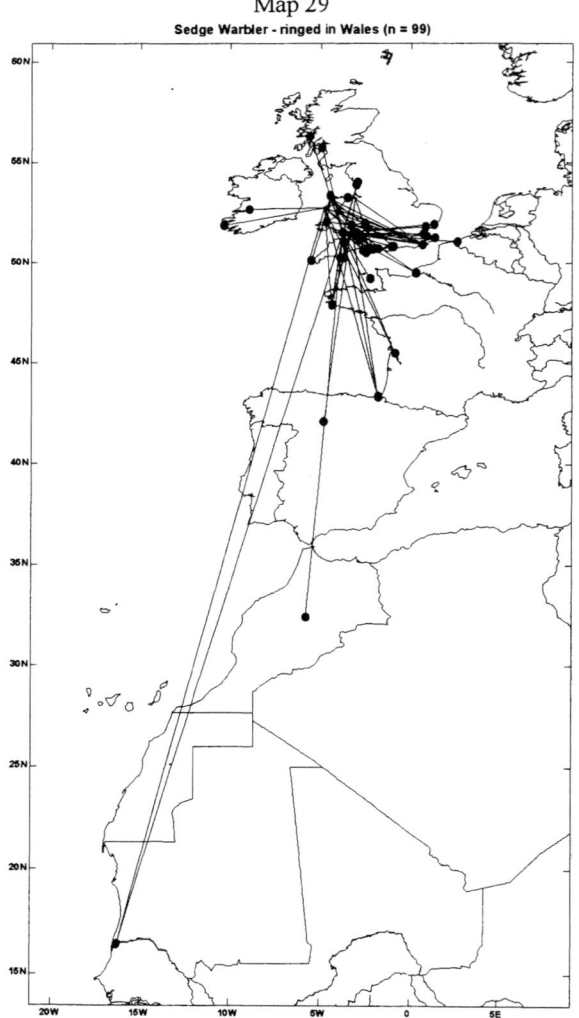

Map 29
Sedge Warbler - ringed in Wales (n = 99)

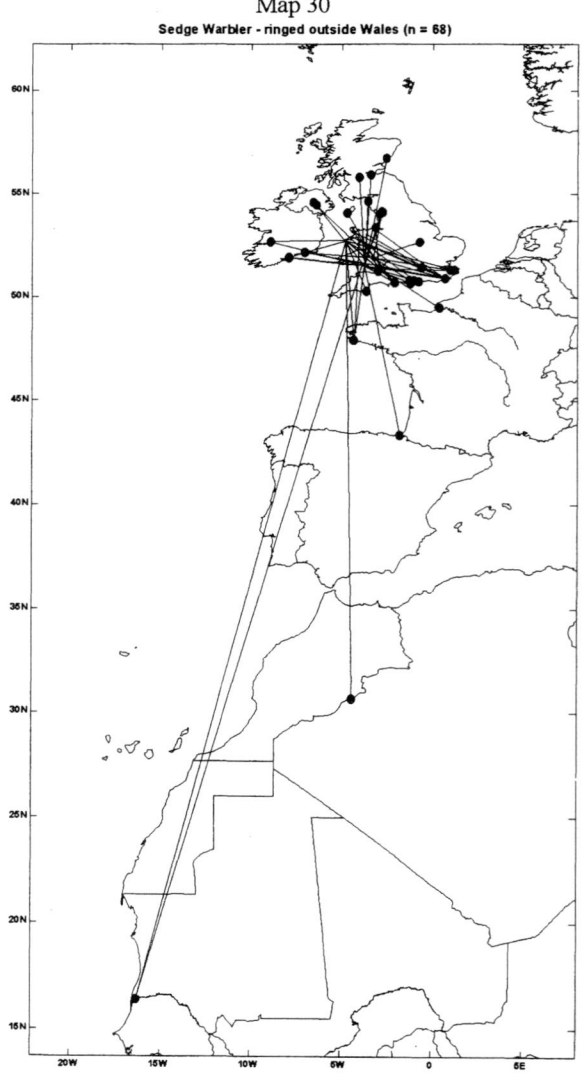

Map 30
Sedge Warbler - ringed outside Wales (n = 68)

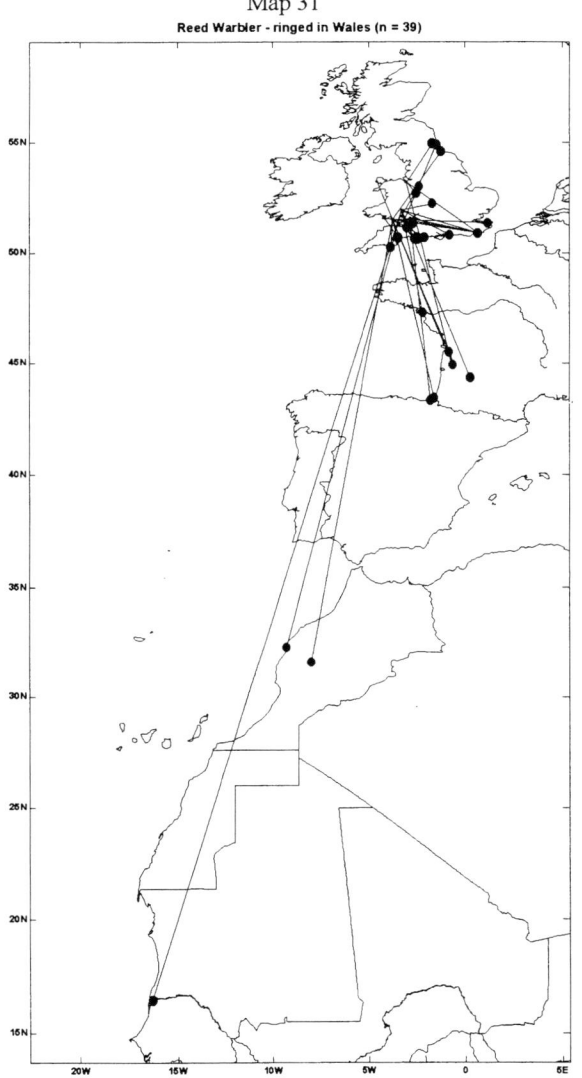

Map 31
Reed Warbler - ringed in Wales (n = 39)

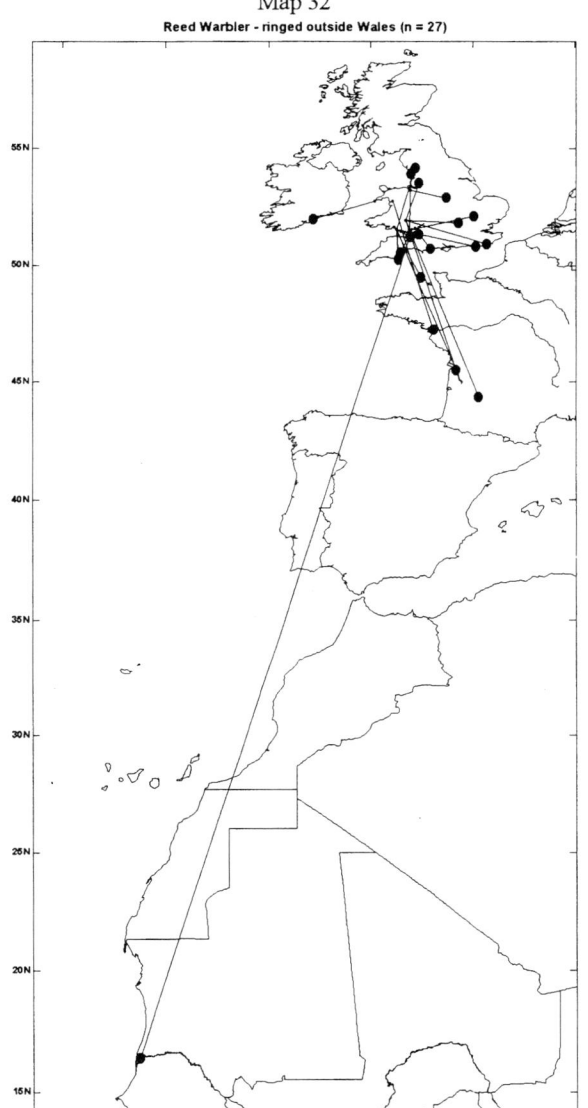

Map 32
Reed Warbler - ringed outside Wales (n = 27)

These recoveries showed which areas were favoured as stopping-off points on their journey south (very little ringing is done in reedbeds in spring so there are very few spring recoveries) and sometimes gave a good indication of how quickly birds could travel. Kenfig is a staging post for birds from Ireland, Scotland and northern England and north and west Wales. They then move on to southern reed beds such as Slapton in Devon and Radipole in Dorset before crossing the Channel often via the Channel Islands. Once in France they move through Brittany and down the western seaboard, in particular using the vast reedbeds on the Loire and Gironde estuaries. Unfortunately there are relatively few ringers in Spain and recoveries are scarce so we have less idea which areas are important. So the next and last stop, as they had reached their winter quarters, was Djoudj in Senegal, West Africa.

Birds in Wales 1992-2000

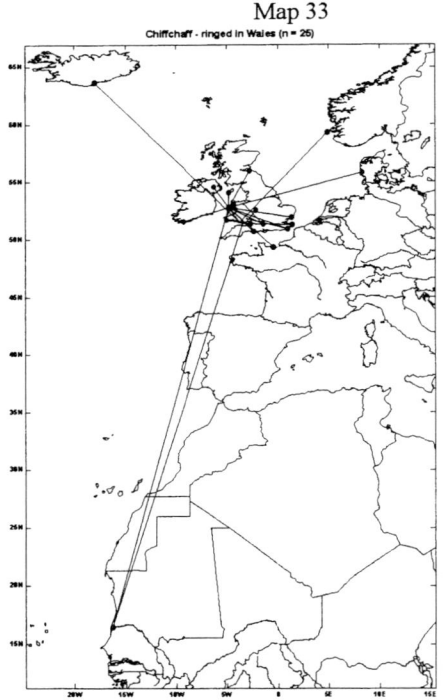

Map 33

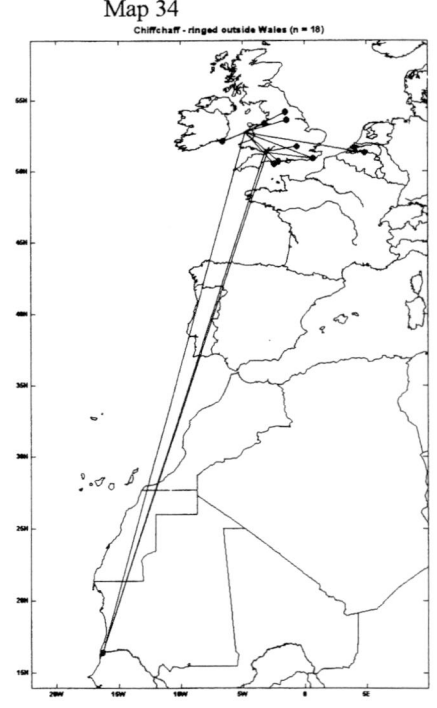

Map 34

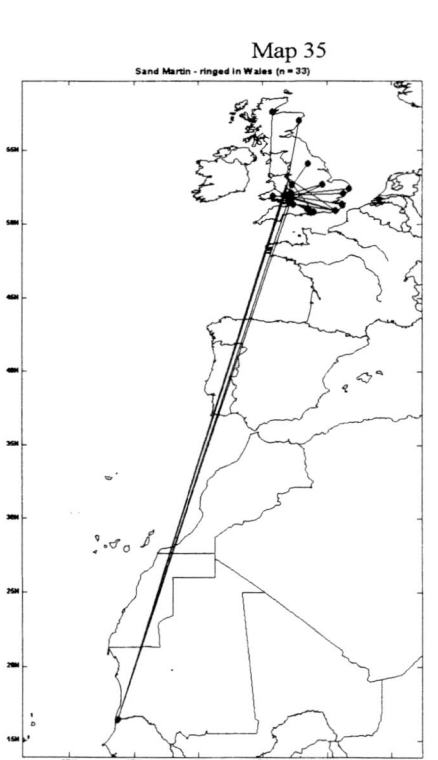

Map 35

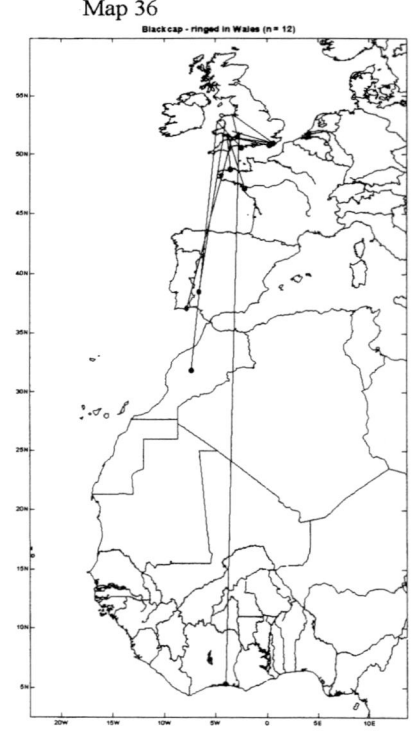

Map 36

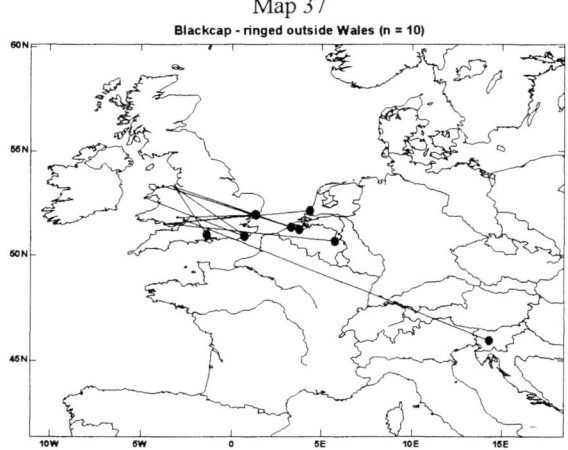

Map 37
Blackcap - ringed outside Wales (n = 10)

The project in Senegal proved that Djoudj was also the winter home of British Chiffchaffs and Sand Martins (maps 33-35). The recovery map for Chiffchaff also gives us a clue to the origin of the birds which have taken to wintering in Britain in the last 20 years or so with some birds showing a more east-west orientation. This same pattern can be seen in the map for recoveries of Blackcaps ringed outside Wales (map 36), another species which is an increasingly common sight in gardens during winter months. Welsh-bred birds show the typical north-south movement expected for a summer visitor (map 37).

Other summer migrants to the British Isles show a similar pattern of recoveries. The recoveries of that classic harbinger of summer, Barn Swallow (map 38), show just how far south they fly to their winter quarters in southern Africa. The recoveries for another common migrant, Willow Warbler (map 39) - one of the most numerous birds during autumn migration in north-west Europe - shows that birds passing through Wales originate mainly from more northerly populations (a pattern visible in many of the other maps). However, it also shows birds of more easterly origin come this far west on their way south.

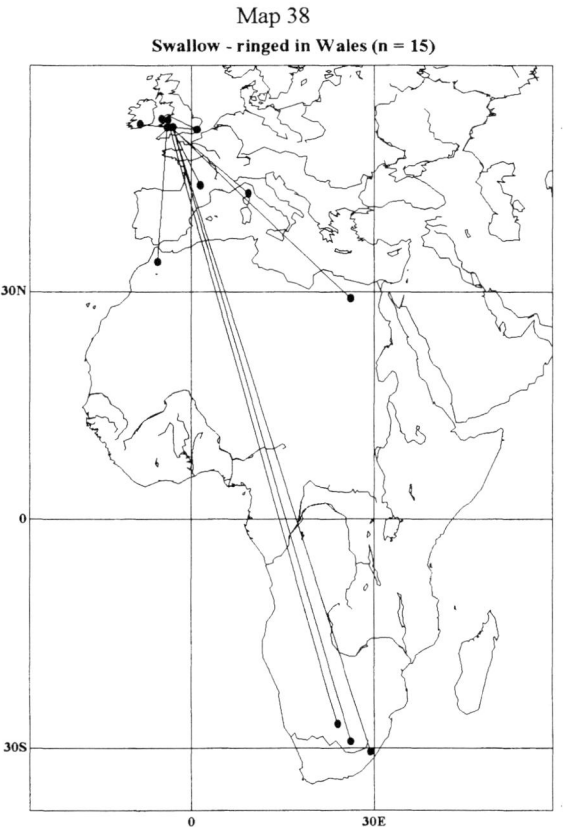

Map 38
Swallow - ringed in Wales (n = 15)

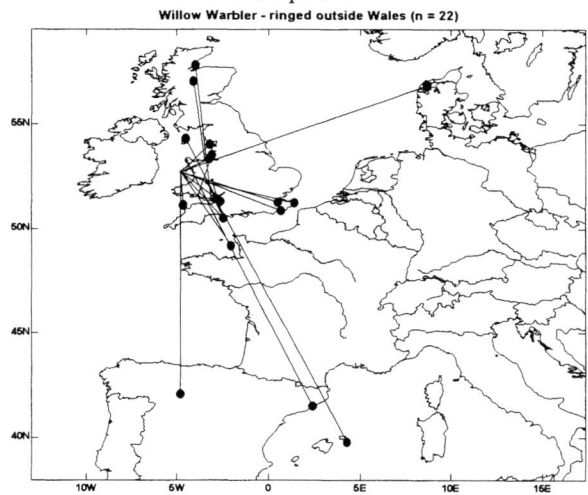

Map 39
Willow Warbler - ringed outside Wales (n = 22)

Birds in Wales 1992-2000

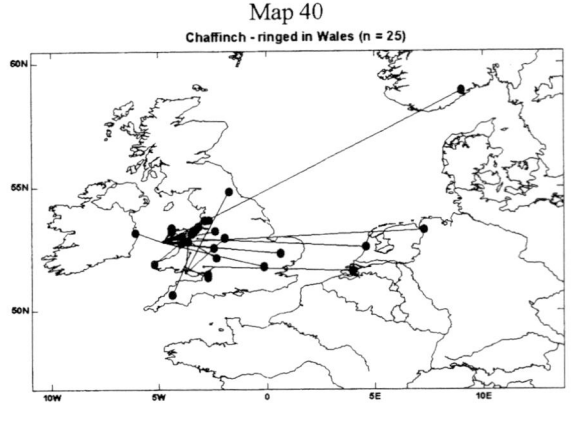

Map 40
Chaffinch - ringed in Wales (n = 25)

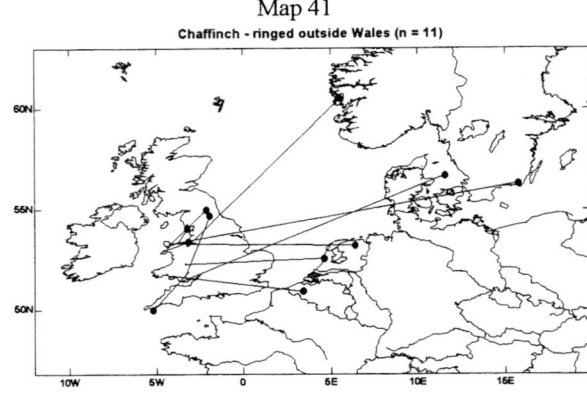

Map 41
Chaffinch - ringed outside Wales (n = 11)

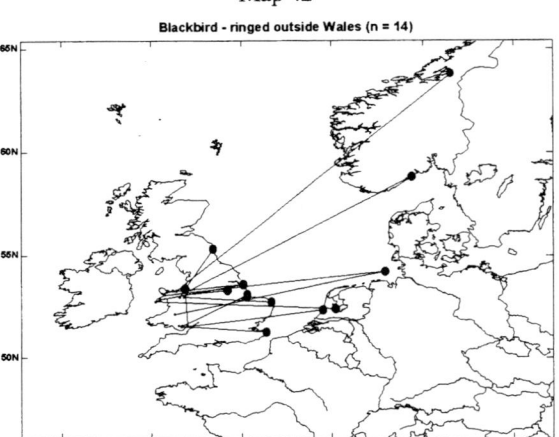

Map 42
Blackbird - ringed outside Wales (n = 14)

Our summer migrants typically move north-south but winter visiting passerines tend to have a more east-west bias. Typical of these are Chaffinch, Blackbird and Song Thrush, recoveries for which indicate Scandinavian and low countries origins (maps 40-44). Recoveries for Redwing ringed outside Wales also shows a similar pattern (map 45), however, the map for Redwing ringed in Wales is rather different (map 46). These recoveries are not from the same winter and indicate birds having wintered in Wales one year then moving much further south into France or even Portugal in subsequent winters.

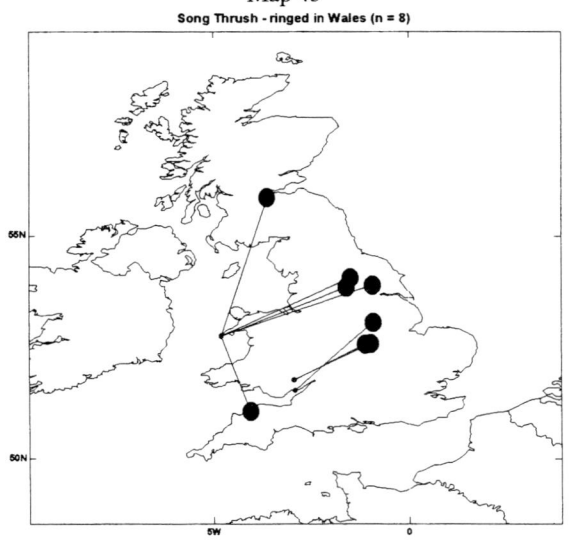

Map 43
Song Thrush - ringed in Wales (n = 8)

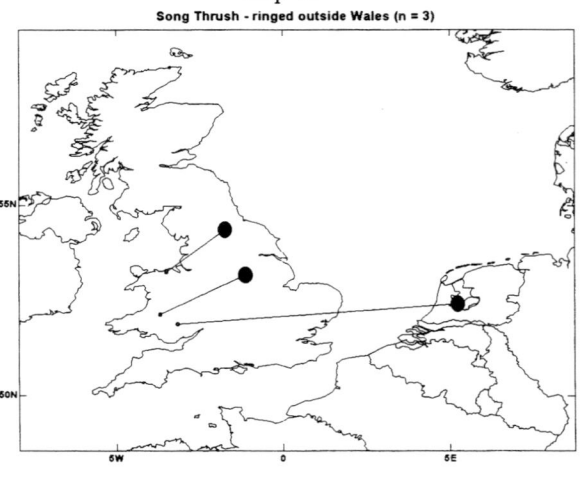

Map 44
Song Thrush - ringed outside Wales (n = 3)

Birds in Wales 1992-2000

Map 45
Redwing - ringed outside Wales (n = 4)

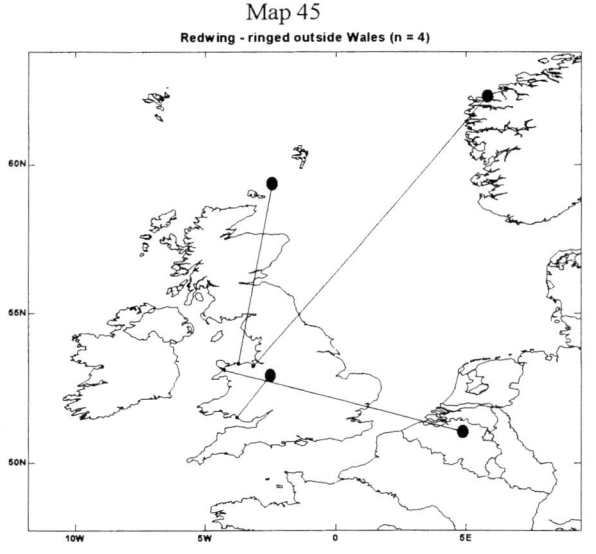

Map 46
Redwing - ringed in Wales (n = 8)

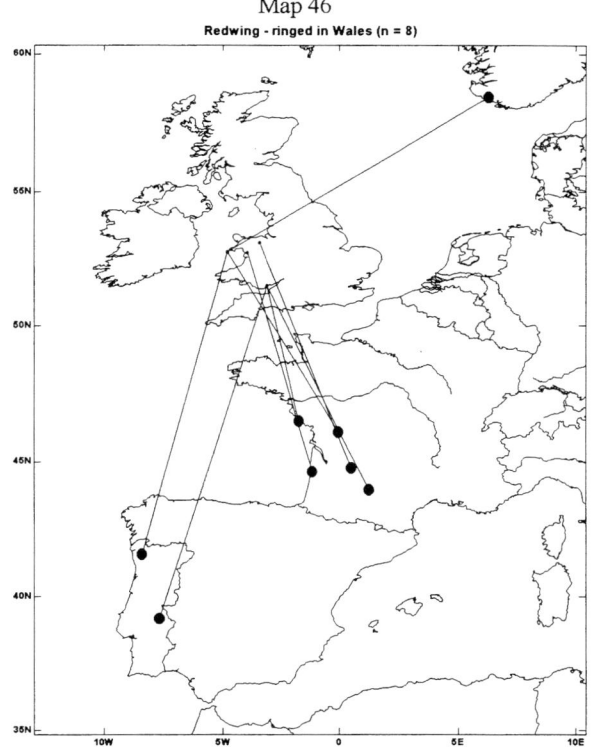

Map 47
Greenfinch - ringed in Wales (n = 22)

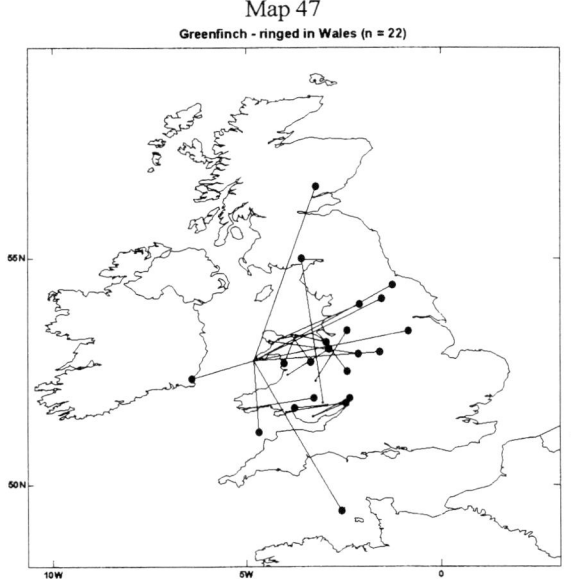

Map 48
Greenfinch - ringed outside Wales (n = 14)

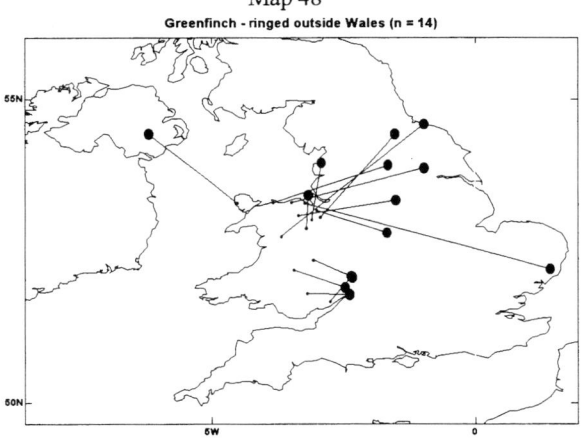

Birds in Wales 1992-2000

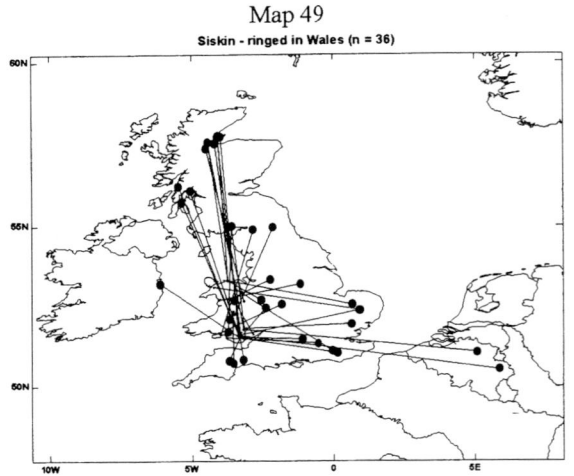

Map 49
Siskin - ringed in Wales (n = 36)

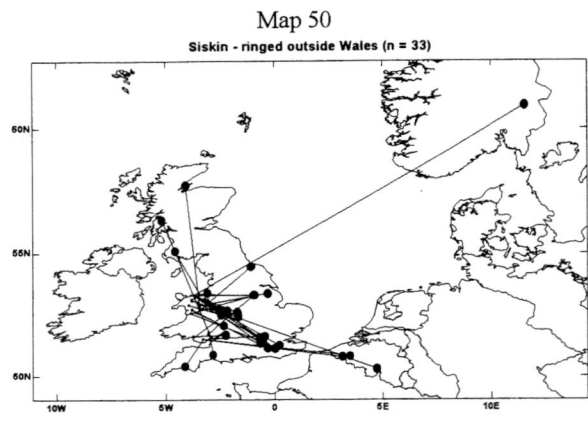

Map 50
Siskin - ringed outside Wales (n = 33)

The maps for Greenfinch and Siskin (maps 47-50) are much less clear-cut as there is a huge amount of fairly random movement within the UK by British breeding birds, as well as immigration of continental birds in the winter months.

There have not really been enough recoveries of Goldcrest (maps 51-52) to give much indication of movements but they suggest that birds breeding in Scotland move south into southern areas and that continental birds also move west into Britain.

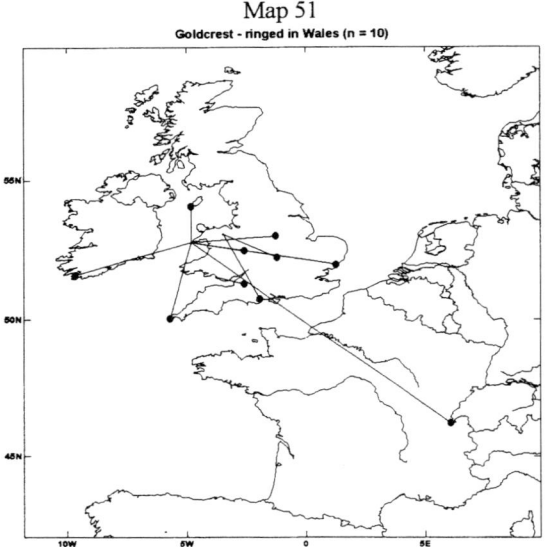

Map 51
Goldcrest - ringed in Wales (n = 10)

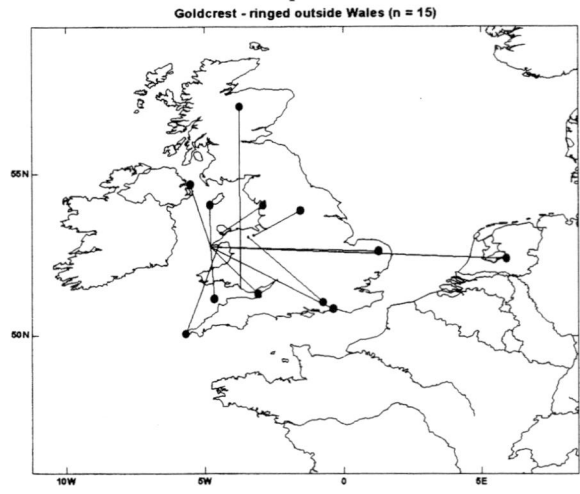

Map 52
Goldcrest - ringed outside Wales (n = 15)

Map 53
Starling - ringed in Wales (n = 12)

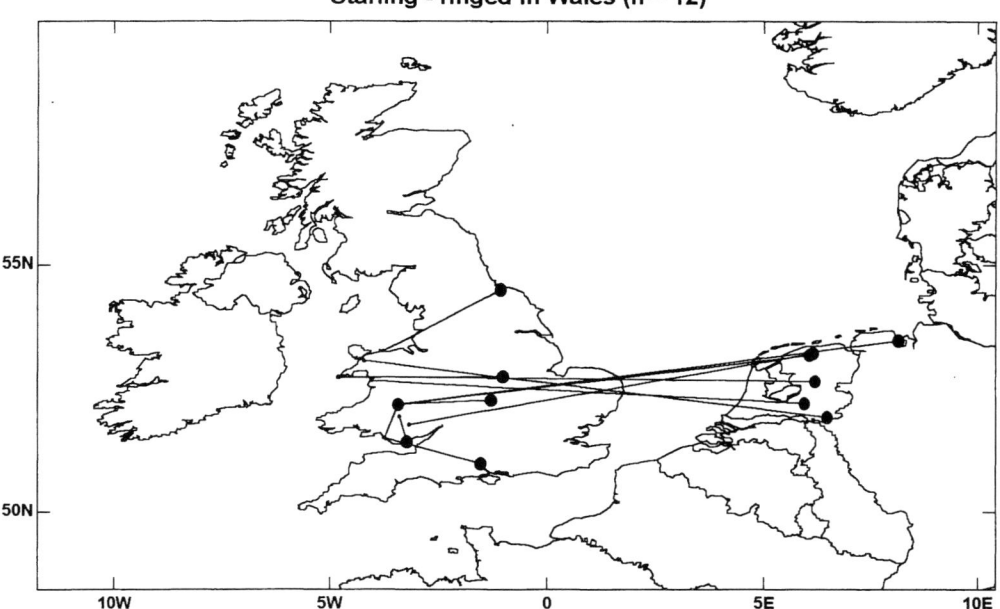

There is one passerine for which many recoveries are received yet relatively few are ringed in Wales, despite their winter roosts being large and often easy to net. Starlings are an important prey item for birds like the Peregrine and Sparrowhawk and if plucking posts are checked regularly an impressive number of rings can be found. The bulk of the Starlings seen here in winter come from eastern Europe, probably as far east as central Russia (maps 53 & 54). The trapping methods employed by ringing stations on the Baltic coast ensures that many thousands are ringed which make the likelihood of them being found much higher.

Map 54

Starling - ringed outside Wales (n = 26)

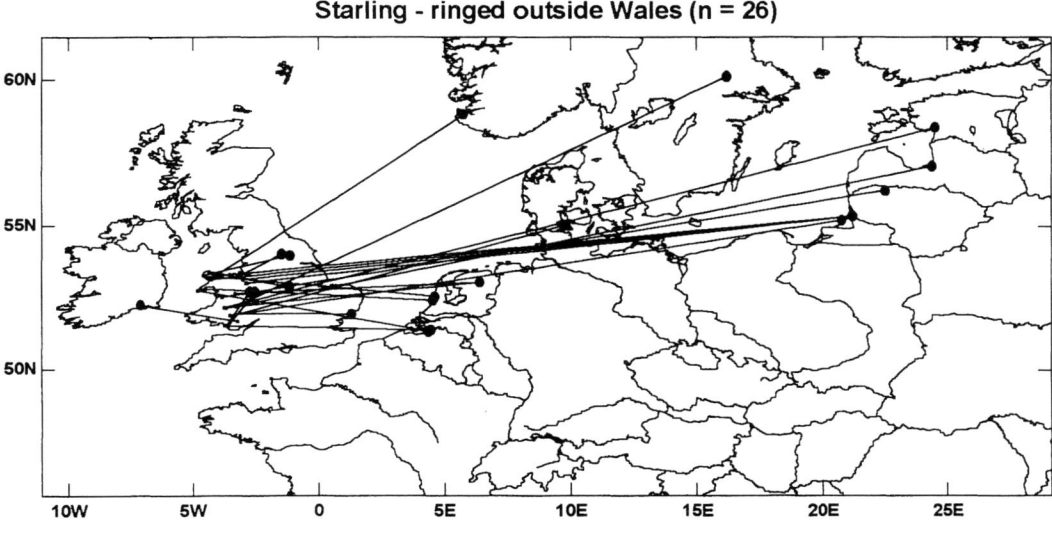

Map 55

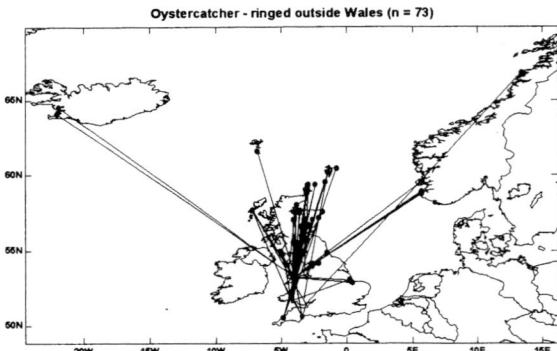

Map 56

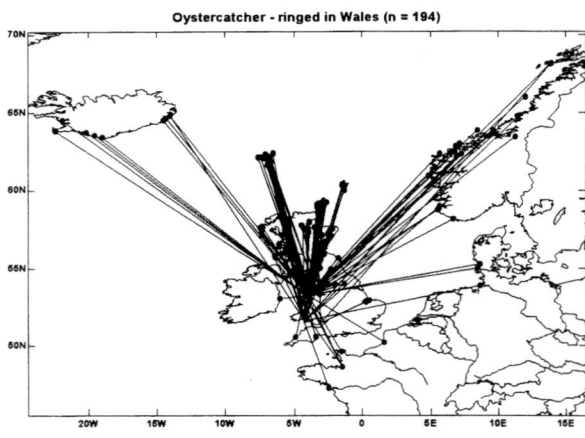

Another group of birds targeted for conservation reasons are waders. Tens of thousands are ringed around the British Isles and other parts of Europe each year, the task made much easier because that gather in large flocks and hundreds can be caught in on go. This concentrated activity means that there are a large number of recoveries for the species most commonly caught in Wales, which are Oystercatcher, Dunlin and Redshank.

Oystercatchers are also commonly ringed as chicks which means we know exactly where they come from if they are recaught. Ringing recoveries show us that the majority of our wintering Oystercatchers come from Scotland, the Faeroe Islands with a fewer from Iceland and Norway (map 55). Correspondingly our first-year birds move south into south-west Britain, France and Portugal, with adult birds less likely to move far from their breeding areas (map 56). Wintering Redshanks are also from the north with a large number of Icelandic birds coming into Wales as well as those from northern England and Scotland (maps 57 & 58). Unfortunately there are too few Redshank chicks ringed to give any information on their movements.

Map 57

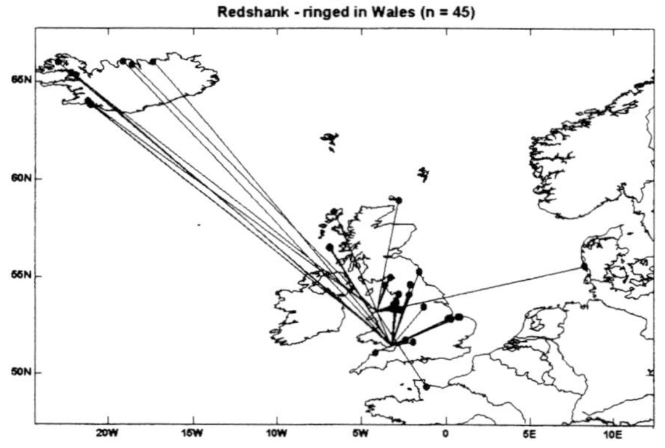

Map 58

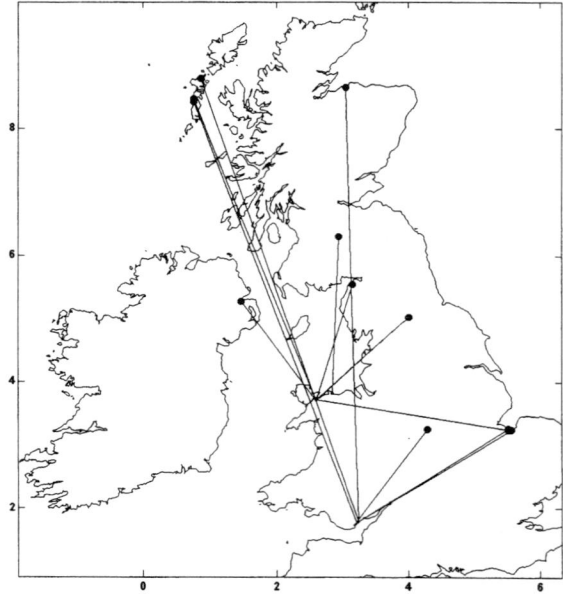

Map 59
Dunlin - ringed in Wales (n = 67)

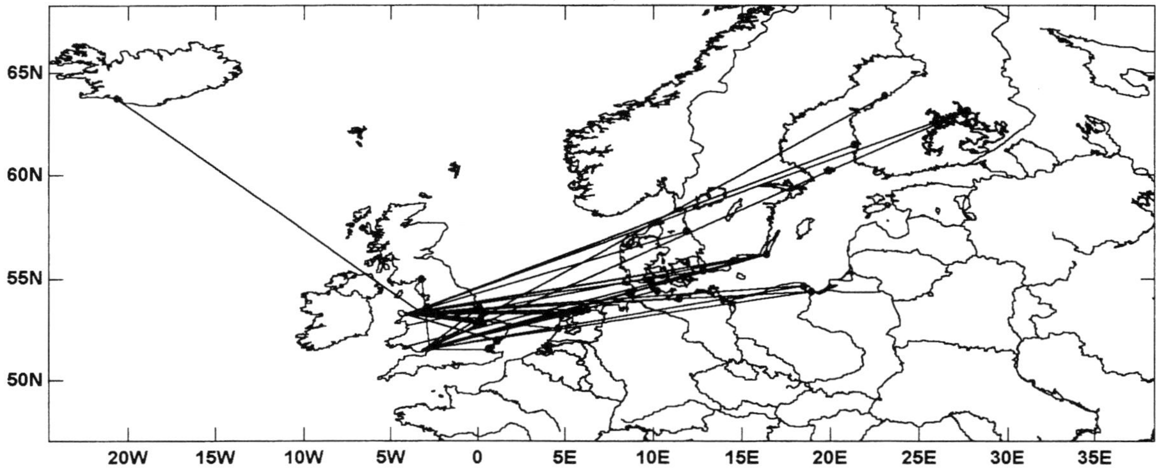

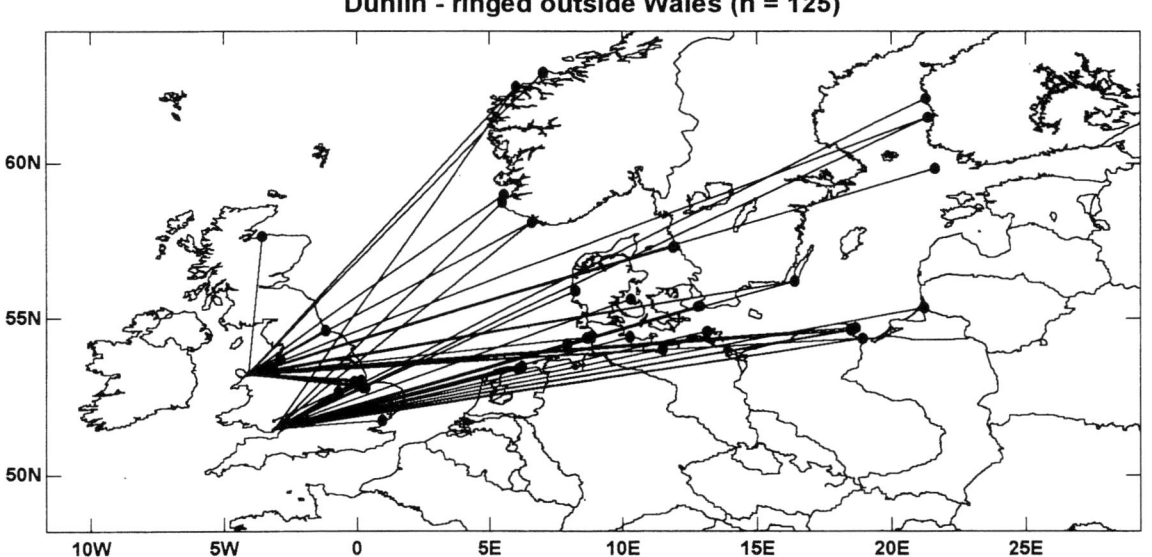

The bulk of wintering Dunlin come from much further afield. Very few are ringed as chicks so exact breeding areas are hard to pin down but they are caught on migration through the Baltic and along the North Sea coast from Denmark to Holland which suggests they breed in Siberia or northern Finland (maps 59 & 60).

Map 60

Dunlin - ringed outside Wales (n = 125)

Birds in Wales 1992-2000

Map 61
Ringed Plover - ringed in Wales (n = 5)

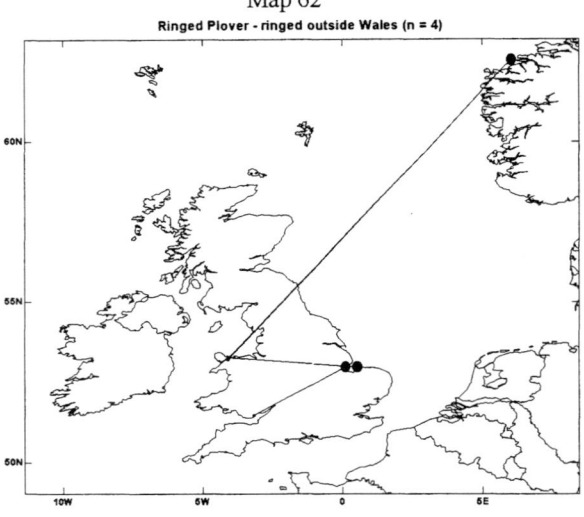

Map 62
Ringed Plover - ringed outside Wales (n = 4)

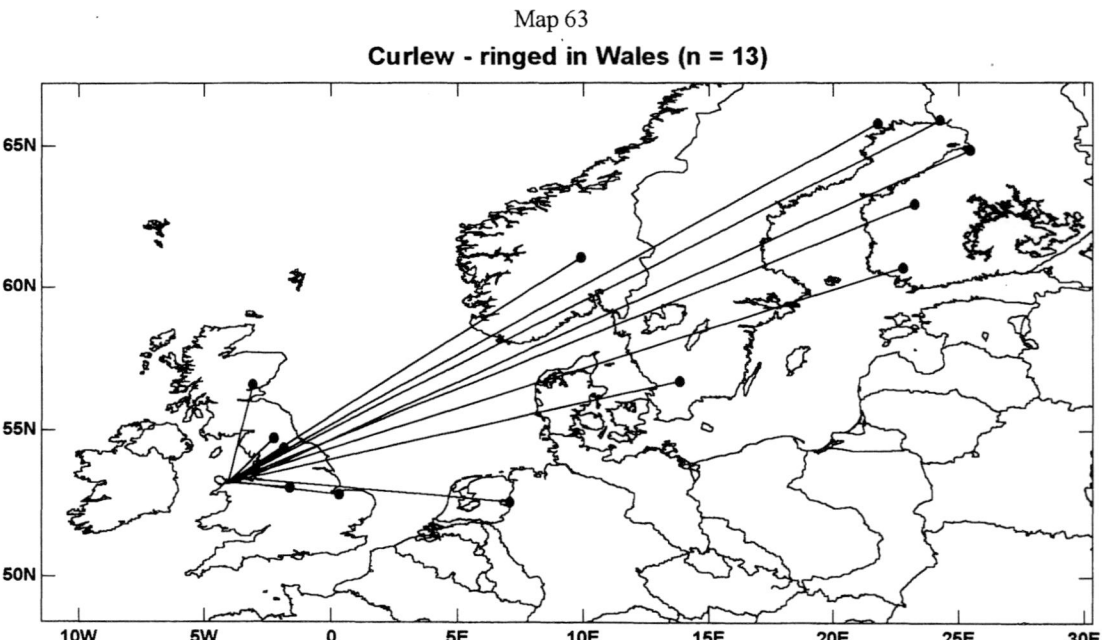

Map 63
Curlew - ringed in Wales (n = 13)

Although far less numerous, and therefore less frequently ringed, our wintering Ringed Plovers show a similar pattern (maps 61 & 62). Unfortunately few Welsh breeders are ringed so there is little information on the movements of locally bred birds.

For a common (although declining) breeding bird it is perhaps surprising to find that many of the Curlew wintering in Wales come from further east and north than almost any other wader. Maps 63 and 64 show that their breeding grounds are in the far north of Finland and east into Russia. They are also long-lived, many of the recoveries from Finland relate to birds around 17-20 years old.

Map 64

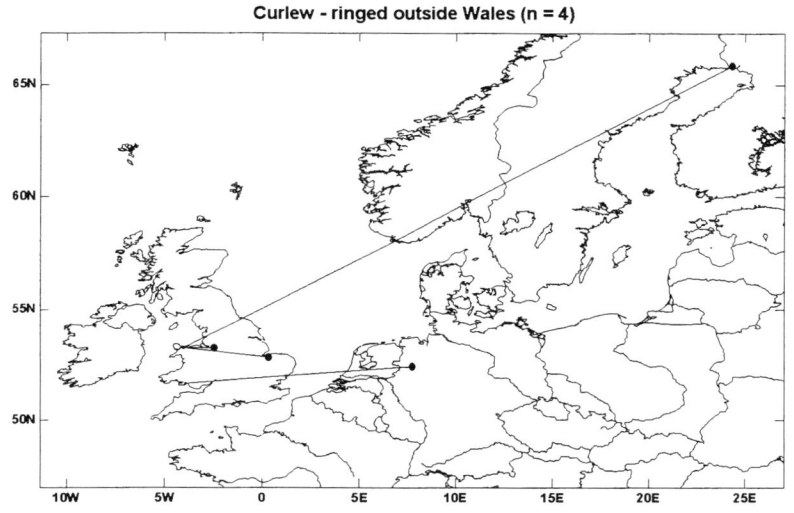

Map 65

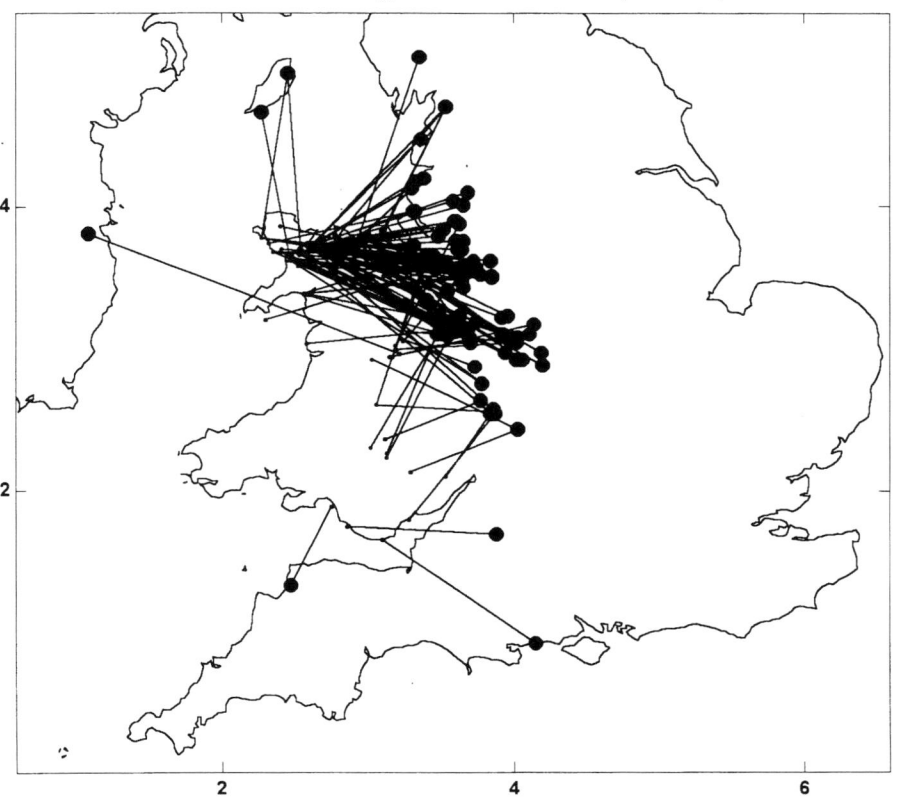

Birds in Wales 1992-2000

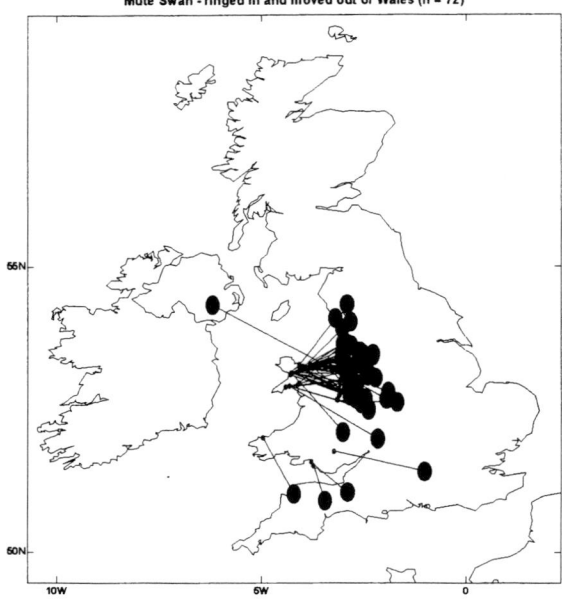

Map 66
Mute Swan - ringed in and moved out of Wales (n = 72)

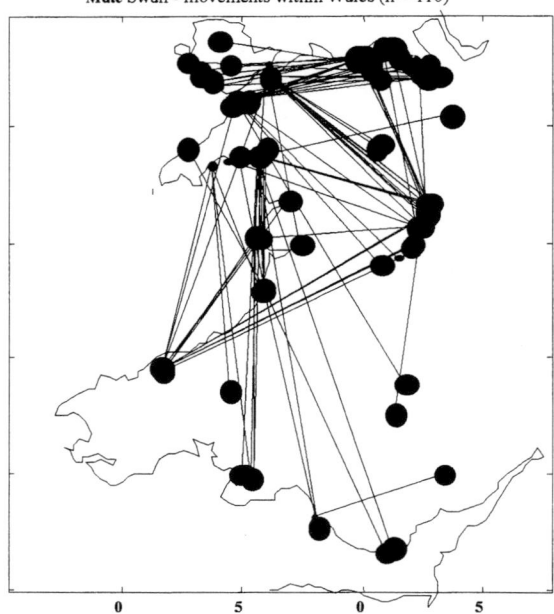

Map 67
Mute Swan - movements within Wales (n = 110)

A handful of ringers put quite some effort into ringing Mute Swans, not surprisingly given their size and behaviour, they have the highest recovery ratio of any British bird. The north Wales coast and areas around Cardigan see a build up of swans during the winter. Ringing, and in particular the use of Darvic rings, has shown that birds in north Wales birds generally move between there and Merseyside, Shropshire and Herefordshire (map 65 & 66). Rather surprisingly a concentrated effort on Mute Swans in Glamorgan for a few years in the 1990s resulted in few if any recoveries outside of the county, which would suggest that all the wintering birds are local to the county. This is in contrast to Mute Swans elsewhere in Wales which seem to be rather more mobile (map 67).

For most species only a handful of recoveries come in each year - indeed for some thousands can be ringed each year and there may only be one recovered every few years - making it difficult to draw any conclusions on movements and breeding or wintering areas. However, these recoveries do shed some light on the amazing feats birds can perform, often over many years and can sometimes be proof that some species can be very long lived.

One such example is a Reed Warbler at Kenfig that was trapped just about every summer for six years, it was an adult when it was ringed which meant it was at least seven or eight years old - a record for a British Reed Warbler. What is perhaps more awe-inspiring is that a bird weighing 12g flew over 6,000km a year between its breeding site at Kenfig and its wintering site in Guinea Bissau, West Africa because this bird was also retrapped twice there as well as being caught at Kenfig.

Ringing has also shown the Storm Petrel to be an amazing bird. It is little bigger than a Barn Swallow but spends most of its life at sea migrating between north-west European breeding grounds and wintering areas off southern Africa - and manages to do this for in excess of 25 years. Other seabirds are also long-lived. A Bardsey Manx Shearwater set a new longevity record for the species - just over 36 years. In these exceptional cases the age is more likely to be limited by when ringing started than by the age of the bird.

These may seem irrelevant facts but the age to which birds live needs to be taken into account if conservation measures are required. Long-lived birds are likely to reproduce less frequently than short-lived birds making small populations less viable and more susceptible to adverse environmental impacts. Ringing is an important tool used by British conservation bodies in their efforts to monitor and conserve our birds, this may be action at home or it may require action elsewhere to conserve an important habitat or reduce other pressures. Although this is frequently achieved through the activities of ringers and scientists the contribution made by members of the public reporting sightings from the field or dead birds found carrying rings is cannot be emphasised enough.

SPECIES ACCOUNTS

Treatment of Records

It is the policy of the Society **NOT TO PUBLISH** in it's Reports or in this book any records of species regarded as British rarities on the list of the British Birds Rarities Committee (BBRC) unless that record has been considered and accepted by BBRC., or likewise those on the Welsh Records Panel List.

For wildfowl and waders monthly maximum counts for the main Welsh sites are tabulated as available. The lack of count figures in tables does not mean that there was no count just that the data was not available at the time of processing. Some overlap may occur between sites, because birds may shift between count areas during the month, and over the winter, with movement between sites. E.g. The number of Scaup at Peterstone Wentloog and the Rhymny estuary, which move up and down the Severn between the two sites.. Nevertheless they show clearly the comparative importance of sites, the scale of populations involved and the seasonal pattern of occurrence. Levels of Importance, either National or International are also included, using the percentage thresholds outlined by BTO / WWT / RSPB / JNCC in The Wetland Birds Survey 1999-2000 – Wildfowl and Wader Counts (Musgrove, A.J. et al 2001).

The arrangement of bird recording areas in Wales follows the Watsonian Vice-Counties as outline in the Chapter Birding in Wales 1992-2000. Records from Vice-county 41 are given as Glamorgan, unless the count only refers to one of Eastern Glamorgan or Gower or where there is a benefit from splitting the totals.

Birds in Wales 1992-2000

For each species there is a code and species comment. The coding system is:

A   A new species to Wales.
B   Rare or scarce in Wales, all or a summary of records are quoted.
C   No change in status but details of important counts and or breeding records are quoted.
D   No change in status and no new useful data from the period.
E   An apparent change in status and a new species comment.

While the species comment is that quoted in *Birds in Wales* unless there has been a change in status during the period under review. For example:

Red-throated Diver: *(C) Winter visitor and passage migrant, regularly recorded off-shore, sometimes in large concentrations. Rare inland.*

Stock Dove: *(E) Breeding resident occurring widely but not numerously up to approximately 305m; most common in lowland areas. Appears to be getting scarcer across most of Wales, probably as a result of a change in farming practices and a decline in arable land.*

Charts and Maps

Using the information in Welsh Bird Reports, County Bird Reports and specific requests from recorders and experts it has been possible to present monthly occurrence charts for the period 1992-2000 for many species.

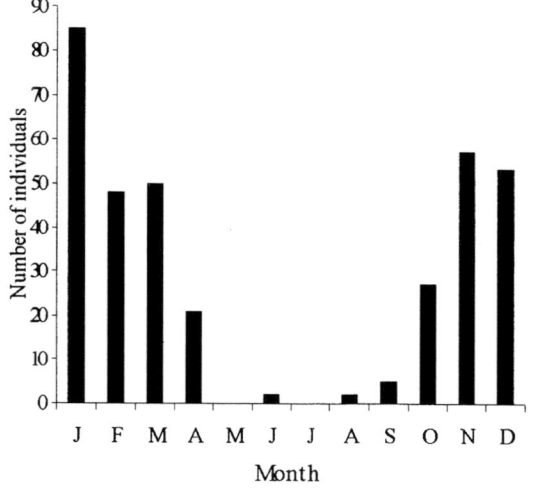

The bars represent the approximate totals of individuals for that species in that month.

e.g. Black-throated Diver occurrence by month, during the period 1992-2000.

For the rarer species it has been possible to analyse all the occurrences in Wales by utilising the Rare Birds in Wales data-base. This data-base has been put together from Welsh Bird Reports, County Bird Reports, British Birds Rarities Reports and the original data sheets using in the production of *Birds in Wales*, which are housed in the National Library of Wales, Aberystwyth.

This has enable the construction of charts showing the monthly and even weekly distribution of all records in Wales.

e.g. the chart below, monthly distribution of all the Rose-coloured Starlings recorded in Wales.

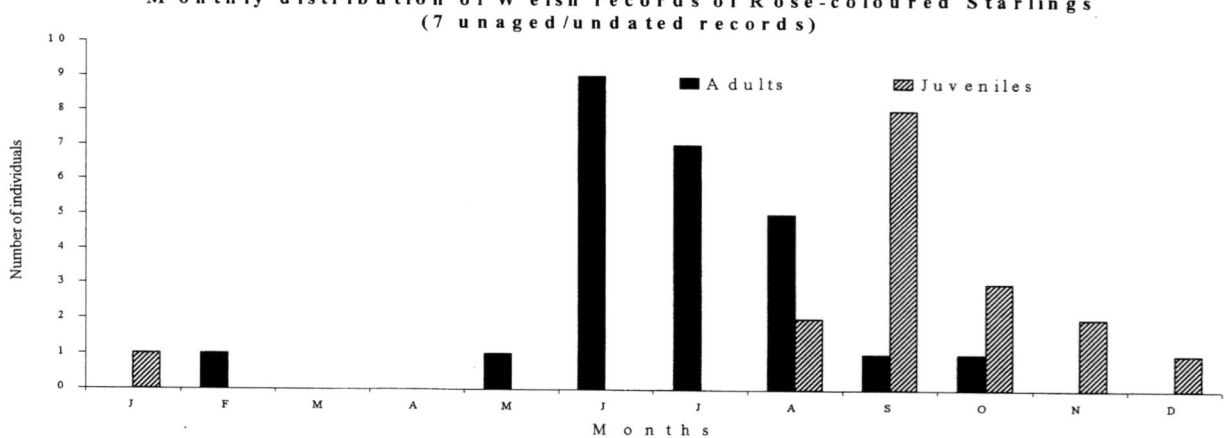

In the case of weekly graphs, e.g. the chart below, showing the weekly distribution of Yellow-browed Warblers in Wales, weeks have been taken as: Week one – days 1 to 7, Week two – days 8 to 14, Week three – days 15 to 21 and Week four – days 22 to the end of the month.

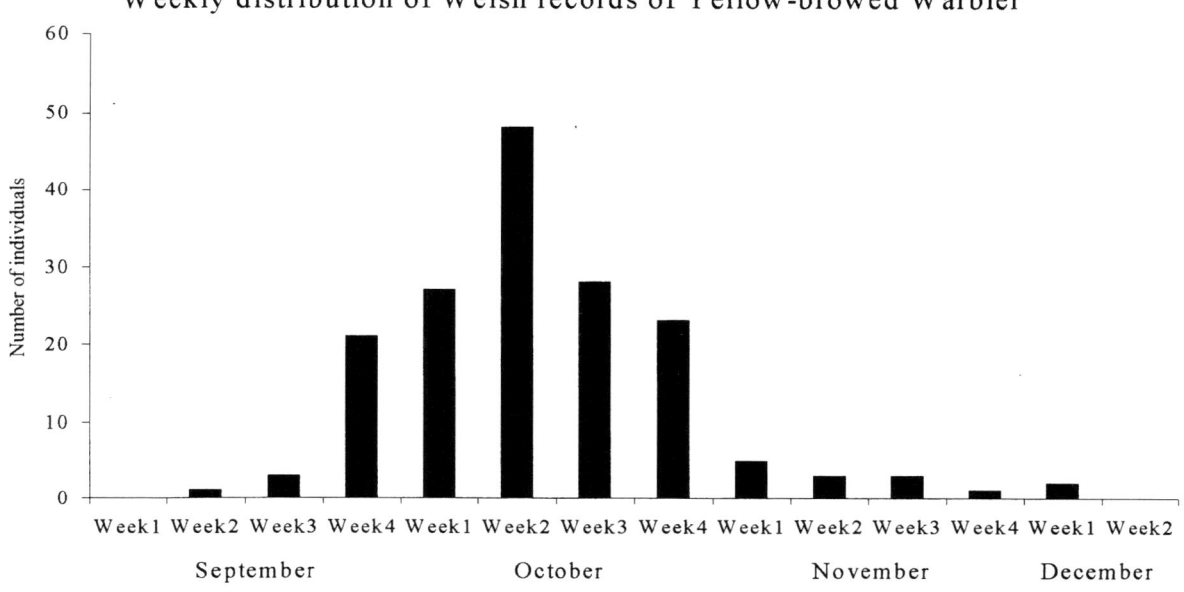

This data-base has also allowed the analysis of records by county and by specific sites. These analyses have been presented as numbers super-imposed over the relevant county.

The number in each county is the total number of individuals in that county. Totals for specific sites within various counties are also given.

e.g. the map opposite shows the distribution of records of Cory's Shearwater in Wales, since 1975. A total of 58 accepted individuals in Pembroke, of which 42 were off Strumble Head, 6 off Skokholm, 7 at Sea and a further 3 off other sites.

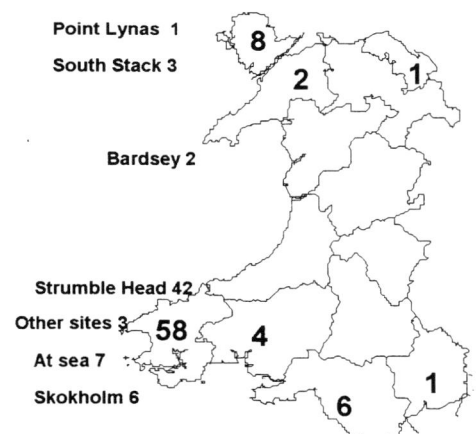

Where ever possible breeding information has also been demonstrated by the use of a map. The figures represent the number of pairs within that county during the period 1992-2000, the data coming from Welsh Bird Reports, County Bird Reports, British Birds Scarce Breeding Bird Reports and from specific requests to the relevant county recorder / expert.

It must be noted that in some cases figures are from full surveys but in many cases the figures are those estimated by county recorders.

e.g. the map shown opposite shows a summary of breeding counts of pairs of Little Grebes for each county during the period.

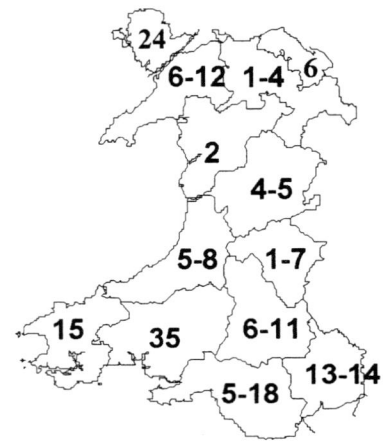

Population estimates

Where ever possible estimates of the current Welsh breeding population are given for a species. For many these were derived by totalling the number of pairs or sites reported in 1999 or 2000. For others the maximum number of pairs / sites within the period was used. For the more common species estimates were derived from three different sources.

1. Common Bird Census – estimates based on the tables of counts of common birds published since 1995 in the Welsh Bird Reports. These figures were then converted to densities, using the highest count in the series for each site as the base, then to the number of pairs. The samples are biased towards uplands by the dominant size of the Epynt area, so for some very numerous garden and woodland birds only the areas that can be categorised as woodland / farmland / garden & sub-urban were used. Estimates were then cross-referenced to estimates derived from the National Atlas, of which Wales forms 9%.

2. National Atlas – estimates derived from the raw data included New Atlas for which Wales was surveyed in 1988-1989. It is important to note that the figures for Wales were calculated by multiplying the UK total by the percentage of occurrence in Wales. These extrapolations may introduce large errors for species that, although show similar distribution across the UK, are more or less numerous in Wales compared to the rest of the UK.

3. Breeding Bird Survey – although a relatively new survey, analysis of the results from the period 1994-1999 allows the calculation of population change over that period. These percentages can then be multiplied by the population estimates from the New National Atlas to produce revised population estimates.

For example: Robin - Using data from the National Atlas produces a Welsh population estimate of 437,000 pairs. Results from BBS, 1994-1999 suggests a population change of 111%, producing a revised population estimate of 485,000 pairs. This is consistent with estimates generated from CBC data.

Abbreviations

The following abbreviations are used in this List: county names are used as shown on the county map, months are abbreviated to the first 3 letters, except June and July (in full) and September (Sept.); S,N,E,W etc, cardinal points of the compass, max. and min., maximum/maxima and minimum/minima; Est. estuary; Res. reservoir; R. river; NNR, LNR, NR, nature reserves of various designations; GC golfcourse; A/F airfield; CP Country Park; GP gravel pits; S/F sewage works; s/p summer plumage; w/p winter plumage; pr/prs pair/pairs; c/s clutch size, b/s brood size.

In wader and wildfowl tables the following abbreviations of place names are used: PW/SF, part of the coastal area of the Wentloog Levels in Gwent, GLWR, the Gwent Levels Wetland Reserve Goldcliff, Burry (S), the south side of the Burry Inlet, Burry (N), the north side of the Burry Inlet; T/T/G, the Taf/Tywi/Gwendraeth complex in Carmarthen; D'Glas., the Dwyryd/Glaslyn estuary in Meirionnydd; PoA, Point of Ayr at the seaward end of the Dee Est.; IMF, Inner Marsh Farm RSPB reserve in Flint.

RED-THROATED DIVER *Gavia stellata*   *TROCHYDD GYDDFGOCH*
(C) Winter visitor and passage migrant, regularly recorded off-shore, sometimes in large concentrations. Rare inland.

The largest winter concentrations of this species is in north Cardigan Bay, Ceredigion / Meirionnydd/ Caernarfon. Numbers here are of national importance. Co-ordinated counts from this area are tabulated below. The count in January 1997 was the largest ever recorded in Wales and was above International Importance Levels.

	1991/92	92/93	93/94	94/5	95/6	96/7	97/8
October	78	42	122	248			
November	427	351	85	252	153	1028	
December	313					1214	
January	362					1916*	370
February	140	390	740	187	404		

Within this area, large concentrations were recorded at three places:

Off Borth / Ynyslas, Ceredigion: 390 in January & 168 in November 1992, 900 in December 1995, 724 in December 1996, 1023 in January 1997, 270 on Jan. 11th and 141 on Nov. 20th 1999, 242 on Jan. 1st and 107 on Dec. 27th 2000.

Off Wallog, Ceredigion: 302 in March 1994, 209 in December 1996 and 122 in December 1998.

Off Aberdysynni, Meirionnydd: 290 in January & 106 in December 1992, 459 in February 1994, 430 in January 1995, 320 in February & 791 in December 1996 and 640 in January 1997.

The only other part of the Welsh coast from which there were co-ordinated counts was in north Wales, where there were 31 between Traeth Lafan, Caernarfon and Mostyn, Flint in January 1998 and 90 off Traeth Lafan in February 2000.

Smaller numbers were reported off all Welsh coasts, large counts (over 60) included: 80 at Worms Head, Glamorgan in December 1992, 200 Amroth, Pembroke February 1993, 75 off Goultrop Roads, Pembroke January / February 1996, with 85 there January 1998 & 60 in December 1998. Off the Caernarfon coast, 74 at Llanfairfechan Mar. 27th 1999, 75 at Pontllyfni Apr. 3rd 1999 and 90 at Morfa Dinlle Apr. 2nd 2000.

Records from freshwaters are quite unusual, during the period 1992-2000 there were 8 records: from Glamorgan at Broad Pool Jan. 16th 1994, Eglwys Nunydd Res. Feb. 8th – 15th 1996 and on a borrowpit lake near Resolven Dec. 19th 2000; at Llangorse, Brecon, Mar. 22nd 1998 and Sept. 28th – Oct. 1st 1999; on Anglesey at Llyn Traffwll Oct. 14th 1993 and at Llyn Coron Dec. 5th 1999; Llyn Helyg, Flint Jan. 28th – Feb. 11th 1995.

Most individuals have left Welsh waters by the end of April but during the period under review there were three summer records of individuals: at Black Rock, Caernarfon July 29th 1999, between Cemaes & Skerries, Anglesey June 23rd 1999 and an individual found dying at Black Rock, Caernarfon July 15th 1994 had been ringed on the Orkneys in 1992.

One at Bardsey, Caernarfon Mar. 11th – 13th 1999 had had its legs severed by monofilament fishing line.

59 were picked up oiled from the Sea Empress disaster in February 1996. At the same time a large passage of birds, totalling 96, was noted at Strumble Head, Pembroke going north into Cardigan Bay on Feb. 25th.

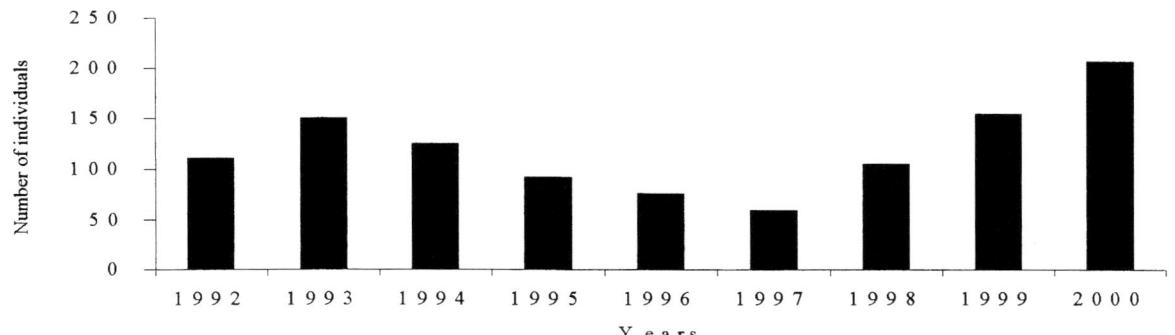

Annual totals of Red-throated Divers passing Strumble Head in Autumn

Large numbers are recorded passing the coast in autumn. Passage off Strumble Head, Pembroke usually starts mid-August and continues until December. Passage was also noted off Point Lynas, Anglesey, but unfortunately no definitive counts are available for 1992 - 1998. In 1999, 142 were logged, between Oct. 3rd and Dec. 15th and 51 were logged in 2000.

BLACK-THROATED DIVER *Gavia arctica* — *TROCHYDD GYDDFDDU*
(C) *Winter visitor and passage migrant, recorded in small numbers annually.*

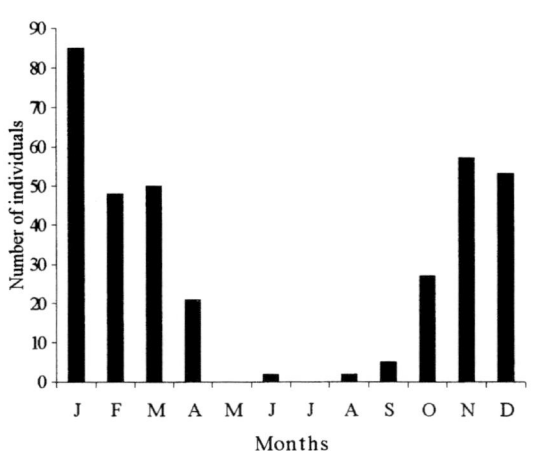

The rarest of the common divers recorded of all coastal counties in Wales but in low numbers. The total number of individuals recorded in each month, during the period 1992-2000 is shown in the chart. The monthly distribution clearly shows autumn peaks relating to passage, January had the most number of individuals, decreasing thereafter. The summer record was of 2 summer plumage birds off Blackpill, Glamorgan on June 17th 1998. The earliest return passage was of a single past Bardsey, Caernarfon on Aug. 8th 2000.

Inland records are rare and during the period 1992-2000 there were 8 records as follows: at Llandegfedd Res., Gwent Jan. 6th – Feb. 21st & Nov. 11th – Dec. 15th 1996, Jan. 25th 1997; at Eglwys Nunydd Res. Oct. 21st 1996 and Dec. 29th 1997 – Mar. 11th 1998; at Llyn Padarn, Caernarfon Jan. 29th & Mar. 15th 1998; at Llyn Traffwll, Anglesey Jan. 22nd – Feb. 19th 1995.

GREAT NORTHERN DIVER *Gavia immer* — *TROCHYDD MAWR*
(C) *Winter visitor and passage migrant, recorded offshore in small numbers annually. Recorded from inland waters more frequently than other diver species.*

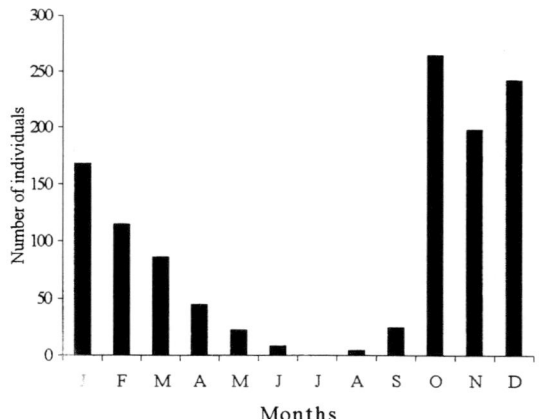

The largest of the 3 commoner diver species, seen off all Welsh coasts and occasionally on freshwater. The total number of individuals recorded in each month, during the period 1992-2000 is shown in the chart. The large numbers in October – December were due to passage birds, particularly off Strumble Head, Pembroke. Generally there is a lack of passage in spring.

Eleven individuals were recorded during the summer months: in Glamorgan: off Port Eynon June 25th 1994 and June 21st 1997, an adult and imm. Blackpill June 9th 1998; 2 off Ceibwr, Pembroke June 22nd 1997, singles off Skomer May 22nd & Ramsey June 3rd 2000; off Aberystwyth, Ceredigion May 31st 1998; at South Stack, Anglesey June 4th 1994 and May 8th 1998.

Autumn passage recorded annually from Strumble Head, Pembroke:

Year	Passage Period		Total	Year	Passage Period		Total
1993	Sept. 19th	- Nov. 30th	19	1997	Oct. 10th	- Dec. 31st	47
1994	Sept. 16th	- Dec. 12th	37	1998	Sept. 13th	- Dec. 31st	27
1995	Sept. 27th	- Nov. 27th	13	1999	Sept. 11th	- Dec. 27th	71
1996	Sept. 20th	- Dec. 31st	30	2000	Aug. 13th	- Dec. 30th	51

Birds in Wales states that this species is more often seen on freshwater than the other diver species. During this period 1992-2000 eighteen individuals were recorded as follows: at Llandegfedd Res., Gwent (1), in Glamorgan at Eglwys Nunydd Res. (7), at Llanishen Res. (1) and at Kenfig (1), Llys y fran Res., Pembroke (4), Craig Goch Res., Radnor (1) and in Denbigh at Rhosllanerchrugog, (1) and at Ty Mawr Res. (1).

WHITE-BILLED DIVER *Gavia adamsii* *TROCHYDD PIGWEN*
(B) A vagrant.
An adult past Strumble Head, Pembroke on Sept. 27th 1999 was the 2nd Welsh record. The only previous record was of a 1st winter in Holyhead Harbour February – May 1991.

PIED BILLED GREBE *Podilymbus podiceps* *GWYACH YLFINFRAITH*
(B) A trans-Atlantic vagrant.
Up to 1991 there had only been three accepted records of this species in Wales. The first at Aber Ogwen, Caernarfon Nov. 13th - Dec. 30th 1984. In 1987 an individual was discovered at Kenfig, Glamorgan, Jan. 31st - Apr. 23rd 1987 and what was presumed to be the same returning bird there, Oct. 31st 1987 - Apr. 1st 1988.

There were no other Welsh records until 1999 when two individuals arrived, the first at Llangorse Lake, Brecon Jan. 9th – Feb. 7th, the second at Cosmeston Lakes, Glamorgan Jan. 31st – Mar. 31st.

LITTLE GREBE *Tachybaptus ruficollis* *GWYACH FACH*
(C) Rather uncommon breeding resident on well vegetated lakes, marshes and water courses; scarcer in the land-locked counties.

Little Grebes are an uncommon breeding resident on small farm ponds, lakes, gravel pits, streams and canals. Much of Wales is unsuitable as breeding habitat due to the lack of vegetation. The map opposite summarises the total number of breeding pairs in each county. It must be noted that breeding data was not always from complete surveys but the total number of breeding pairs as suggested by these figures (derived from Welsh Bird Reports), is within the 120 – 150 pairs quoted in *Birds in Wales*.

The largest congregations recorded in Wales during the period were 40 at Llyn Cefni, Anglesey Aug. 20th 1993, 40 at Dwyryd / Glaslyn, Meirionnydd Nov. 26th 1993 with 41 there in November 1999.

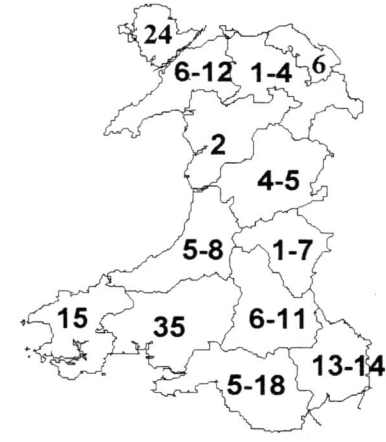

GREAT CRESTED GREBE *Podiceps cristatus* *GWYACH FAWR GOPOG*
(C) Resident breeding species in all counties except Pembroke, breeding on shallow, vegetated lakes. Shows a more coastal distribution in winter and recorded on passage from several counties.
The largest wintering concentrations, of National Importance, are in north Cardigan Bay. Results from co-ordinated counts are tabulated below.

	/92	92/93	93/94	94/5	95/6	96/7	97/8
October		66	225	154			
November			160	110	119	198	
December		322				228	
January	97					412	177
February	120	287	229	341	133		

Individual counts of more than 100 birds were noted from 2 locations within this area, off Borth and off Glaslyn / Harlech.
Off Borth, Ceredigion: 295 Jan. 17th and 158 on Dec. 26th 1992, 278 on Dec. 26th 1993, 252 on Jan. 17th and 200 in December 1994, 292 on Dec. 29th 1996, 396 on Jan. 21st 1997, 121 on Dec.1st 1998 & 100 Feb. 1st 1999 and 92 on Dec. 1st 2000.
Off Dwyryd/Glaslyn/Harlech, Meirionnydd / Caernarfon: 116 February 1992, 118 in October 1993, 190 in January 1995, 266 on Feb. 2nd 1995 and 100 in November 1996.

The only other co-ordinated count in Wales was of the north coast in January 1998, when 32 were counted between Traeth Lafan, Caernarfon and Llandulas, Denbigh with a further 72 between Abergele and Mostyn, Flint.

Smaller numbers were recorded off all Welsh counties during the winter months. Counts of over 50 were:
in Glamorgan: off Blackpill 65 in December 1996 and 95 in January 1997, off Aberavon 80 in December 1996, 56 on Feb. 13th & 92 on Dec. 13th 1997, 66 in November & 60 in December 1998, 85 on Jan. 11th 80 in February & 102 on Nov. 21st 1999 and 149 in Baglan Bay in December 2000; Caernarfon: 70 at Traeth Lafan February 1997; Denbigh: 67 in Pensarn / Kinmel Bay Feb. 4th 1996; off Flint: 75 off Connah's Quay Nov. 24th 1992, 97 in the Dee estuary October / November 1993, 109 there on Nov. 3rd 1995.Smaller numbers were recorded on freshwaters, max. counts were: 48 at Llandegfedd Res. Gwent in July 1999, 42 at Eglwys Nunydd Res. Glamorgan April 1994 and 35 in March 1995, 31 at Llangorse, Brecon April 1998, 40 there on Nov. 4th 1999 and 46 during March 2000.

Birds in Wales reported an increase in the Welsh breeding population from 42-46 birds in 1931, to 71 birds in 1965, to 163 birds in 1975 and by 1991 the total Welsh population was estimated at around 100 pairs. The map opposite summarises the numbers of breeding pairs in each county during the period 1992-2000. Breeding data was not always from complete surveys,
however the total number of breeding pairs, as suggested by these figures, is probably still at the 100-120 pairs as suggested by *Birds in Wales*.

RED-NECKED GREBE *Podiceps grisegena*
(C) Scarce but regular winter visitor.
The rarest of the five grebes which regularly winter around the shores of Wales, predominately on coastal waters. The map below shows the total number of wintering individuals recorded in each county during the period 1992-2000. Although recorded all along the Welsh coast, there were 13 preferred sites, data from which is also tabulated (none were recorded in the winter of 2000/01).
The total number of individuals recorded in each month, during the period 1992-2000 is shown in the chart. The

The first ever-recorded breeding attempts in Pembroke were in 1995-7.

During August to October large numbers of post-breeders / non-breeders congregate off the north Wales coast in the Traeth Lafan to Llanfairfechan, Caernarfon, area to moult. Numbers are of National Importance and are tabulated below:
1992 1993 1994 1995 1996 1997 1998 1999 2000
 273 275 508 283 244 360 389 165 388
Another important gathering (perhaps the same, just moved) was in the Dee estuary, Flint. Counts from there were: 56 off Bagillt Bank Aug. 8th increasing to 223 by Oct. 21st 1996, 51 there August 1997 increasing to 242 in September, with 220 in October, 164 in November but only 30 in December and 147 in October 1999.

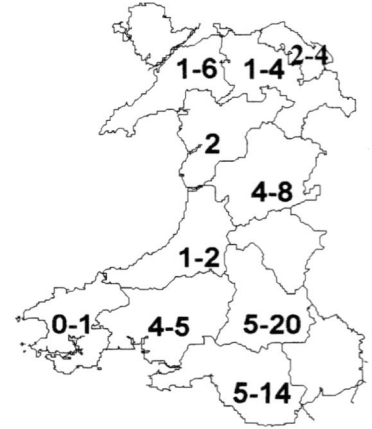

GWYACH YDDFGOCH

monthly distribution shows that February was the most favoured month but this is exaggerated by the 1996 influx of 22 individuals. The Sea Empress disaster in February of that year brought an unprecedented influx of 14 birds to Pembroke with birds remaining into early April. There were up to 4 in Fishguard Harbour, 3 at the Gann, and 3 oiled birds picked up in the clean-up. There were also three inland records, singles at Llys y fran Res., Penberi Res. and Heathfield Gravel Pits.

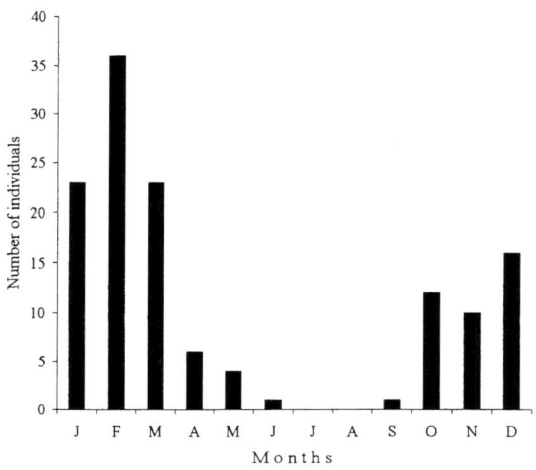 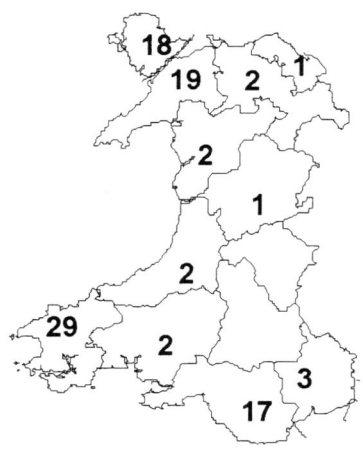

Breakdown of the wintering, October – March, sites with more than one record, none in 2000/01:

		/1992	92/3	93/4	94/5	95/6	96/7	97/8	98/9	99/00
Glam.	Whiteford / Rhossili	1	1						1	1
	Kenfig		1			2				
	Eglwys Nunydd Res.				1			1	1	
	Queen's Dock, Swansea			1	1					
Pemb.	Angle Bay						1		1	
	Fishguard Harbour					4	1			
	St. Bride's Bay						1		1	
	Gann				2	3	1			
Cere. / Meiri	Dyfi mouth / Borth							1	1	1
Caern.	Borth y Gest				1	1		1		
	Bangor Pier				1	1		1		1
	Traeth Lafan/Beaumaris		1					3	2	2
Angle.	Beddmanarch Bay			1				1	2	1

Records from freshwaters are rare and during the period ten individuals were recorded as follows: at Llandegfedd Res., Gwent (1), in Glamorgan at Kenfig (4), at Eglwys Nunydd Res. (2) and at Llanishen (1), at Llys y fran Res., Pembroke (1) and at Trawsfynydd, Meirionnydd (1).
As the monthly distribution chart shows, most individuals have departed Welsh waters by the end of March. There were however six spring/summer records: at Eglwys Nunydd Res., Glamorgan Apr. 25th – May 1st 1993, at Llandegfedd Res., Gwent May 14th 1995, at Skokholm, Pembroke May 21st 1997. In 2000 at Llanfairfechan, Caernarfon Apr. 15th, at Shotwick Lake, Flint Apr. 27th and at a site in Denbigh May 9th – June 6th, where a nest was started to be built.

SLAVONIAN GREBE *Podiceps auritus*   GWYACH GORNIOG

(C) A scarce and local winter visitor, chiefly to coastal waters, occasionally recorded on passage. Rare inland.

Birds in Wales states that this species is a more common visitor to the Welsh coast than Red-necked Grebes, visiting traditional sites each winter. The 6 main wintering areas are off Whiteford Point, Glamorgan; Angle Bay & Gann in Pembroke; in north Cardigan Bay, off Morfa Harlech / Black Rock Caernarfon; the north Wales coast, Traeth Lafan, Caernarfon to Llandulas, Denbigh and at Beddmanarch Bay, Anglesey. A breakdown of records for each site is shown in the following table.

Birds in Wales 1992-2000

	/1992	92/3	93/4	94/5	95/6	96/7	97/8	98/9	99/00	2000
Whiteford Point	6		5	6	3	11	5	14	9	3
Angle Bay		1	1	1	2	4	5	1	2	1
Gann	1		1	1	2		1		3	
Morfa Harlech	13	13	21	12	16	12	19	5	11	6
Traeth Lafan / Llandulas						9	5	5	4	5
Beddmanarch Bay					4		4		3	3

The only other large counts elsewhere were: 4 off Blackpill, Glamorgan Dec. 2nd 1998 and 4 off Newgale, Pembroke January – February 1997.

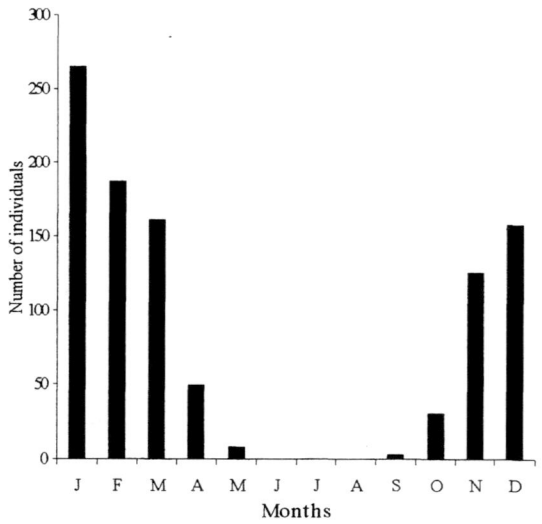

The total number of individuals recorded in each month, during the period 1992-2000 is shown in the chart opposite. A distinct peak in January with a decline afterwards and no summer records.

Inland, this species is quite often seen at a number of sites: in Gwent at Llandegfedd Res. (6) and Ynysfro Res. (1); in Glamorgan at Kenfig (2) Hensol Lake (1); in Pembroke at Llys y Fran Res. (1) and Carew Mill Pond (1); at Llyn Eiddwen, Ceredigion(1); in Brecon at Usk Res. (1) and Llangorse Lake (2-3); at Llandrindod Wells Lake, Radnor (1); on Anglesey at Llyn Alaw (4); at Rhyl brickwork's, Flint (1).

An individual at Llyn Brenig, Denbigh on Apr. 5th was thought to be the same bird that was seen at Llyn Helyg, Flint on May 31st.

BLACK-NECKED GREBE *Podiceps nigricollis*   *GWYACH YDDFDDU*
(C)   Rare winter visitor to coastal waters, occasionally recorded on passage. A regular visitor in small numbers only to Anglesey and Caernarfon. Has bred on Anglesey in the past.

A rarer winter visitor than Slavonian Grebe, with smaller groups recorded at a handful of sites, mainly in north Cardigan Bay around Morfa Harlech, Meirionnydd / Caernarfon and on the north Wales coast around Traeth Lafan, Caernarfon. Fewer records from the south of the country. The table below shows the number of wintering individuals at each of the main 9 sites (those with more than one record).

Birds in Wales states that from 1953/4, this species was recorded regularly from Whiteford Point, Glamorgan, max. 8 in 1970 but has declined since and by 1980's rarely seen there. Data from this period, 1992-2000 suggest that this species is usually present but in smaller numbers.

The map opposite gives the total number of individuals recorded for each county during the period 1992-2000.

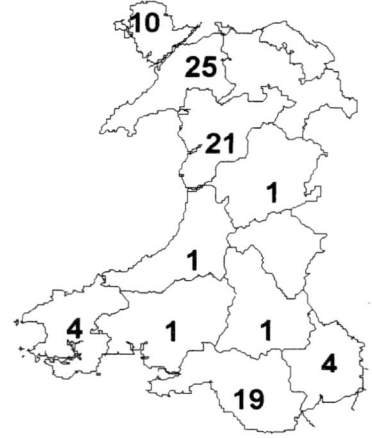

		92/3	93/4	94/5	95/6	96/7	97/8	98/9	99/00	00/01
Gwent	Llandegfedd Res.				1			1-2		
Glam.	Kenfig / Eglwys Nunydd Res.	2	1-3						1	
	Port Eynon / Whiteford Pt.		2	1	1	1			2	1
Pemb.	Gann / Angle Bay						1			2
Meiri./ Caern.	Morfa Harlech / Dwy. / Glas.	3	4	5	2	6	2	2	1	1
Caern.	Pwllheli				1		1			
Caern./ Denbigh	Traeth Lafan / Llandulas	3			1	2	4	2	2	
Angle.	Beddmanarch Bay				1		1		1	
	Anglesey Lakes		2	1						

The largest gathering was in Conwy Bay, Caernarfon in March 1996. A single was present on the 6th, which was joined by several others, max. 11 on the 19th.

This species was not proved to have bred in Wales in the period under review but single individuals in suitable habitat were recorded in two years. Spring/summer records were: in Gwent at Llandegfedd Res. 27th May 27th – June 2nd 1994, at Ynysfro Res. Apr. 25th – 26th 1996; in Glamorgan at Kenfig one Apr. 6th – 21st with a second individual Apr. 21st 1997, one Apr. 3rd – 12th 1998; at Llanbwchyllyn, Radnor May 19th 1997; on Anglesey at Llyn Penrhyn 2 May 6th-8th 1995, 2 Mar. 29th – Apr. 3rd with display noted on the 1st, then a single until Apr. 6th 1997, a single Mar. 30th 1999 and 2 on Mar. 18th 2000.

Recorded nearly annually in autumn, records included: in Gwent at Llandegfedd Res. (8), and at GLWR Goldcliff (1); in Glamorgan at Kenfig (3), Eglwys Nunydd Res. (3) and at Llanishen/Lisvane Res. (1); at Penclacwydd, Carmarthen (1); at Gann, Pembroke (1); at Conwy RSPB, Caernarfon (2) including a juvenile in Sept. 23rd – Nov. 6th 1999; on Anglesey at Llyn Alaw (2), Llyn Traffwll (2) and at Llyn Cefni (1).

FULMAR *Fulmarus glacialis*   *ADERYN-DRYCIN Y GRAIG*

(C) A numerous species, breeding in all coastal counties except Gwent. Frequent offshore in most months.

		1992	1993	1994	1995	1996	1997	1998	1999	2000
Pembroke	Stackpole – Giltar Pt.					56	85	83	89	93
	Castle Martin Coast	103	98	118	117	133	132	109	130	120
	Skokholm	110	119	124	130	137	170	147	147	155
	Skomer	692	722	696	715	678	662	638	581	691
	Ramsey	197	216	226	225	295	321	288	285	215
Ceredigion	New Quay/Cardigan I.				270		257	159		380
Caernarfon	Bardsey			51		38	35		28	37
	Great Orme	83	48		64	46	53	35	47	18

The expansion along the Welsh coast, as described in *Birds in Wales* appears to have come to a halt, with very little change detected during the period 1992-2000. In some parts there may have even been a slight decline. Breeding data from the main colonies are tabulated above, figures are for "apparently occupied sites".

Very little change was detected in Ceredigion in 1996 compared to 1986/7, when a county total of 437 occupied sites counted. Similarly in Pembroke where 2254 apparently occupied sites were counted in 1996, 2591 counted in the following year. There was no noticeable effect of the Sea Empress disaster on this species.

Inland records are very rare and during the period the only record was of one at the top of the Dyfi estuary at the William Condry Reserve, Montgomery Oct. 30th 2000.

The only large movements were noted at Strumble Head, Pembroke 560 in 3 hrs. Nov. 15th & 500 on Dec. 3rd 1998 & 1,300 in 3.25 hrs. on Feb. 8th 2000, 211 from Bardsey, Caernarfon on Sept. 8th & 238 on the 9th 1999 and at Point Lynas, Anglesey 320 in 3.5 hrs. on July 22nd 1999.

Blue morphs were reported from Pembroke: off Strumble Head Dec. 5th 1992, Mar. 27th, Apr. 1st & Oct. 3rd 1994, Sept. 10th 1998 & Feb. 8th 2000 and at Druidston Feb. 26th 1994; from Flint: Prestatyn Quarry during the breeding season 1992 and at Point of Air July 12th 1992.

"SOFT PLUMAGED PETREL" *Pterodroma madeira/feae/mollis*   *PEDRYN MWYTHBLU*
(A) A vagrant.
The first accepted Welsh record came from Bardsey, Caernarfon on Sept. 10th 1994 this was followed by one off Strumble Head, Pembroke on Oct. 4th 1996.

CORY'S SHEARWATER *Calonectris diomedea*   *ADERYN-DRYCIN CORY*
(B) A scarce passage species off the Welsh coast, chiefly recorded from west coast localities in late summer.

Birds in Wales quotes 34 individuals since 1975, with the increase in sea-watching, this total has more than doubled. Annual totals for 1992-2000 are shown in the chart below. 1999 was a bumper year for the occurrence of this species in Wales, probably linked to unusual warm currents coming north from Spain. The November record from Strumble Head was the latest Welsh record and was seen after a "good" gale preceded by two days of strong winds – presumably pushing the bird up from a long way south.

Breakdown of records during the period 1992-2000:
Carmarthen: the first county record was of an extremely early individual off Telpyn Point Feb. 29th 1992. Since then 2 were seen off Cefn Sidan Aug. 1st 1993 and another there Aug. 5th 1997.

Anglesey: a single at Point Lynas, Aug. 30th 1999 was the 1st record for this site.

Flint: a single at Gronant, June 21st 1999 was the first county record.

Pembroke: singles at Skokholm Sept. 17th & 25th 1995, at least one off Grassholm July 31st 1993, "Celtic Deep pelagic trips" from Pembroke in 1999 counted 2 on Aug. 15th & 16th, one on the 25th and at least one on Sept. 26th. The majority of records came from Strumble Head, where this species was recorded passing in every year except 1992, 1994 and 1995. In total of 32 individuals were recorded during the period, mainly singles but at least 6 on Aug. 15th 1999. Also in that year a very late individual was seen on Nov. 29th 1999.

As with Great Shearwater, the increase in pelagics is providing more information on the true status of this species in Welsh waters. The February record in 1992 off Telpyn Point was the earliest yet recorded in Wales. A breakdown of all the Welsh records, since 1975 of this species is shown on the map.

Totals (since 1975) for each county are given and where possible data totalled for specific sites, e.g. 58 in Pembroke of which 42 were seen off Strumble Head, 6 off Skokholm, 7 seen at sea and a further 3 from other sites.

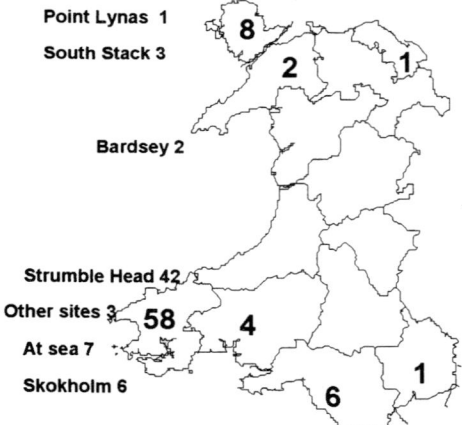

Birds in Wales 1992-2000

GREAT SHEARWATER *Puffinus gravis*   *ADERYN-DRYCIN MAWR*

(B) A scarce passage species occasionally seen inshore Welsh waters in summer.

A more frequent visitor to Welsh waters than the other large shearwater.

Birds in Wales quote 58 individuals up to 1991, since when there have been a further 97 in Welsh waters. Annual totals for 1992-2000 are shown in the chart opposite. All of these subsequent records except one, at Point Lynas in 1998 and one off Bardsey in 2000, came from Pembroke, indeed of the 153 accepted Welsh individuals 142 have been off Pembroke. With the increase in pelagic trips from this county in autumn, the numbers of records will no doubt increase dramatically.

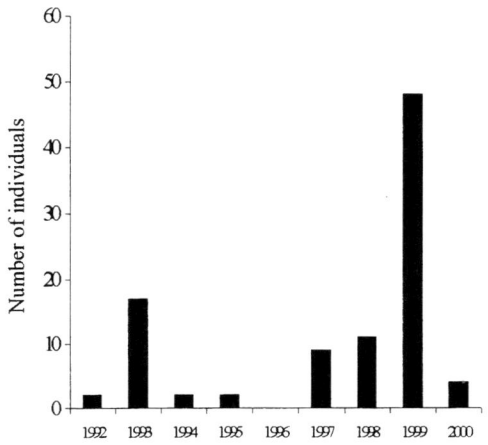

All records were:

Caernarfon: a single off Bardsey Sept. 8th 2000.

Anglesey: Single Point Lynas Oct. 25th 1998.

Pembroke: 2 off Skokholm Sept. 9th 1993 & a single Aug. 9th 2000, a single off Skomer Oct. 29th 1994. Celtic Deep pelagic trips recorded one Aug. 10th 1998 but in 1999 there were 20 on Aug. 15th, 6 on the 16th, 5 on the 25th, one on Sept. 14th, 4 on the 21st. 3 from the Fishguard / Rosslare Ferry Sept. 10th 1999.

As with the other large shearwater, the majority of records came from Strumble Head, where there were records in every year except 1996. Most records involved just ones or twos in a day but 9 were logged on Aug. 21st 1998 and 5 on Sept. 4th 1999.

A breakdown of all Welsh records by county and where possible by specific sites is shown on the map, e.g. 142 in Pembroke, 88 seen off Strumble Head, 2 off Skomer, 3 off Skokholm, 40 seen at sea and 9 from other sites.

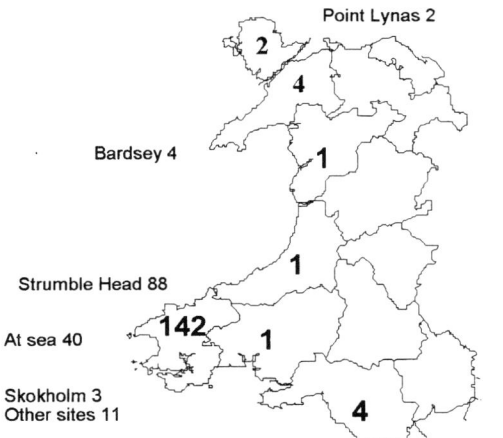

SOOTY SHEARWATER *Puffinus griseus*   *ADERYN-DRYCIN DU*

(C) A fairly common passage migrant to western coastal areas, principally in the period late August to early October.

A regular autumn passage migrant off the coast of Wales. Passage numbers in the period under review have not approached the maxima of the 1980's, e.g. 397 off Strumble Head, Pembroke in September 1983, 126 off Bardsey, Caernarfon in September 1980.

Main passage noted from Strumble Head, Pembroke, data from the other main watch points, Skokholm in Pembroke, Bardsey in Caernarfon and South Stack & Point Lynas on Anglesey also included;

Strumble Head	1992	1993	1994	1995	1996	1997	1998	1999	2000
Total	47	14	119	72	45	52	99	51	67
Day Max.	11	9	27	14	8	14	29	16	25
Peak date	Aug. 30th	Sept. 10th	Sept. 29th	Sept. 26th	Aug. 24th & Oct. 16th	Sept. 5th	Sept. 10th	Oct. 2nd	Sept. 6th
Passage period	Aug. 9th - Oct. 19th	Aug. 30th - Sept. 11th	July 27th - Nov. 11th	July 30th - Oct. 27th	Aug. 6th - Nov. 7th	Aug. 27th - Oct. 11th	July 31st - Oct. 29th	July 21st - Nov. 16th	Aug. 2nd - Nov. 11th

ELSEWHERE	1992	1993	1994	1995	1996	1997	1998	1999	2000
Skokholm	9	7	9	8	6	3	8	1	1
Bardsey			55	1	8	8	4	14	11
South Stack		1	4	3			3		
Point Lynas		1	1	12	4		3	1	

55 off Bardsey, Caernarfon in 1994, including a count of 39 on Sept. 10th. Elsewhere 5 records from Glamorgan, one from Carmarthen, 12 from other Pembroke sites, 2 from Ceredigion, 3 from Caernarfon mainland and two from Flint. A late individual off Traeth Llugwy, Anglesey Dec. 5th 1999 and one off Whitesands, Pembroke on Dec. 28th 2000.

MANX SHEARWATER *Puffinus puffinus*   *ADERYN-DRYCIN MANAW*

(C) *Breeding on Pembrokeshire islands and also in Caernarfon. Large feeding movements and autumn passage are regular off western coasts and even well up into the Bristol Channel.*

The total world population was estimated in 1991 at 260,000 – 330,000 of which approximately half breed on 6 Welsh islands. Below is a summary of all the estimates in the period, a.o.b. represents apparently occupied burrow:

Skomer	Pemb.	Population thought to be stable in 1994. Major survey in 1998, total of 101,000 a. o. b. - from responses to taped calls.
Skokholm	Pemb.	Survey in 1998, total of 45,000 a. o. b. - from responses to taped calls.
Middleholm	Pemb.	Survey in 1998, total of 2,800 a. o. b. - from responses to taped calls.
Ramsey	Pemb.	Estimates of 500 - 1,000 pairs in 1993, over 1,000 pairs in 1995, c900 pairs in 1999.
Bardsey	Caern.	404 apparently occupied burrows on the SE side of island in 1994, similar figures to 1985. In 1995 a total of 1,424 a.o.b. on the lowland, with on average 0.68 chicks / burrow, while in the following year a total of 6,927 a.o.b., higher than all previous estimates. An average of 0.79 chicks / burrow.

Changes in methodology of breeding surveys on the four main islands have altered population estimates. The Skomer figures are lower than those estimated in 1990 by S.J. Sutcliffe, of 135,000-185,000 pairs. Skokholm's population was estimated at 30-40,000 in 1969, a slight improvement there, as on Middleholm where there were an estimated 2,000 pairs in 1983. Overall the population is thought to be stable in Pembroke.

Increase in population at Bardsey where the population was estimated at 4,500 pairs in 1986 although this may be due to changes in census methodology.

Birds landed Cardigan Island, Ceredigion summer 1992 & 1993 but breeding not proved.

An individual ringed as a chick on Skokholm in 1967 was found breeding on Skomer in 1999 – 32 years old.

Summer movements up the Bristol Channel noted annually, usually up to 7,000 recorded but a max. 18,000 off Port Eynon, Glamorgan June 9th 1998 and on the same date in 1999. Similar movements along Cardigan Bay, up to 5,000 recorded.

MEDITERRANEAN SHEARWATER *Puffinus mauretanicus*   ADERYN-DRYCIN MOR Y CANOLDIR

(C) *A fairly common passage migrant to western coastal areas, principally in the period late August to early October.*

A regular occurrence in the daily logs off Strumble Head, Pembroke, with smaller numbers recorded off the other Welsh sea-watching hotspots. A summary of passage off Strumble Head, Pembroke, data from the other main watch points, Skokholm in Pembroke, Bardsey in Caernarfon and South Stack & Point Lynas on Anglesey also included:

	1992	1993	1994	1995	1996	1997	1998	1999	2000
Strumble Head									
Total logged	24	18	86	115	145	210	65	402	124
Max in a day	4	3		16	33	21	12	36	17
Peak date	Aug. 30th & Sept. 6th	Oct. 2nd		Sept. 24th	Oct. 4th	Sept. 1st	Oct. 17th	Oct. 2nd	Sept. 6th
Passage period	July 25th- Oct. 25th	Aug. 12th- Oct. 25th	Aug. 12th- Nov. 1st	July 20th- Nov. 26th	June 30th- Nov. 7th	July 13th- Oct. 10th	June 30th- Nov. 12th	July 22nd- Dec.12th	Aug. 2nd - Nov. 11th
ELSEWHERE									
Skokholm	9	7	15	11	1	5		2	3
Bardsey	6		9	4	14	32	3	91*	12
South Stack			2	15-19			2		
Point Lynas		1	2		13		16	2	5

Small numbers were seen off Glamorgan: Port Eynon and Worms Head; Pembroke at Skomer, Ramsey and Grassholm; Ceredigion at Mwnt, Borth and Ynyslas; Flint off Point of Air.

1999 was an exceptional year with 567 individuals logged passing off the Welsh coast – the 402 off Strumble Head was much higher than the previous highest annual count of 274. This included 9 off Glamorgan, 7 off other sites in Pembroke, 8 in Ceredigion, 16 off other Caernarfon sites and 33 off Anglesey. As with the two large shearwaters, this influx was thought to be influenced by warm currents from the SW, throughout August and September.

LITTLE SHEARWATER *Puffinus assimilis*   *ADERYN DRYCIN BACH*
(B) A vagrant.

An additional record to the 13 individuals listed in *Birds in Wales*, of a single seen passing Strumble Head, Pembroke on Oct. 7th 1988 (the same day that one was seen off Bardsey – quite possibly the same bird). Since then only one further record, again off Strumble Head, on Sept. 12th 1997.

WILSON'S PETREL *Oceanites oceanicus*   *PEDRYN WILSON*
(B) A vagrant.

Additional record to those in *Birds in Wales* of an individual off Strumble Head, Pembroke Sept. 16th 1990 – the third Welsh record (the other two being in St. George's Channel Sept. 12th 1980 and from Strumble Sept. 3rd 1986). Since then that site has recorded two others, on Sept. 5th 1997 and Sept. 11th 1998. With the initiation of pelagic trips visiting the Celtic Deep channel off Pembroke in August 1999 and the use of "chum", a single was seen on the 16th and at least 3 on the 25th (although up to 10 birds were thought to be feeding around the boat, only 3 could be confirmed by plumage differences). In 2000 2 were seen on Aug. 6th with a single (presumed to be of the same) the following day.

STORM PETREL *Hydrobates pelagicus*　　　　　　　　　　　　　　　　　　　　*PEDRYN DRYCIN*
(C) Breeds regularly on several island sites off the west coast, mainly in very small colonies. Skokholm holds the only population over 1,000 pairs.

Breeding data:
Skomer	Pembroke	35 pairs in 1993, c100 apparently occupied sites in 1996 but only 55 in 1998.
Skokholm	Pembroke	2,700 – 5,400 pairs in 1995.
Ramsey	Pembroke	2 colonies totalling 100+ pairs in 1995.
Bardsey	Caernarfon	Suspected to have bred 1996-1998.

Birds in Wales quotes c6,000 pairs on Skokholm in 1990, a decrease since 1967-9 when there were an estimated 6,200 pairs. Recent work suggests that the population in the island's wall systems has decreased although much of this may be as a result of errors in earlier censuses. A similar story at Skomer where 100 – 200 pairs quoted in *Birds in Wales*. The Ramsey population, only established in 1980, continues to increase, from 3-5 a.o.s. in 1990.

Tape luring took place at Wooltack Point, Pembroke were 100-200 were caught regularly, at Strumble Head, with 50 caught 1993 & 97 in 1994 and at Bardsey, with 21 in 1994, 25 in 1996, 66 in 1999 and 87 in 2000. One caught in 1998 on Skokholm in 1998 was first ringed as a breeding adult in 1969 – a 29 year gap, the second longest recovery gap.

Max. counts of birds passing headlands were from Glamorgan: off Tutt Head, 40 on May 31st 1993, 66 on June 21st 1994, 46 on May 29th 1995, 31 on June 25th, 34 on July 13th 1998 and 32 on May 17th 2000; off Port Eynon, 31 on June 21st 1994 and 31 from Porthcawl July 12th 1998, 3 from Porthcawl Aug. 7th and 15 from Pwll Du Head Aug. 26th 1999.

Recorded off Strumble Head, Pembroke until Dec. 12th in 1999. Passage of 16 past Point Lynas, Anglesey July 21st 1999 with 11 off there the next day.

Inland there was a single record of an individual picked up on the River Neath, Glamorgan Oct. 31st 1998.

LEACH'S PETREL *Oceanodroma leucorhoa*　　　　　　　　　　　　　　　　　　　*PEDRYN GYNFFON-FFORCHOG*
(C) A scarce autumn visitor, commonest off the north Wales coast.

Birds in Wales states that the majority of records came from north Wales, principally Denbigh and Flint, where wind-blown birds tend to gather in Liverpool Bay after north west gales in October and November. Within the period 1992-2000 there have been far fewer records from the north Wales coast, with the exception of 1997. There have been far more recorded off Strumble Head, Pembroke than quoted in *Birds in Wales* and about the same (1-10 annually) from Point Lynas, Anglesey. Some of these differences are likely to be due to changes in observer coverage rather than any change in birds distribution. The tables below summarises the counts from these headlands, off Strumble Head, Point Lynas and Point of Air, Flint:

	1992	1993	1994	1995	1996	1997	1998	1999	2000
Strumble Head									
Total logged	41	-	19	64	53	99	83	38	19
Max in a day	23	-	5	47	26	79	33	31	11
Peak date	Sept. 4th	-	Sept. 8th	Sept. 24th	Oct. 29th	Sept. 13th	Oct. 17th	Oct. 3rd	Oct. 31st
Passage period	Aug. 30th - Oct. 25th	-	Aug. 29th - Oct. 15th	Sept. 4th - Oct. 1st	Oct. 5th - Oct. 29th	Sept. 5th – Oct. 10th	Aug. 3rd - Oct. 30th	Oct. 2nd - 4th	Sept. 8th - Nov. 3rd
Point Lynas									
Total logged	3		4		5	1	35	13	7
Max in a day	1		3		3		29	7	4
Peak date			Aug. 29th		Aug. 30th		Sept. 13th	Oct. 4th	Nov. 2nd
Passage period	Sept. 12th – Oct. 15th		Aug. 29th – Sept. 16th		Aug. 30th – Oct. 5th	Sept. 8th	Aug. 24th – Sept. 14th	Oct. 2nd – 4th	Oct. 2nd & Nov. 4th

Point of Air	1992	1993	1994	1995	1996	1997
Total logged	61	3	-	42	10	488
Max in a day	41	-	-	29	5	258
Peak date	Sept. 4th	-	-	Sept. 25th	Oct. 29th	Sept. 8th
Passage period	Sept. 1st – 15th	Sept. 14th	-	Aug. 29th - Sept. 25th	Sept. 30th - Oct. 29th	Sept. 7th – 17th

Recorded annually passing Bardsey, Caernarfon in small numbers. Elsewhere ones or twos from Sluice Farm in Gwent, Mumbles in Glamorgan, Burry Port in Carmarthen, in Pembroke at St. Govan's (3); Ceredigion: Mwnt (2); Caernarfon: Criccieth and Dinas Dinlle;

Anglesey: Holyhead Harbour, Cemlyn (22 on Sept. 16th 1998 and 3 on Oct. 4th 1999); Flint: Oakenholt 92 on Sept. 14th & 3 on 16th 1998). One inland at Llanishen / Lisvane, Glamorgan Sept. 9th 1998.

GANNET *Sula bassana*   HUGAN
(C) A common bird offshore throughout the year, most plentiful during the extended breeding season, March - October. One major breeding colony on Grassholm, Pembrokeshire.

The single Welsh colony on Grassholm, Pembroke has continued to increase in size, from 28,535 pairs in 1987 to 30,688 apparently occupied sites in 1999. This represents a population increase of 7.5% over the 12 years.

A single pair nested on St. Margaret's island, Pembroke from the mid 1990s, with other birds being seen there as well – possibly the start of a new colony.

Three inland records, of singles on a farm pond at Llanafan, Brecon June 23rd 1993, at Rhayader, Radnor Oct. 10th 2000 and in rhododendron bushes at Abergwynant, Meirionnydd on Nov. 27th 2000.

CORMORANT *Phalacrocorax carbo*   MULFRAN
(C) Breeding resident in western counties from Pembroke to Anglesey. Frequently seen offshore in other counties. A regular visitor to inland freshwaters, mainly between late summer and early spring.

The Seabird Colony Register, 1985-87 estimated the Britain and Ireland breeding population as 11,700 pairs, of which 1,700 were in Wales (an increase of 19% since the original survey in 1969-70), some 15% of the British and Ireland population. The only complete survey since was in 1994, which revealed 2291 nests, a 37% increase since 1985-87 [the results of Seabird 2000 have yet to be analysed].

In 1996 a complete Pembroke survey counted 315 pairs in 7 colonies (St. Margaret's showing a 10% decrease since 1995) and 303 pairs in 7 colonies in 1997 (a further 8.8% decline at St. Margarets) - the decrease is thought to be due to winter mortality and probably not due to the Sea Empress disaster. A complete Ceredigion survey in 1996 found 164 nests (cf. 271 in 1987) a 39% decline.

The first definite breeding in Gwent occurred in 1999 at Denny Island.

Breeding data, number of pairs tabulated:

		1992	1993	1994	1995	1996	1997	1998	1999	2000
Pembroke	St. Margarets	208	320	260	229	207	187	198	172	
	Green Scar			40		33	36	32		
	Thorn Island						34	41		
	Skomer	17	16	14	12	12	13	16	10	
Ceredigion	Penderi	91	86	133	94	96	73	64	69	52
	N/Q - Penbryn			32	43		39	34		34
Meirionnydd	Craig yr Aderyn	85	95							
Caernarfon	Great Orme	38	48	78	60	60	35		13	46
	Little Orme	217	208	354	396	347	296	268	171	183
	Ynys Gwylan F.				60		25			163
	Ynys Gwylan B.						47			
Anglesey	Puffin Island			640					700	

Webs counts at selected sites show variations of importance between different sites at different times of the year. The largest spring congregations, March to April, were at Queen's Dock, Glamorgan, counts there were:

1993	1994	1995	1996	1997	1998	1999	2000
94	152	125	92	126	114	122	100

In autumn, August to October, the largest counts were:

		1992	1993	1994	1995	1996	1997	1998	1999	2000
Glamorgan	Port Talbot			62	46	82	71	79	27	49
	Whiteford Pt.	70	106	95	68	90	80	96	89	93
	Blue Pool Bay		67	76		76	88	68	65	75
Carmarthen	T/T/G	129	200	413		151	120	107		57
Meirionnydd	Dwyryd/Glas.	98	173	108	310	257	174			140
	Dysynni		245	141	250	214	83	184		176
Denbigh	Clwyd Est.			302				120		

Large autumn congregations elsewhere included: 171 at Aberdyfi/Borth, Ceredigion/Meirionnydd Sept.12th 1995, 127 at Broadwater, Meirionnydd Aug. 13th 1999; 250 off Penmon Point, Anglesey August 1998; off Rhyl, Denbigh there were 272 in September 1996, 237 in June 1998 and 220 in October 1999; in Flint, 219 at Gronant July 1996 with 228 there in September 1998.

In winter, November to February, the largest count were:

		/92	92/93	93/94	94/5	95/6	96/7	97/8	98/9	99/00	00/01
Glam.	Port Talbot		50	87	38	53	43	79	67	24	20
	Whiteford Pt.	136	32	140	106	62	36	49	78	70	66
Carm.	T/T/G	30	83	89	134	137	39	108	48	40	19
Flint	Oakenholt		89	95	88	130	102	94	105		

Large congregations elsewhere: 278 Rhos Point Nov. 5th 1995 and 280 off Rhyl in November 1998, both Denbigh.

The largest inland winter counts were, at Llangorse & Talybont in Brecon (thought to involve the same birds) and Craig y Aderyn, Meirionnydd:

	1992	92/3	93/4	94/5	95/6	96/7	97/8	98/9	99/00	00/01
Llangorse	78	90	64	71	82	50	56	66	70	30
Talybont			48	81	49					
Craig y Aderyn			75							

North Cardigan Bay, Ceredigion to Caernarfon, was counted systematically on an annual basis, counts from there included:

	/1992	92/93	93/94	94/95	95/96
October		148	152	252	
November			222	50	79
December		53			
January	38				
February	28	60	45	61	

Records of the "white-headed", *sinensis* race are not infrequent along the Welsh coast in winter. *Birds in Wales* quotes a total of 41 individuals in the period 1981-1990, compared to just 18 in the period of this review. All records were: at Llanishen/Lisvane Res. Glamorgan Feb. 6th 1993, Llanrhidian, Gower Apr. 6th 1996, 2 at Dysynni Mar. 19th 1995 and a single Porthmadog Mar. 7th 1993, both Meirionnydd.

Relatively a large number of records in 2000: in Glamorgan: Cosmeston Jan. 30th & Dec. 26th, Cardiff Bay January & February, with 2 there Jan. 20th, Roath Park Lake Jan. 6th & Feb. 17th, Rhymney Est. Feb. 11th, Llwyn On Res. Feb. 20th and Llan/Lisvane Res. Dec. 4th; at Penclacwydd, Carmarthen Apr. 3rd; in Pembroke at Fishguard Harbour Jan. 2nd and at Llys y fran Res. Nov. 11th; at Cemlyn, Anglesey Mar. 22nd.

SHAG *Phalacrocorax aristotelis*   MULFRAN WERDD

(C) *A regular breeding species on rocky coasts from Gower northwards to Little Orme.*

		1992	1993	1994	1995	1996	1997	1998	1999	2000
Pemb.	Middleholm		25	20	35	37	33	20	29	37
	Caldey/St. Margarets	7	13	11	16	8	4+	11	12	
Cere.	Cardigan Isl.		10	8	10			7		12
	Penderi		10	7	5		8	10	7	
	New Quay/Penbryn				10	8	19	15		20
Caern.	Bardsey			12	17	18	18	18	33	44
	Ynysoedd Gwylan			68	68		82			133

The Britain and Ireland breeding population was estimated in 1991 as 47,300 pairs, some 38% of the total European total. The Welsh population is very modest in comparison, approximately 770 pairs. Unfortunately there has not been any further all Wales survey work until Seabird 2000, the results of which are still to be analysed. The only regular surveys of breeding colonies, data as number of pairs, are summarised in the above table.

A complete survey of Pembroke coast found 115 pairs at 19 sites in 1996, a 50% decline along the coast from Milford Haven to Caldey compared to 1995 - presumably as a direct result of the Sea Empress disaster (when 22 oiled individuals were recovered). A similar survey in 1997 found 120 pairs. A complete survey of Ceredigion in 1996 found only 35 nests compared to 61 in 1987. A minimum of 31 pairs were found on Puffin Island, Anglesey in 1998.

Co-ordinated counts of Cardigan Bay;

	/92	92/93	93/94	94/5	95/6	96/7	97/8
October		11	39	104			
November			48	28	150	187	
December		427				209	
January	62					58	170
February	14	105	129	142	157		

Within this area, large congregations off Ynyslas / Aberdyfi, Ceredigion / Meirionnydd: 45 in January & 43 in December 1992, 70 in January 1993, 68 in January & 60 in December 1995, 100 on Jan. 17th & 74 on Dec. 31st 1999. Elsewhere in Cardigan Bay, 62 were off Pwllheli, Caernarfon in January 1998 and 78 off Aberystwyth, Ceredigion Dec. 2nd 1998.

A survey of the north Wales coast, from Traeth Lafan, Caernarfon - Mostyn, Flint in January 1998 found 233 individuals, 166 of which were off Penmaenmawr.
70 were counted at Goultrop Road, Pembroke on Jan. 6th 2000, with 40 there on Dec. 21st.

FRIGATEBIRD *Fregata sp.* *FFRIGAD SP.*
(A) A vagrant.
A female Frigatebird, probably Magnificent, over Skomer, Pembroke June 4th 1995, was the first of this genus to be recorded in Wales.

BITTERN *Botaurus stellaris* *ADERYN Y BWN*
(C) Scarce but regular winter visitor chiefly to Anglesey, Ceredigion, Carmarthen, Glamorgan and Pembroke.

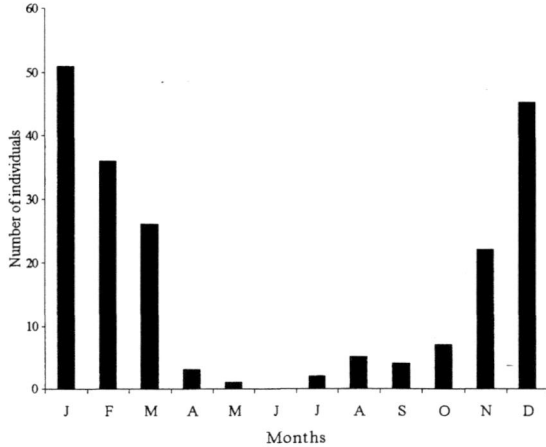
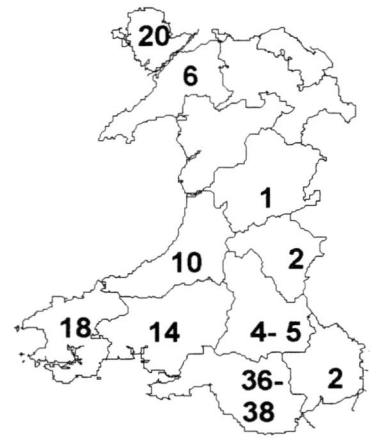

The total number of individuals recorded in each month during the period 1992-2000 is shown in the chart. The number of individuals in each county is shown on the map. During January 1997 there was an influx into Pembroke, a total of 8 individuals were seen. Oddly there was no similar influx into any other county during that month.
The majority of wintering Bitterns were recorded in Glamorgan (at three key sites) and on Anglesey (at 5 key sites). The table below summarises the number of wintering individuals at the main sites, those with more than one record.

There was not any breeding attempts in Wales during this period but a booming male was heard at Llangorse, Brecon on Jan. 28th 1999. This was the first record of a booming male in Wales since 1990.

		/1992	92/3	93/4	94/5	95/6	96/7	97/8	98/9	99/00	00/01
Glam.	Kenfig	1	1	2	1	3	2	2	1	2	2-3
	Cosmeston					1	1	2	2	1	
	Oxwich	1	1	2	1	1	1	1			
Carm.	Penclacwydd		1	1		1				1	
	Witchett	1		1	2	2	2				1
Pemb.	Bosherston	1					2			1	
	Dowrog					1	1				
Pemb. /Cere.	Teifi Marshes		1	1		1	1				
Cere.	Ynyshir				1	1	1	1	1		
Brecon	Llangorse		1	1-2					1	1	
Caern.	Conwy								1	1	
Angle.	Malltraeth					1	1		1	1	
	Llyn Alaw			1				1			
	Llyn Maelog					1	1	1	1		
	Other Valley Lakes						1	1	1		

LITTLE BITTERN *Ixobrychus minutus*   *ADERYN-BWN LEIAF*
(B)   A rare visitor.

Following a review of Glamorgan records (Lansdown and Hurford 1995) there are only 15 acceptable records in Glamorgan up to 1991. Since the publication of *Birds in Wales* there have been an additional 4 accepted records in Wales. One at Newport, Gwent May 15th 1994, otherwise all records came from Pembroke, where a male was found dead at Whitesands Apr. 12th 1993, a wing from a male bird taken by a raptor was found at Castle Martin Apr. 2nd 1995. A female seen in the Cilgerran Gorge May 6th 1995 on both sides of the Pembroke / Ceredigion border*.

Breakdown of all Welsh records by month (12 undated records) in the chart and by county on the map.

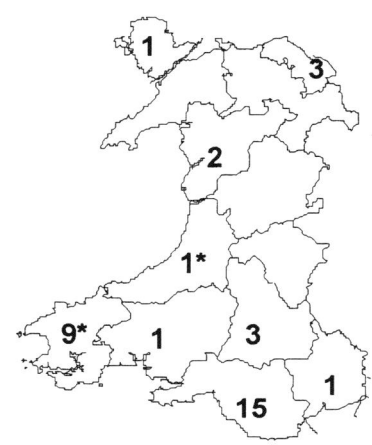

NIGHT HERON *Nycticorax nycticorax*   *CREYR Y NOS*
(B)   A rare visitor to Wales.

Re-analysis of the *Birds in Wales* archive shows that all the records in Wales up to 1991 only involved 31 individuals. Since then there have been a further 7 records.
Gwent: An adult at Newport May 6th – 23rd and a 1st summer bird there 6th – 25th 1994, a juvenile at Llandegfedd Res. July 7th – 30th 1997.
Ceredigion: An adult at Cors Caron Apr. 24th 2000.

Anglesey: Adults at Cefni Res. June 2nd 1994 and at Pentre Berw Apr. 16th - 18th 1998.
Flint: Immature at Ddol Uchaf Caerwys June 18th 1993.

Breakdown of all Welsh records of Night Heron, by month (arrival dates, 4 undated records) in the chart and by county on the map.

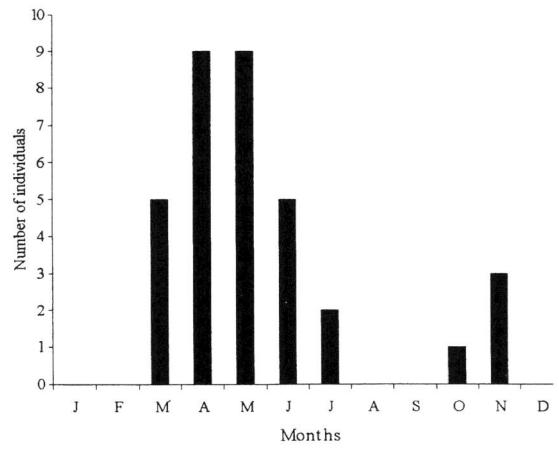
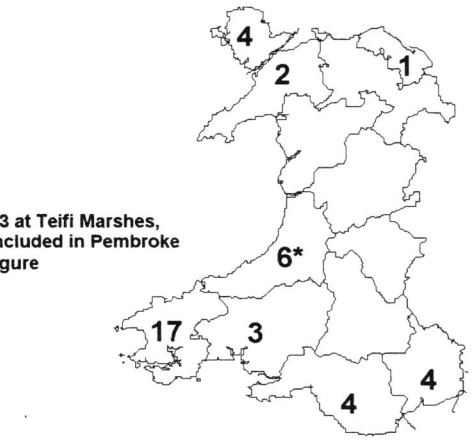

* 3 at Teifi Marshes, included in Pembroke figure

SQUACCO HERON *Ardeola ralloides* *CREYR MELYN*
(B) A vagrant.
The only record was from Kenfig, Glamorgan, on June 28th 1994. This was the 6th record for Wales.
Of the previous 5, 3 were shot: at River Conwy, Denbigh 1828, R. Wye, Radnor 1867 and near Montgomery 1875.
The only other living individuals were at Porthcawl, Glamorgan May 17th – 30th 1954 and at Cemlyn, Anglesey June 11th 1988.

CATTLE EGRET *Bubulcus ibis* *CREYR Y GWARTHEG*
(B) A vagrant.
Three records in the period all of singles, two from Caernarfon: at Aberdaron May 2nd - 4th 1992 and Foryd Bay Apr. 19th 1998, the other from Skomer, Pembroke Apr. 30th 1996.

All the previous records came in the December 1980 – May 1981 influx, 2 to Pembroke and singles to Gwent, Caernarfon and Anglesey.

GREAT WHITE EGRET *Ardea alba* *CREYR MAWR GWYN*
(B) A vagrant.
Five accepted records in *Birds in Wales* (all post 1981), since then there have been a further 7 individuals at: Kenfig, Glamorgan July 22nd 1997, two individuals at Penclacwydd, Carmarthen the first on June 27th 1995 (individual seen the following day at Peterstone Wentloog, Gwent), the second on Sept. 30th 1996, Llangorse, Brecon Apr. 21st 1999 and at Inner Marsh Farm, Flint July 25th – Aug. 25th 1995.
One at the Braint est., Anglesey on July 8th 2000 was later seen at the Alaw est. July 25th – Sept. 29th. A ringed individual was seen at the latter site on Aug. 14th before moving to Llyn Alaw until the 30th.

Breakdown of all Welsh records by county:

Gwent	Glamorgan	Carmarthen	Pembroke	Brecon	Meirionnydd	Caernarfon	Anglesey	Flint
1*	1	2*	1	1	1	2	2	2

* the 1995 bird at Penclacwydd & Peterstone Wentloog.

LITTLE EGRET *Egretta garzetta* CREYR BACH

(E) Now a fairly common autumn migrant and winter visitor. Has attempted to breed.

	1992	1993	1994	1995	1996	1997	1998	1999	2000
Gwent		5-10	4		2-3	5	15	10+	35+
Glamorgan	2	8-13	27+	27	4	5	8	11	17+
G/C Burry Inlet	2	17-21	18-20	39	42	46	53+	81+	127+
Carmarthen	2			4	4-6		7	6	12+
Pembroke	4-5	19	22-24	30	52	45	50	45+	53+
Ceredigion		2-7	8	9	6	8-10	4-6	5	15+
Brecon		1		1	1				
Radnor			1					1	1
Montgomery		5	2	2	1		2	5	6
Meirionnydd		4	2		4	1		4	4
Meir/Caern Glaslyn			6	4				2	
Caernarfon		2	4	2	5	6	5	13+	15+
Anglesey		3-7	1	2-3	2	2	1	8	8+
Denbigh		3-4	1	7	2				
Flint					2	9	4	6+	11+

Birds in Wales states that this species is an uncommon but increasing visitor to Wales with about 70 records of 78 individuals up to 1991, principally in the period from 1970 onwards, reflecting an increase and spread in its breeding range northwards in France. This increase has continued dramatically, with this species being a familiar autumn and winter visitor to many estuaries in south Wales. In 2000 alone there were an estimated 304 individuals recorded in Wales. There is a distinct influx at the end of July / early August. The table above is far from complete but summarises the number of individuals recorded in each county in each year. It must be noted that as numbers increase the total number of individuals in Wales in any year is confused by many individuals crossing over the county boundaries. Figures for Glamorgan and Carmarthen can be broken down to birds in or away from the Burry Inlet, the main wintering area for this species in Wales.

The largest congregations were at Penclacwydd (Burry inlet) which reached its maxa. in August. The other large congregation was on the Cleddau estuary in Pembroke, where birds started arriving also in August but maxa. usually over the winter period. The development of the Gwent Levels Wetland Reserve, Goldcliff has also started to attract large numbers with 30 in August 2000.

Although this species has not bred successfully in Wales yet, there have been three occasions when nesting / displaying have been noted. The first in 1995 in Pembroke when birds were seen displaying in a heronry in March but nesting was not attempted. The following year 6 birds were seen displaying, 2 pairs appeared to have built nests, one of which may have incubated eggs. In the same year a single bird was observed collecting nest material in Carmarthen.

Unfortunately these pairs were unsuccessful but surely it is only a matter time before this species breeds successfully in Wales (this species has bred in recent years in Dorset, Devon, Kent and southern Ireland).

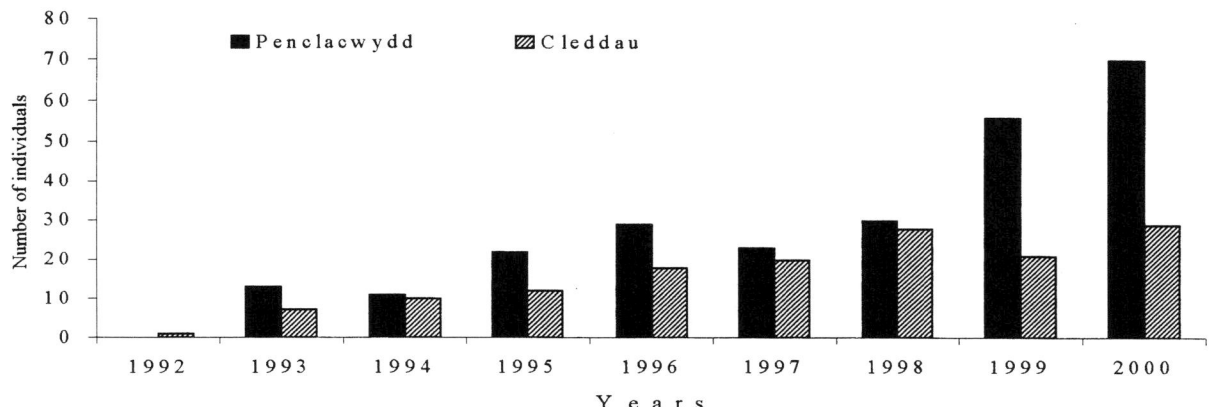

Maximum counts of Little Egrets 1992-2000

Birds in Wales 1992-2000

GREY HERON *Ardea cinerea* — CREYR GLAS

(E) Resident breeding species occurring widely around estuaries, lakes and inland watercourses; although the population is increasing in some counties, overall in Wales, it appears to be stable.

Birds in Wales quoted 55 nests at Dryslwyn, Carmarthen in 1985, this is now thought to be an over-estimate. Although incomplete, the data suggests a total Welsh population of 600-700 pairs. Table of heronry counts, the first number is the total number of nests, the second the number of heronries counted:

	1992	1993	1994	1995	1996	1997	1998	1999	2000
Gwent	114-5/5	102/5	103-7/6	159-161/8	156/8	127-9/7	162/	152/9	129/7
Gower	43/8	38/8	38/7	46/7	52/8	58/8	63/9	70/9	64/9
Carm.	140-3/15		22/5	54-63/14	53-58/15	27-30/6	42-7/8	65/16	
Pemb.	38/5	53/6	24/4	40/6	28/5	59/7	41/5	26/4	
Cere.	93/11	86/11	90/11	95/11	79/11	85/11	84/11	102/12	108/12
Brecon	10-12/2	23-43/2	43-7/5	80/5	53/5	60/5	44-50/5	82/5	85/7
Radnor						6/1		11/2	12-13/2
Mont.	31/4	21/3	29/4	38/5	40/5	41/6		59/7	34/4
Meiri.	38/8	38/6							
Caern.	10/2					10/1	7/1	66/8	59-78/7
Anglesey					61/10	8/2	43/	62/9	52/9
Denbigh						13/1	26/1		
Flint						6/1			

PURPLE HERON *Ardea purpurea* — CREYR PORFFOR

(B) A scarce visitor.

At least 8 records in the period, all of singles: in Gwent: at Caerleon May 3rd then presumably the same bird at St. Brides on the 24th 1992 and Goldcliff Sept. 26th 1993; in Glamorgan at Kenfig, a first summer bird May 27th 1994, Apr. 27th 1997 and at Baglan Pool Sept. 24th, Oct. 1st & 23rd 2000; at Gorsgoch Ceredigion May 16th 1998; in Caernarfon: at Bardsey Nov. 27th 1996 and at another location Apr. 12th - 15th 1998. Breakdown of all Welsh records by arrival date in the chart and by county on the map (one undated record).

Birds in Wales quote a record of 3, one of which was shot, at Talybont, Brecon in 1882. The shot individual is now thought to be the only reliable record.

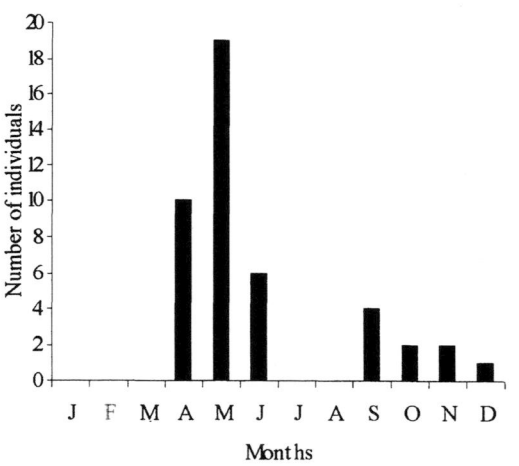

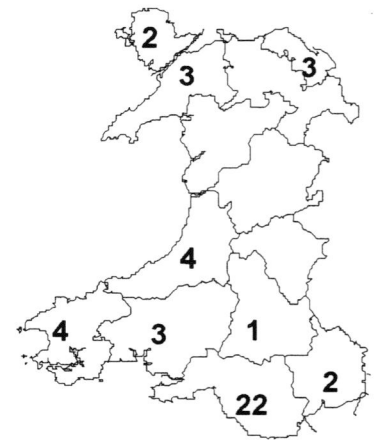

WHITE STORK *Ciconia ciconia* — CICONIA GWYN

(B) A rare visitor but there has been a dramatic increase in sightings in the past 28 years.

At least 7 individuals described in the period, although the possibility of escapes was never ruled out.

The first was from Llanfair Caereinion, Montgomery April 1993 until 1994, which was highly mobile being seen at Penrhyncoch Ceredigion May 26th - 28th. On Apr. 30th 1995 one was seen flying west over Rhos-on-Sea, Denbigh then at Bardsey Caernarfon the next day. An individual at Peterstone Wentloog, Gwent Nov. 9th then at Sker, Glamorgan 10th - 12th 1996 may have originated from Bristol Zoo. In 1997 a bird was observed over the M4 at Bridgend on Mar. 1st, while in 1998 there were three sightings probably only relating to two birds: at Llywernog Mine Museum, Ceredigion May 8th, at Llandrillo, Meirionnydd June 6th and later over Brecon on the 10th. Finally a single at Danescourt, Cardiff, Glamorgan May 29th 2000.

GLOSSY IBIS *Plegadis falcinellus*   CRYMANBIG DDU
(B) A vagrant to Wales, with only 12 records, three of them dating back to the 19th century.
A single at Marloes Mere, Pembroke June 16th 1996 was the only record since 1991.

SPOONBILL *Platalea leucorodia*   LLWYBIG
(E) A uncommon visitor to Wales, recorded mainly from estuaries and coastal lagoons and occasionally over-wintering and summering.

Dramatic increase in records of this species over the eight years. Five immatures wintered 1997/8 on the Dee estuary, with at least 3 remaining right through the summer. These birds were extremely mobile, accounting for many of that year's records. Four frequented the Dee, Flint/England March – October 1999 (presumably from the original five) and more than likely account for all that year's records in north Wales (e.g. 4 at Foryd Bay in September). Two new immatures arrived in September, remaining until October.

In the south, Penclacwydd, Carmarthen has entertained 3 individuals together on three occasions since 1991, birds departing from this site probably account for the majority of birds in south Wales.

A colour-ringed bird from Holland was on the Gann, Pembroke May 10th - 15th 1992.

The map summarises the distribution of individuals by county, although records often refer to the same roving individuals. The table below shows the minimum number of individuals in each year:

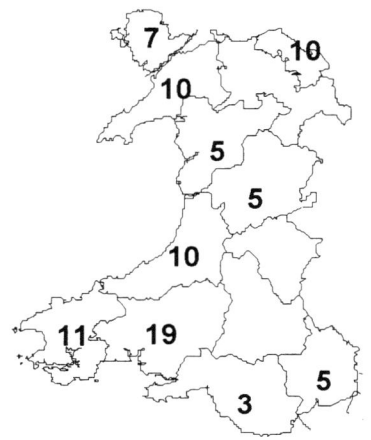

	1992	1993	1994	1995	1996	1997	1998	1999	2000
Min. number of birds	2	2	2	8	11	11	9	9	6+

An adult female of the Mauritanian race *P. l. bulsaci* that travelled to the UK via a boat, was present at Inner Marsh Farm, Flint from May 29th - June 4th 1998 and Nov. 25th - Dec. 2nd, at Inland Sea, Anglesey Dec. 16th – Jan. 5th 1999 then at Aber Ogwen, Caernarfon Jan. 9th – Mar. 5th before returning to Inner Marsh Farm Mar. 14th – May 12th and again on Sept. 23rd 1999 until May 28th 2000.

MUTE SWAN *Cygnus olor*   ALARCH DOF
(C) Resident breeding species, now increasing in most counties.

Birds in Wales states that due to the lack of waters large enough to allow the take-off runs, much of Wales is unsuitable for this swan. The main breeding areas are in Gwent, Glamorgan, Brecon and Montgomery. The table below summarises the number of breeding pairs in each county. Few counts were from complete surveys except from Brecon, the absence of figures for various counties does not indicate that this species does not breed there, just that there are no accurate census details available. Their large size, white plumage and huge nests makes this species one of the easiest to survey and as such this species is crying out for a comprehensive and accurate all Wales survey.

	1992	1993	1994	1995	1996	1997	1998	1999	2000
Gwent	10-11	10				16		12	23
E. Glamorgan	7		16	15		19	10	10	16
Gower		7			11		10	8	9
Carmarthen	13	12	13	18	7	8	11	11	11
Ceredigion		8+	9+			8	9		
Brecon	9	4	11	10	11	12	11	11	17
Montgomery		2	2	4	5	16	9	17	5
Caernarfon				21				7+	8
Anglesey			5+	5+	4+	5+	5+	8+	7
Denbigh								4+	
Flint - River Dee	5	5	3+	2+	3+	10	5	10	4

Birds in Wales 1992-2000

The breeding population in Carmarthen is about 25 pairs and is increasing due to newly created water-bodies.
Although incomplete, the data suggests a total Welsh population of between 100 and 200 pairs, little change from the last survey in 1983.
Large congregations recorded at many places, are tabulated in the table below. The max. counts however can be split into different times of the year.

Large counts at other sites included the flock at Dryslwyn, Carmarthen which numbered 42 in November 1994, 44 in December 1995 and 48 in June 1996. 55 were at Cilsan Bridge, Carmarthen in February 1996. 5 colour-ringed individuals from Llandrindod Wells Lake in 1998 were subsequently seen at the Ogmore estuary on June 5th 1999 and at the Knapp in November 1999, both Glamorgan.

Winter	1992	1993	1994	1995	1996	1997	1998	1999	2000
Glamorgan Knapp	76		107	117				85	102
Brecon Glasbury/Wye				73		130		155	103
Montgomery Pool Quay						54		50	
Flint Inner Marsh Farm						110		14	
Flint Shotton				86					

Summer / Autumn	1992	1993	1994	1995	1996	1997	1998	1999	2000
Glamorgan Ogmore			68					43	
Ceredigion Teifi estuary	40	36	48	42			49	47	40
Brecon Llangorse	75			42		85		105	103
Brecon Glasbury / Wye	82	68	67	100+				82	70
Caernarfon Aber Ogwen							69	72	84
Anglesey Llyn Alaw						54	51		

BEWICK'S SWAN *Cygnus columbianus*   *ALARCH BEWICK*
(C)   Regular winter visitor in small numbers to Brecon, Carmarthen, Gwent and Radnor. Elsewhere recorded in larger flocks on spring passage.

Birds in Wales states that this species is found predominantly in the south of the country in comparison to the Whoopers north west distribution. The table below summarises the number of individuals recorded in each winter period in each county. There appears to be little change in their distribution over the period except for the large numbers on the Dee / Inner Marsh Farm. Influx of birds into Pembroke during January 1997, when 21 at Bosherston, 16 at St. David's and 19 at Bicton Res.

Spring passage noted through the Dyfi estuary during March, 12 increasing to 19 on the 15th 1992, 50 on the 6th 1996, 21 on the 14th 1997, 38 on the 5th 1998.
Two birds wearing blue-neck rings recorded at Inner Marsh Farm, bird 409P present 1996/7 winter, returning October 1997 and January - Feb. 8th 1998, bird 38U, ringed in Siberia as a nestling on Aug. 15th 1996, present Nov. 29th - 30th 1998.

	1992		1993		1994		1995		1996		1997		1998		1999		2000	
Gwent			54	41	26	18					63	28			59	13	13	20
Glam.		1						3		9								
Carm.	24	15	11	1		1	1	17-20	14	29	4	14	3	2				
Pemb.		1		5		1		12	1	6	56	1	1	7	2			2
Cere.	19	7		1		2	2	15	59		36		38			3	20	
Brecon				5				13	20		9	5	1		1			2
Radnor				9	6					3	6			2	2			
Mont.				1			4									3		
Meir.	20	16	7		2	4	11		3	9	5	1	1	4		4*		
Caern.							6	6								4*		
Angle.	6						11	5	4	10								
Denbigh								3										
Flint	53	70		5	10	4	13	60	55	4	108	66	102	37	44	40	60	84

* 4 individuals on the Dwyryd / Glaslyn estuary in both Caernarfon & Meirionnydd.

The only large wintering flocks were at Llandegfedd Res. & Nedern in Gwent and Inner Marsh Farm in Flint:

	/92	92/3	93/4	94/5	95/6	96/7	97/8	98/9	99/00	00/01
Llandegfedd Res.		47	30				27	2	4	168
Nedern								36	2	4
Shotwick/Inner Marsh Farm	53	70			60	108	102	44	60	84

WHOOPER SWAN *Cygnus cygnus*   *ALARCH Y GOGLEDD*

(E) Regular winter visitor in small numbers to several Welsh counties, notably Anglesey, Caernarfon, Montgomery and Meirionnydd..

The main wintering grounds in Wales are Cors Caron in Ceredigion, Aberhafesp in Montgomery, Dwyryd / Glaslyn estuary in Meirionnydd and at Llanerchymedd on Anglesey. The max. counts for each winter at these sites is tabulated below:

	/92	92/3	93/4	94/5	95/6	96/7	97/8	98/9	99/00	00/01
Cors Caron	21	13	20	15	19	9	17	7	20	16
Aberhafesp		24	13	14	15	25	29	20	23	
Dwyryd / Glaslyn	69	79	56	37	41	35	70	43	59	43
Llanerchymedd	44	37	32	56	48	42	24	60	43	34

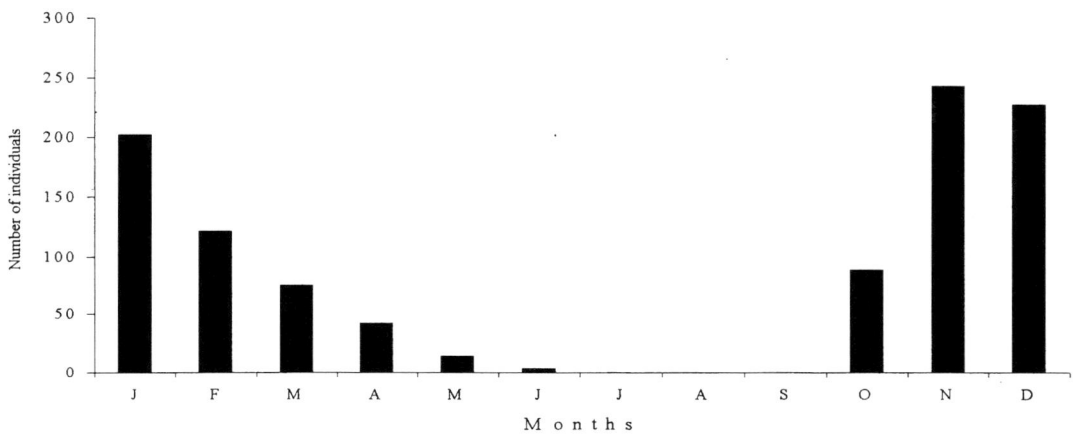

Monthly distribution of Whooper Swans (excluding the four main flocks) 1992-2000

Spring birds reported from Pembroke, at Skomer, 4 on May 8th & at Bosherston June 5th 1993, 5 at Ramsey May 30th 1994 and 3 at Skokholm / Skomer May 13th, then at Treleddyn from the 20th until June 4th 1995. A single turned up at Llyn Alaw, Anglesey on July 5th 2000.

An Icelandic ringed individual reported on Anglesey at Llyn Rhos Du in October & at Llyn Alaw Nov. 1st, then at Llyn Eiddwen, Ceredigion Nov. 6th, then Cilsarn Bridge, Carmarthen Dec. 11th 1994 until 1995. Another at Ro-fawr, Carmarthen Jan. 19th 2000, then at Dinefwr Pools on Mar. 18th, returning to Dryslwyn on Dec. 4th.

BEAN GOOSE *Anser fabalis*   *GWYDD Y LLAFUR*

(B) Rare and irregular winter visitor.

As stated in *Birds in Wales*, there have been numerous records of birds of uncertain origin. Since 1991 there have been 9 individuals recorded in Wales that were thought to be of wild origin, at: Llyn Traffwll & Llyn Alaw, Anglesey January 1992 and a *rossicus* there December 1996 – January 1997, a *fabalis* at Nedern, Gwent Jan. 31st 1996 then at Mathern on Feb. 10th and a *fabalis* at Ffairfach, Carmarthen December 1996 – January 1997. The only multiple record was of 5 *fabalis* at Trefeidan, St. David's, Pembroke Dec. 27th 1997.

PINK-FOOTED GOOSE *Anser brachyrhynchus*   *GWYDD DROED-BINC*

(C) A scarce and irregular winter visitor, chiefly to north Wales. Subject to hard weather influxes from the Ribble estuary.

The majority of records of this species come from north Wales, Anglesey and the Dee. Max. counts from these sites is shown in the table below. The only large wintering flocks were at Llyn Alaw, Anglesey and on the Dee at Inner Marsh Farm, Flint/ Cheshire.

	1995/6	1996/7	1997/8	1998/9	1999/00	2000/01
Llyn Alaw	136	104	41		36	
Dee		26	8	70	90	2

Large movements noted at: Dyfi est., Ceredigion 50 on Jan. 15th 2000, 25 – 30 at Llyn Alaw, Anglesey Sept. 19th – Oct. 22nd 1994, 115 over Penmynydd, Anglesey Jan. 9th 1993, 68 at Brynteg, Anglesey Dec. 13th 2000, 39 Talacre, Flint Jan. 1st 1996, 160 flew north over Inner Marsh Farm, Flint Jan. 1st 1997, 95 there in March 1997.

Elsewhere this species is a scarce and irregular visitor, with the numbers of individuals confused by birds of uncertain origin. The total number of individuals recorded in each county is tabulated below:

	1992	1993	1994	1995	1996	1997	1998	1999	2000
Gwent		6				1	1		
Glamorgan	1		1						
Carmarthen		6	5						1
Pembroke	10	2				2	1		1
Ceredigion	4	4	3	6	4	2	1	1	50
Radnor							1	2	
Montgomery									3
Meirionnydd	1	1		20				3	2
Caernarfon						1	2	10	2
Anglesey	10	6	29-34	15	186	124	65	11	108
Denbigh		1			2			3	
Flint	19-24	5	12	25	41	262	14	182+	80

WHITE-FRONTED GOOSE *Anser albifrons* GWYDD DALCEN-WEN

(E) Scarce winter visitor; regular flock of albifrons frequented Dryslwyn, Carmarthen but this ceased 1996, whilst Ceredigion supports the only regular flock of flavirostris although a new flock developed on Anglesey for a few years. Rare elsewhere, except occasionally during hard weather.

Main wintering flock of *flavirostris* on the Dyfi with a smaller flock on Anglesey at Llyn Alaw / Traffwll / Treflesg / Porth Cwyfan, while the only flock of *albifrons* was at Dryslwyn, Carmarthen. Here numbers dropped dramatically by 1995 & 1996 when birds only arrived in January. In 1997 birds only present on one day in March. There have been no records since.

Peak winter counts on the Dyfi, Anglesey and Dryslwyn are tabulated below.

Larger flocks noted elsewhere included: 28 at Ginst Point, Carmarthen Mar. 1st - 3rd 1994 and Trefenter, Ceredigion 70 on Nov. 24th 1993.

Influx of 77 birds into Pembroke October / November 1992, including 20 on the Dowrog and 47 at Treginnis / Porth Liski. Similar movement in October 2000 involving at least 24 birds: 10 passed Strumble Head on Oct. 31st, 10 the following day, 14 at Caerfai on the 31st and up to 7 at Treleddyn October – Nov. 27th. All those that were sub-specifically identified were *flavirostris*.

Passage observed over Bardsey in October 1998, 31 on the 19th, 23 the next day, one on the 23rd and 4 on the 25th. Similarly in 2000, 3 on Oct. 13th, 9 on the 14th and 13 on the 31st.

flavirostris	/92	92/3	93/4	94/5	95/6	96/7	97/8	98/9	99/00	00/01
Dyfi	179	134	141	155	147	124	97	167	116	134
Anglesey	2		3	20	61	3	3	2	2	
albifrons										
Dryslwyn	110	64	47	22	18	12				

Annual totals for the other records were:

	1992	1993	1994	1995	1996	1997	1998	1999	2000
flavirostris	1/2	27	30	22	22	1	21	12	24-42
albifrons			3		5-6	26	32		1
unknown		193	88	32	152	14		12	57

LESSER WHITEFRONTED GOOSE *Anser erythropus* GWYDD DALCEN-WEN LEIAF

(B) A vagrant.
A bird that had been collar-marked from the Finland re-introduction scheme was at Dryslwyn, Carmarthen Feb. 2nd - 10th 1991 before moving on to Slimbridge.

GREYLAG GOOSE *Anser anser*   *GWYDD WYLLT*

(E) *A rare winter visitor. Majority of records relating to feral birds mainly in Caernarfon, Anglesey and Denbigh.*

Most records appear to refer to birds of feral origin. Flocks over 100 were:

1994: 190 Aber Ogwen, Caernarfon in October, 100 River Clwyd, Denbigh in July.

1995: 104 Ynyshir, Ceredigion in October, 840 Llyn Alaw, Anglesey in September, 300 Llyn Coron, Anglesey in December, 100 at Pont Glan y Wern in March and 149 at Llyweni Hall in August, both Denbigh.

1997: 245 at Pont Glan y Wern.

1998: 268 at Aber Ogwen increasing to 632 in September and on Anglesey 1,000+ Llyn Maelog, 300 Llyn Alaw, 249 Llyn Coron, 215 Braint Estuary and 219 River Clwyd.

1999: 497 Aber Ogwen Aug. 27th, 114 Llyn Maelog in June, 205 in August and 223 in November, 100 at Llyweni Hall Jan. 23rd.

2000: 150 on the River Taf, Glamorgan January – February & 156 in October, 903 at Traeth Lafan, Caernarfon in September and on Anglesey, 22 at Llyn Maelog in January & 279 in November and 181 at Llyn Coron in September.

Two adjoining populations have now been established on the Carmarthen coast, at Taf, Tywi & Gwendraeth and since 1993 in the SE, based at Penclacwydd – Machynys, where winter numbers have recently peaked at 100 individuals.

Report of possible wild birds were: 9 at Olway Meadow, Gwent Feb. 13th 1993 (which were assigned to the race *rubirostris*). A spate of records in January 1996: in Pembroke one at Newgale on the 1st, 2 at Ramsey on the 29th, 2 also at Llangorse, Brecon on the 3rd and 13 over Bardsey, Caernarfon on the 27th.

CANADA GOOSE *Branta canadensis*   *GWYDD CANADA*

(C) *A locally common feral breeding resident in most counties, with populations continuing to expand.*

Max. winter counts from the main sites, Cleddau estuary in Pembroke, Newcastle Emlyn Ceredigion/Carmarthen, Dyfi in Ceredigion/Meirionnydd, Llangorse in Brecon, Welshpool in Montgomery, Dwyryd/Glaslyn in Meirionnydd and in Flint, Shotwick/Oakenholt and Inner Marsh Farm:

	/92	92/3	93/4	94/5	95/6	96/7	97/8	98/9	99/00	00/01
Cleddau		69	216	293	250	482	469	469	760	1080
Newcastle Emlyn	96	104	119	95	160	160	161		216	
Dyfi	175	280	453	520	681	682	1020	899	1707	1683
Llangorse	200	350	260	460	430	450	520	720	560	400
Welshpool						400	400	470		
Dwyryd/Glaslyn	17	6				120		115	178	247
Shotwick/Oakenholt	250	450	262	330	665	746	900	1300	1200	
Inner Marsh Farm									1200	1450

Two flocks on the Teifi in September 1998 (223 and 201), one of which was thought to have come from the Dyfi. Other large counts were: 200 Glasbury, Brecon Oct. 13th, 280 Pwll Patti, Radnor Nov. 13th 1999 & 250 there Nov. 11th 2000, 208 at Trawsfynydd, Meirionnydd and on Anglesey, 300 at Llyn Alaw/Traffwll in December 1998, 200 at Llyswen, Brecon Aug. 15th 1999.

BARNACLE GOOSE *Branta leucopsis* *   *GWYDD WYRAN*

(C) *A rare winter visitor. Majority of records relating to escapes from captivity.*

As with Greylag, very difficult to separate the wild from the feral. The only wintering flock of wild origin was on the Dyfi. In recent years individuals have started to arrive on the Dyfi with the Whitefronts in October. The monthly distribution is distorted by feral individuals, particularly in the summer months. Max. winter counts on the Dyfi were:

92/3	93/4	94/5	95/6	96/7	97/8	98/9	99/00	00/01
7	17	21	20	27	37	36	46	60

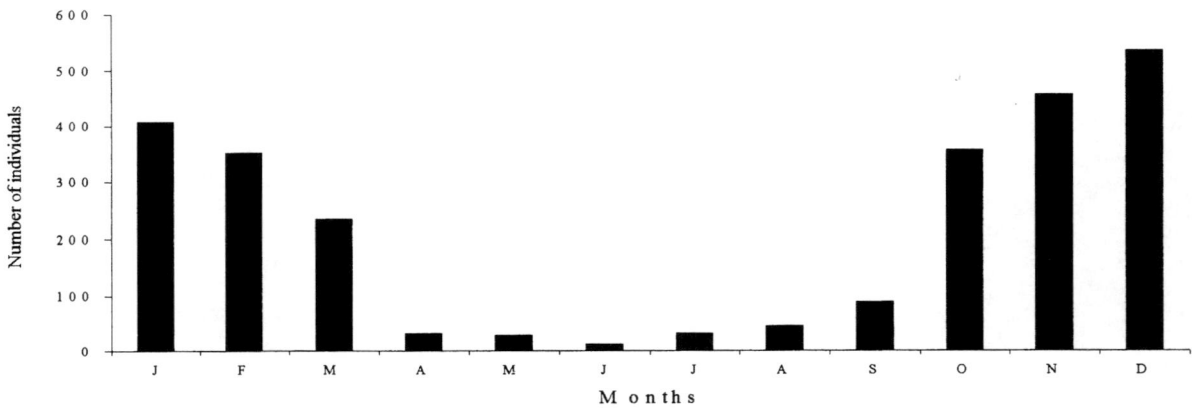

Monthly distribution of Barnacle Geese 1992-2000

BRENT GOOSE *Branta bernicla*   GWYDD DDU

(C) A regular wintering population is found in the Burry Inlet. Elsewhere, small numbers are recorded in winter and on passage.

Main wintering flock of *B. b.bernicla* in Wales is on the Burry Inlet, max. monthly counts were:

Year	Jan.	Feb.	Mar.	Apr.	Sept.	Oct.	Nov.	Dec.
1992	1360	1435	1030	85	20	510	555	975
1993	835	990	955	15	10	685	955	995
1994	825	1165	855	10	30	455	690	930
1995	845	980	695	5	10	165	485	820
1996	970	750	895	140	215	655	790	755
1997	765	1045	1015	5		665	815	89
1998	850	1165	670	20	5	775	1440	835
1999	970	1520	535		5	765	1095	750
2000	1210	975	695	5	85	195	1155	1105

Other large flocks of *B. b. bernicla* at the Menai Straits, Caernarfon 47 on Dec. 31st 1994, 19 in February, 21 in March & 15 in December 1999 and 58 in December 2000. 88 in Broughton Bay, Glamorgan winter 1995/6, with 80 the following November and off Penclacwydd, Carmarthen, 400 in March and 288 in December 2000.
One summered in Angle Bay, Pembroke 1998, 1999 & 2000.

The light-bellied race *hrota* is a scarce visitor in small numbers. The largest counts recorded were: 30 at Blackpill, Glamorgan on Oct. 8th 1992, 26 on Mar. 31st 1995 and 56 on Apr. 20th 1998; 23 on the Gann Est. Pembroke, Sept. 13th 1998 and 24 there Sept. 7th 1999; 36 at Foryd Bay, Caernarfon Jan. 29th with 31 there on Dec. 11th 1999, 48 in January 2000 and 13 in November; 63 at Abermenai, Anglesey Jan. 21st 1999, 31 there in January 2000 but only 2 in December, on Anglesey: 52 at Beddmanarch Bay Jan. 11th 1997, 33 there in March 1999, 29 in January 2000 and 90 in December.

Inland records at Llandegfedd Res. Gwent Oct. 27th 1994, in Brecon at: Llangorse: Feb. 13th 1993, Dec. 26th 1993 until Jan. 16th 1994, Nov. 12th 1994, 3 on the 19th and 2 on Nov. 25th 1995, Mar. 21st 1997 and Nov. 23rd 1998, at Glasbury Mar. 19th 1997, Ffordd Fawr Apr. 6th 1998 and at Talybont Oct. 23rd 2000; at Pwll Patti, Radnor Nov. 13th 1999 and at Dolydd Hafren, Montgomery Jan. 2nd – 17th 1999.

A single "Black Brant", *B. b.nigrans* Mar. 23rd - 30th 1997 at Beddmanarch Bay, Anglesey was the 1st Welsh record.

RED-BREASTED GOOSE *Branta ruficollis*   GWYDD FRONGOCH

(D) Feral.

Single, presumed escape, Dyfi Estuary, Ceredigion Oct. 3rd – 15th 1999, then on Anglesey to the end of the year.
There have only been two genuine wild individuals recorded in Wales. The first was shot at Milford Haven, Pembroke in 1935, the other at Camlad Meadows, Montgomery in 1950.

EGYPTIAN GOOSE *Alopochen aegyptiacus*   *GWYDD YR AIFFT*
(D) Feral.
A single at Inner Marsh Farm, Flint Aug. 3rd – 7th and Sept. 3rd – 7th 2000 and a pair at Celtic Manor Golf Course, Gwent in April 2000.

RUDDY SHELDUCK *Tadorna ferruginea*   *HWYADEN GOCH YR EITHIN*
(D) Feral.
A female at Inner Marsh Farm, Flint Jan. 1st – May 25th 2000 and July 19th until the end of the year.

SHELDUCK *Tadorna tadorna*   *HWYADEN YR EITHIN*
(C) A locally common breeding resident in coastal areas, but rare passage migrant to inland counties.

The Shelduck's winter distribution in Wales is similar to its breeding distribution – almost exclusively coastal. The most important sites in Wales are the Dee estuary (the 2nd most important site in the UK) and the Severn estuary. Max. winter counts, November – March, are shown in the table below. Those with an average over 750 are of National Importance.

Mean peak winter counts during the period 1996-2000 for the Burry (as a whole, where the total monthly count is used not necessarily the largest shown in the table) was 1712, the Cleddau 910 and for the Dyfi 430. Interestingly counts from the Dee peak in October while all other sites tend to peak in January or February. A record count on the Dee in October 1997 of 10,418 birds mainly on the English side.

Other large winter counts were: in Glamorgan, 430 at Taff / Ely, in February 1992, 1000 at Cardiff docks in December 1994 and 500 at Cardiff Heliport Feb. 15th 2000; 350 at Malltraeth, Anglesey in January 1993, 460 in January 1998 and 604 on Dec. 26th 2000.

		1992	92/3	93/4	94/5	95/6	96/7	97/8	98/9	99/00	00/01
Gwent	St. Brides/W		850	1350	500		500		1055	1370	
Glam.	Rhymney	1500	930	922	900	100	750	823	800	1200	800
	Burry S.	1465	1210	860	1225	1155	1325	1395	1250	1400	780
Carm.	Burry N.		473	520	592		1064	549	626	618	60
Pemb.	Cleddau	923	470	877	1178	976	1023	939	921	690	509
Cere.	Dyfi	430	345	420	301	356	398	457	435	502	262
Caern.	Traeth Lafan									506	406
Flint	Point of Air	1000	346	590	1316	807	660	1163	1500	1900	
	CQ/O'holt	600					713				

In 1992 a major breeding survey was conducted by the BTO / WWT, the results of which are tabulated below:

	Spring		Summer	
	Adults	Territories	Adults	Juveniles
Gwent	327	90	441	79
S. Glamorgan	171	71	12	23
M. Glamorgan	26	0	4	15
W. Glamorgan	131	0	27	14
Dyfed	692	177	193	241
Powys	0	0	0	0
Gwynedd	1940	776	232	609
Clwyd	1016	203	3368	213
TOTAL	4303	1317	4277	1194

There have been few comprehensive surveys since, the table below gives the number of ducklings counted each year at some of the main breeding grounds.

		1992	1993	1994	1995	1996	1997	1998	1999	2000
Gwent	Severn shore		53						20	
Glamorgan	Burry S.	18	14	18	7	50	38	10	20	
Carmarthen	Penclacwydd				4	30	38		32	
Pembroke	Cleddau	91	75	64	122	217	186	205	102	112
Ceredigion	Dyfi	103		53						
	Teifi	23			13		3	29	15	14
Meirionnydd	Dysynni		22							33

The dramatic increase in the number of ducklings on the Cleddau in 1996 was thought to be influenced by a fall in the fox population as a result of mange.

Inland breeding is rare, mainly on the Wye, where 8 pairs were found between Chepstow and Wyncliff in 1999. Breeding was also recorded in Gwent, a pair with 5 ducklings on the River Usk at Caerleon in 1993 (also bred in 2000) and 3 pairs bred at Llandegfedd Res. in 1993 (also bred in 2000).

The first ever breeding on Bardsey took place in 1999, 8 young were fledged.

The Welsh population is estimated at between 500 and 800 pairs.

MANDARIN DUCK *Aix galericulata*   HWYADEN GRIBOG
(C) Scarce breeding resident, recorded, but not breeding in the majority of Welsh counties.

The small breeding population in Montgomery declined during the period under review and no birds were seen on the Mochdre brook, a former stronghold, since 1995. No birds were seen on the Severn near Llandinam since the early 1990s. This species may now be extinct in the county.

A new breeding population was discovered in 1999 in Caernarfon and Meirionnydd and in 2000 pairs bred at Nantmor, 4-5 birds at Porthmadog and a number of broods were seen on the Glaslyn. A female with 7 ducklings was seen on the River Clwyd, Denbigh in 2000, in the same year a pair was also present at Tintern, Gwent in April and individuals were regular at Margam Park, Glamorgan.

The current Welsh population is likely to be fewer than 20 pairs.

WIGEON *Anas penelope*   CHWIWELL
(E) Common passage migrant and locally abundant winter visitor, chiefly to coastal counties.

		/1992	92/3	93/4	94/5	95/6	96/7	97/8	98/9	99/00	00/01
Gwent	Llandeg. R.		610	500				425	265	300	140
Glam.	Burry S.	1670	2370	1510	3140	2585	5180	2675	3020	2735	2080
Carm.	Burry N.	400	383	810	707	615	1050	791	566	855	388
	Dryslwyn			600	450	500			600		
Pemb.	Cleddau	1739	2461	2088	2403	3455	3351	3058	4007	4078	3611
Cere.	Dyfi	4003	3860	4770	3665	4363	4681	2585	2366	2984	1453
Meiri.	D/Glas	2060	1198	1385	1810	1400	1500	1117	1200	1693	781
	Mawddach	779	717	933	338	676	630	25			
	Dysynni	857	1300	1600	2000	2007	953	2000	1000	800	750
Caern.	Traeth Lafan				600		1218		597	1165	1621
	Foryd								400	1640	2790
Anglesey	Llyn Alaw	604		457	304	1085	1012				
Flint	CQ/O'holt	800		410	1650	220	1500	1000			
	Bagillt Bank						6200		900		
	Inner M. Fm				1500		3000	3250	1400	1250	1800

Predominantly a coastal wintering species, with the major concentrations on estuaries. The table above gives the winter maximum counts at the main sites. Mean peak winter counts from the top 3 estuaries during the period 1996-2000 were 3,772 on the Burry Inlet (peak counts for the whole site were used), 3590 on the Cleddau and 3,396 on the Dyfi. These are over National Importance Level of 2,800.

Other large counts were of 400 at Usk Mouth, Gwent in January 1994, in Caernarfon 1700 at Foryd Bay in October 1995 and 1740 in October, 850 at Aber Ogwen in January - March & 450 in October 1998; on Anglesey 600 at the Braint Estuary in December 1994, 450 there in October 1998, 874 on the Inland Sea in November 1997 with 840 there in January 1998; 400 in Kinmel Bay, Denbigh in November 1997. Max. counts on the Dee in 2000 were 1158 in the 1st winter period and 3283 in the 2nd.

Although individual birds often summer at many places there has not been a confirmed breeding record since 1934.

Summering records include one at Dinefwr Ponds, Carmarthen June 2000, on Anglesey: 3 males at Llyn Coron in 1994, 2 at Malltraeth in 1995 & 2000, a single at Llyn Alaw in 1996 & 2000, 2 at Llyn Traffwll in 1998 with a single the following year and one at Cemlyn in 1999 & 2000; single males at the Teifi Marshes, Ceredigion / Pembroke 1996 & 1999 and a single in Caernarfon in 1998 which may have bred. 6 males + 2 females summered at Inner Marsh Farm, Flint in 2000 but there was no evidence of breeding.

AMERICAN WIGEON *Anas americana*
(B) A vagrant from North America.

CHWIWELL AMERICA

There is a lot of confusion over the true number of Welsh records of this species due to the possibility of birds returning to the same sites for a number of years. It is highly probable that all the records relate to just 12 individuals.

The first Welsh record was of a male at Llyn Llywenan, Anglesey, 21st & 23rd June 1910. The next record wasn't until 1975 of an immature male at Kenfig 19th Oct - 2nd Nov. A male was present at Llyn Bodgylched, Anglesey on 30th - 31st Jan & 11th Feb 1977 and again on 21st Jan - 25th Feb & 1st Mar 1979.

A female was at Ynyshir, Ceredigion, on 2nd November 1981 whilst Kenfig pool, Glamorgan played host to an immature 29th Oct - 9th Nov 1985.

A record published by BBRC of a male at Inner Marsh Farm, Cheshire/Flint on Apr. 26th 1990 and therefore was included in *Birds in Wales* was in fact a female and was only recorded on the Cheshire side of the reserve.

A male was recorded at the same location on 30th Nov - 26th December of that year.

Records of a male in north Wales since 1993 probably relate to the same individual. It was first recorded at Foryd Bay on 7th - 9th Mar and again on 3rd - 4th July 1993. It was then found on the Conwy estuary on 20th Dec 1994 until 12th Feb 1995 before moving to Inner Marsh Farm on 18th - 26th Feb. Favouring the Conwy estuary again on 6th Nov 1995 until 15th Mar 1996 and again on 7th Oct until 28th Dec 1996, moving to Llanfairfechan 29th Dec until 21st Feb 1997, moving to Inner Marsh Farm on 14th - 19th Mar. Returning again to Llanfairfechan 13th Dec until 10th Jan 1998 before moving to Foryd Bay 29th March and 9th - 18th April (also being seen at the Braint estuary, Anglesey on Apr. 7th). It returned to the Conwy estuary on 25th September staying until Dec. 27th before moving to Inner Marsh Farm then to Morfa Madryn until Jan. 20th 1999. This bird was probably the same individual that was seen at Inner Marsh Farm May 7th – 25th 1999. This bird did not return the following autumn but did the year after at RSPB Conwy Aug. 23rd – 26th, then at Foryd Bay Aug. 29th – Oct. 22nd.

Another male was found at Porthmadog, Caernarfon Aug. 21st – Sept. 15th 2000.

In south Wales a female was at Dingestow, Gwent, 26th Aug - 10th Sept. 1995 and a male turned up at Lawrenny in Pembroke, 27th Dec 1996 - 15th Jan 1997, a male at Penclacwydd, Carmarthen Oct. 3rd - 9th 1998, a 1st winter male arrived at Peterstone Wentloog, Gwent Oct. 31st – Nov. 27th, then at Goldcliff on the 28th 1999 and a male at Cors Caron, Ceredigion Dec. 25th 1999 until Mar. 7th 2000.

GADWALL *Anas strepera*   HWYADEN LWYD

(C) Regular breeding bird in small numbers on Anglesey, and has bred in at least two other counties. Elsewhere a regular winter visitor to most counties in small numbers.

		/1992	92/3	93/4	94/5	95/6	96/7	97/8	98/9	99/00	00/01
Gwent	GLWR									10	22
Glam.	Kenfig	49	23	47	38	32	18	36	43	34	17
	Eglwys Nun.R.					12		20			
Carm.	Penclacwydd		22		28	41	45		50		26
	Witchett Pool	18	13				14				
	Machynys Pd.			40	28	28	21	29			
Pemb.	Herbrandston				15		23	48		26	4
	Marloes		22				16	11		18	11
Brecon	Llangorse			15				32	16	20	18
Angle	Llyn Alaw		20				23	62			
	Llyn Traffwll					18		29	10		
	Malltraeth									11	48
Flint	Shotwick Lake					39	33				
	Inner M. Fm.					25	25		10	12	3

Concentrations of over 20 individuals are rare in Wales, except on Anglesey. The table above gives the maximum winter, November – March, counts form the main sites.

Large counts elsewhere were from Pembroke in 1997, during January - March 21 at Bicton Res. and 25 at Bosherston and 66 at the latter in November - December; 51 on Valley Lakes Anglesey in November 1997. A total of 65 recorded in Pembroke in January 1999, with 53 in December. 24 were at Llyn Helyg, Flint Dec. 10th 1999 increasing to 33 by Jan. 27th 2000.

Breeding records have been sporadic since the first breeding record in Wales in 1969. *Birds in Wales* states that the main breeding areas are on Anglesey and that the island's population was approximately 15 pairs in 1991. This population has more than doubled in the last 9 years. Data from this period, 1992-2000 are incomplete but the best estimate of the current number of breeding "pairs" in Wales is from 1998 when there was a minimum of 56 pairs.

Breeding only reported at a handful of sites, some more regularly than others. The figures in brackets refer to the max. number of pairs recorded during 1992-2000): Machynys Pond (1) & Penclacwydd (4) in Carmarthen, Skomer Island in Pembroke (2), RSPB Conwy in Caernarfon (a male hybridised with a female Mallard, producing 3 young), on Anglesey at the Valley lakes of Llyn Penrhyn, Treflesg and Dinam (18+) with up to 35prs. on all other Anglesey sites (including Llyn Alaw) and at Inner Marsh Farm in Flint (2).

The breeding record on Skomer in 1999 was the first confirmed case for Pembroke and although 6 eggs hatched only one young eventually fledged. In 2000 pairs were also seen at GLWR & Corus Llanwern in Gwent, Marloes in Pembroke and at Malltraeth on Anglesey.

Summering birds also at Green Moor/ Llanwern, Gwent in 1999, Teifi Marshes in 1992, Ynyshir & Afon Leri in 1995 all Ceredigion, Llangorse, Brecon in 1998, Glaslyn, Meirionnydd in 1992 and Conwy RSPB in 1999.

TEAL *Anas crecca*   CORHWYADEN

(C) Small numbers breed in most counties, otherwise a widespread and abundant winter visitor and passage migrant.

Birds in Wales states that the wintering population favours low-lying lakes, marshes and estuaries (the reverse of the breeding distribution). During the period under review the Dee and Cleddau estuaries continued to be the most important wintering areas, although the Burry Inlet is close behind. The table below gives the max. winter counts, November – March, at sites that recorded over 500 once. Mean Winter Peaks during the years 1996-2000 for the top 2 sites were 2,071 at Burry Inlet (peak counts for the whole site were used) and 2,446 on the Cleddau. These are above the National level of importance of 1,400.

		/1992	92/3	93/4	94/5	95/6	96/7	97/8	98/9	99/00	00/01
Gwent	PW/St. Brides			700			650	550	400	636	540
	Goldcliff/Usk			700			500		115	593	1066
Glam.	Burry S.	385	260	255	505	760	1820	1660	1300	980	1050
Carm.	Burry N.	500	570	497	360	657	1015	1115	1755	990	223
Pemb.	Cleddau	1930	1555	1787	1200	2948	2217	2594	2085	2384	1619
Cere.	Dyfi								417	978	1174
Meiri.	D/Glas	350	486	200	576	604	266	360	397	600	400
Angle	Llyn Alaw	412		409	450	1370		664	268	183	412
	Malltraeth				1000			765	900	609	1229
Flint	CQ/O'holt	2500	2500	750	1000	1000	2500	1800		433	
	Inner M. Fm.				1550	4200	2500	3000	770	3500	3200

Teal is the second most widespread but not the second most numerous breeding duck in Wales. It is a rare breeder in many parts, often overlooked. Although the data is incomplete, what there is suggests a Welsh breeding population of less than 100 pairs compared to the 125-150 pairs as quoted in *Birds in Wales*.

During the period under review breeding were recorded in (figures in brackets are the max. number of pairs): at Oxwich, Glamorgan (2); Pembroke: on Skomer (2), Skokholm (1+) and Ramsey (1+); Ceredigion: at Cors Fochno (20), Ynyshir (10), Cors Caron (5+) and Llyn Rhuddnant (1); Brecon: on the Epynt (unknown), Brechfa pool (2), Pentrebach (1), Pwll Gwy-rhoc (1), Talybont (1), Llyn Traeth Bach (1); Radnor: on the Elenydd - Elan Valley (8); Montgomery: on Llyn Newydd (4), Dyfi saltings [also in Meirionnydd] (2+); Caernarfon: at Conwy RSPB (4); Anglesey: on Malltraeth (14). Birds recorded summering at Kenfig, Glamorgan, Marloes Mere, Pembroke and Teifi Marshes, Ceredigion / Pembroke and Inner Marsh Farm, Flint.

GREEN-WINGED TEAL *Anas carolinensis*
(B) Vagrant from North America.

Correction to *Birds in Wales*, an individual at Llyn Traffwll, Anglesey Dec. $2^{nd} - 3^{rd}$ 1990 not 1991 as published.

Since then there have been 8 other records of males: at Penclacwydd, Carmarthen Mar. 20th 1999; in Pembroke: at Skokholm Apr. $15^{th} - 17^{th}$ 1996, moving to Marloes Mere May $3^{rd} - 11^{th}$, presumed the same returning bird the following year at Skokholm Apr. 17th then at Skomer Apr. 18th – May 1st, an immature male at Skomer Nov. $17^{th} - 20^{th}$ 2000; Teifi Marshes, Ceredigion Apr. 19th 2000; at Llyn Coed y Dinas,

CORHWYADEN ASGELL-WERDD

Montgomery April – May 1997; in Caernarfon: at Conwy RSPB December 1996 – Apr. 21st 1997 & Feb. 8th 1999 and 2 at Morfa Madryn / Spinnies Feb. 22nd – Mar. 26th 2000.

Breakdown of Welsh Green-winged Teal records, graphed by month (not arrival dates, as many were long-stayers / multiple occurrences) and individuals by county on the map.

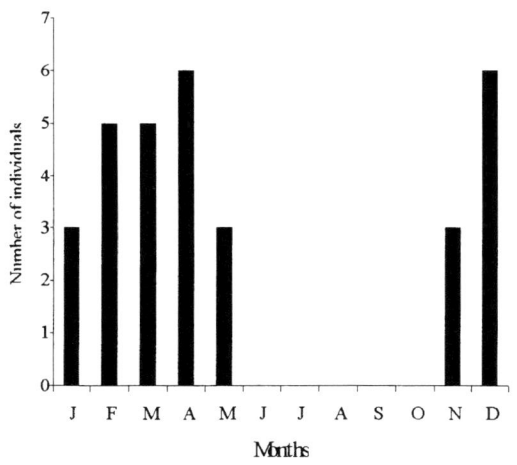

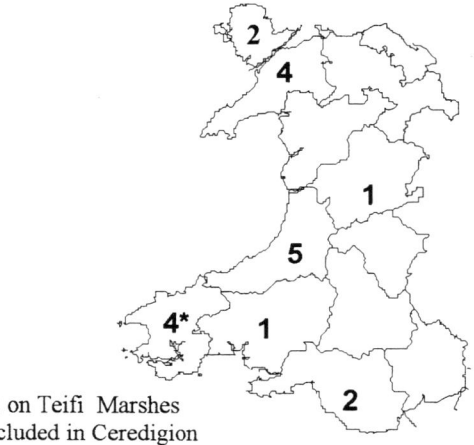

*2 on Teifi Marshes included in Ceredigion

MALLARD *Anas platyrhynchos*
(C) Common and widespread breeding resident, abundant passage migrant and winter visitor to all counties.

Mallard is by far the most widespread and numerous breeding wildfowl species in Wales. Despite this, its is by no means common in all areas and its true distribution is modified by releases by wildfowlers. Using data from the National Atlas, the Welsh population is estimated at 7,200 pairs. This ties in with estimates in *Birds in Wales* of 7-8,000 pairs, with the main concentration on Anglesey. BBS data 1994-1999 however suggests a population change of 51%.

HWYADEN WYLLT

Max. winter counts, November – March, at the main sites, that recorded over 500 once, are tabulated below. Figures during this period are generally lower than those quoted in *Birds in Wales*, e.g. up to 1,000 on Llyn Alaw and regular counts of over 1,000 on the Dyfi.

No more than 18 were reported in Cardiff Bay, Glamorgan in 2000 compared to a pre-barrage max. in 1999 of 326.

Birds in Wales 1992-2000

		/1992	92/3	93/4	94/5	95/6	96/7	97/8	98/9	99/00	00/01
Gwent	PW/St. Brides			710					400	560*	815
Carm.	T/T/G				906	539	664			153	86
Pemb.	Cleddau	291	520			574			564	353	508
Cere.	Dyfi	866	1719	1416	913	953	569	421	398	578	467
Meiri.	Dwyryd/Glaslyn	445	703	942		850	535				
Caern.	Traeth Lafan									292	614
Angle	Llyn Alaw	527	753			600					
Flint	Point of Air			595							

* 620 counted at St. Brides in January are not included in this figure.

PINTAIL *Anas acuta*   HWYADEN LOSTFAIN
(C) Rare breeder, passage migrant and locally abundant winter visitor to some counties.

		/1992	92/3	93/4	94/5	95/6	96/7	97/8	98/9	99/00	00/01
Gwent	PW/SF								300	433	48
Glamorgan	Rhymney	300	260	292	330	256	419	300	415	500	254
	Burry S.	2410	1385	1800	1820	3540	3490	1605	3265	3600	4275
Carmarthen	Burry N.	480	475	804	898	664	1502	793	1342	2075	570
Ceredigion	Dyfi	254	474	181	187	330	265	173	169	266	158
Meirionnydd	D/Glas	301	224	222	181	362	300	310	162	230	150
Flint	CQ/O'holt	3000	2500	2000	2000	7500	8000	4200	5600		
	Point of Air	300	182	383	486	312	460	235	360		

Birds in Wales states that traditionally the most important estuary in Britain is the Dee, with numbers up to 12,000, while numbers on the Burry Inlet have increased during the last century, reaching 3,000 in 1991. The max. winter counts, November – March, at sites that recorded over 300 once are tabulated below. The data from the Dee, Connah's Quay – Oakenholt only tells half of the story but numbers on the Burry have increased during this period. Mean Peak Winter counts during 1996-2000 were 378 for the Rhymney estuary, above National Importance Level (280) and 4,290 for the Burry Inlet (whole site counts were used) which is above International Importance Levels (600) and higher than the max. count quoted in *Birds in Wales* of 3,000 in February 1991. Mean Peak Winter counts for the Dyfi, 241 and the Dwyryd / Glaslyn, 273 are below the National Importance Levels.

Large flocks recorded elsewhere: 300 at Uskmouth, Gwent in January 1999, 400 at Cardiff Heliport, Glamorgan Nov. 17th 1998, 390 in Foryd Bay, Caernarfon Feb. 1st 1996, 400 at the Braint estuary, Anglesey Dec. 24th 1996, 600 on flooded meadows at Holt, Denbigh Jan. 14th 1998 and at Inner Marsh Farm, Flint 300 in February, 800 in September and 75 in December 1999, 310 in November 2000. The Dee count of October 2000, numbered 4216.

A small spring passage recorded through the Cleddau, Pembroke in 1996, 34 on Feb. 24th, 44 on Mar. 3rd with 36 on the 17th.

Breeding in Wales was not proved until 1988, when a pair bred on Skomer. During 1992-2000 the only successful breeding was on the islands of Skomer and Skokholm. Summering individuals were also recorded at Inner Marsh Farm, Flint in 1998, 1999 and 2000.

1992: 2 pairs at Skomer, both hatched young but probably didn't successfully rear them.
1993: a pair on both Skokholm and Skomer, both failed.
1994: 4 – 5 pairs on Skomer fledged 7 young.
1995: 2 pairs Skomer fledged 2 young.
1996: 2 pairs on Skomer, both failed.
1997: a pair on Skomer raised one young. The female was a Mallard hybrid.

GARGANEY *Anas querquedula*   HWYADEN ADDFAIN
(E) Recorded in small numbers on passage. Although probable breeding is more frequent, there have been only 5 confirmed Welsh breeding attempts and only one since 1991.

Birds in Wales states that the first confirmed breeding in Wales was in 1936. It listed only a handful of subsequent breeding records from Gwent, Glamorgan, Ceredigion, Caernarfon, Anglesey and Flint, with the last confirmed breeding in Wales was in 1980 at Llyn Dinam, Montgomery. Since then there the only confirmed breeding record was on the Dyfi NNR, Ceredigion in 1999, although breeding was thought probable at Llyn Llywenan, Anglesey in 1992. A pair were seen on the Dyfi in 2000 but there was no evidence of breeding and in the same year a male at the Teifi Marshes, Ceredigion / Pembroke was paired up with what was thought to be a Teal x Garganey hybrid.

In 2000 individuals were present all summer at GLWR, Goldcliff, Gwent (with up to 6 there in late summer), at Penclacwydd, Carmarthen (females / immatures seen in the late summer) and at RSPB Conwy, Caernarfon.

The following table shows the number of individuals recorded in spring. In light of the difficulty in proving breeding, this gives some indication of the size of the breeding population.

	1992	1993	1994	1995	1996	1997	1998	1999	2000
Number in spring	16	25	16	20	17-18	17-19	14	14	12

Spring and autumn individuals were recorded from all Welsh counties, except Radnor. The totals in each county are shown in the table below.

	1992	1993	1994	1995	1996	1997	1998	1999	2000
Gwent	4	7 - 8	5	9	6	1	3	5	7+
Glamorgan	4	2	5	4	4	2	2	5-6	2
Carmarthen	5	11	3	3	6	7 - 10	4	7	3-5
Pembroke	4	4	4	6	2	1	1	4	1*
Ceredigion	2	3 - 4	5		1	1	2	1	3*
Brecon	1			2					
Montgomery					2 - 3			1	
Meirionnydd				1		2			
Caernarfon					3	2	4	5	3
Anglesey			4	up to 18	6	2	3	12	2
Denbigh				1					
Flint	2				1	3		2	

* single male at Teifi Marshes in both counties.

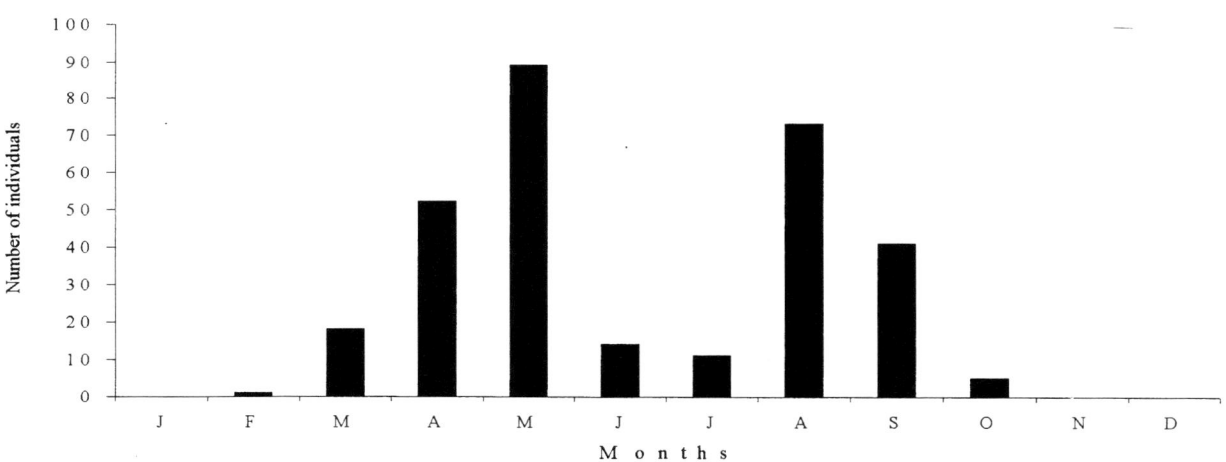

Monthly distribution of Garganeys 1992-2000

BLUE –WINGED TEAL *Anas discors*   *CORHWYADEN ASGELL-LAS*

(B) A vagrant from North America.

Male at Talacre, Point of Air, Flint Oct. 2nd – 12th 1997 and a female at Penclacwydd, Carmarthen Mar. 12th – 26th 2000 were the 5th and 6th Welsh records.

Four other records quoted in *Birds in Wales*, the first of a male shot on Anglesey 1919, a female / immature on Skokholm, Pembroke September 1960, a male at Cemlyn, Anglesey March 1983 and a pair at Shotwick, Flint in June 1983.

SHOVELER *Anas clypeata*   HWYADEN LYDANBIG

(E) Locally common winter visitor and passage migrant. Annually attempts to breed on the Pembrokeshire islands and on Anglesey, and occasionally elsewhere in Wales.

		/1992	92/3	93/4	94/5	95/6	96/7	97/8	98/9	99/00	00/01
Gwent	PW/St. Brides		102	110	80		150	100	100	160	131
Glam.	Burry S.	140	316	268	355	735	575	665	550	520	205
Carm.	Burry N.	70	182	144	400	496	550	630	348	349	180
Angle.	Malltraeth									191	152
Flint	Shotton				154						
	IMF					109		98	70	139	

Birds in Wales states that winter gatherings of more than a few hundred are rare in NW Europe, with only about a dozen sites in the whole of the UK holding more than 250 birds. The only site that supports large numbers of wintering Shoveler in Wales is the Burry Inlet, with the max. ever recorded there of 449 in December 1988. Since then the max. winter counts at this site have been higher than this (except in 2000). The table above shows the max. winter, November – March, counts at sites that recorded over 100 once (National Level of importance, 400 is the International Level). The Mean Peak Winter count for the whole of the Burry Inlet during 1996-2000 was 1,034, more than twice the highest count quoted in *Birds in Wales* on the Burry of 449 in Dec. 1988.

Large counts on Anglesey: in 1999, 109 at Malltraeth in October and 200 at Plas Bog, Bryngwran on Feb. 15th; in 2000, 100 at Llanfaelog Nov. 23rd, 150 at Llyn Penrhyn Dec. 15th and 99 at Llyn Cefni on Dec. 18th.

In Wales, this species has been a regular breeding bird for many years on Anglesey but elsewhere breeding records are sparse. During the period under review breeding was reported from a total of 12 sites in 4 counties. The numbers at the main sites (i.e. those holding more than 3 pairs in any one year) are tabulated below, but otherwise 1-2 pairs bred, in at least one year at: Llyn Alaw and Llyn Cefni (Anglesey), Cors Fochno and Ynyshir RSPB (Ceredigion), and Llangorse Lake and Brechfa Pool (Brecon). The counts for the whole Valley Lakes complex are difficult to interpret and those for 1994 and 1995 could well be over estimates compared to later years. This is due to the drake count information on file not distinguishing group size. However, counts in 1999 (and 2000) suggest that groups of more than 5 drakes would be unusual at this site.

3 prs. were seen throughout the season in 2000 at GLWR Goldcliff, Gwent, display and even a distraction display was observed but breeding was never confirmed. Birds were also present all year at Inner Marsh Farm, Flint.

Total Welsh population estimated at less than 90 pairs, an increase from the 40 pairs quoted in *Birds in Wales*.

		1992	1993	1994	1995	1996	1997	1998	1999	2000
Carmarthen	Penclacwydd WWT							6	3	1
Pembs	Marloes Mere	1	1	3	0	1	0			
	Skokholm	3	3	3	3	1	0		1	
	Skomer	2	2-3	2-4	2-3	3-5	3	2	1	2
Ceredigion	Aber Leri / Cors Fochno	1-2	2	3	1	1	0	0	0	
Anglesey	Llyn Penrhyn, Treflesg and Dinam			35	29+	7+	8	20	13	Pres.
	Other Valley Wetlands complex sites			NC	NC	NC	NC	8	10	Pres.
	Malltraeth RSPB			4	3	4	6	5	5	11

+ Llyn Dinam not counted in 1995 or 1996

RED-CRESTED POCHARD *Netta rufina*   HWYADEN GRIBGOCH

(E) An increase in the feral population in England has resulted in an increase in records in Gwent and Glamorgan.

The true status of this species in Wales is unclear due to uncertainty over the bird's origin. Although there were 19 accepted individuals 1992-1999 the vast majority of sightings went undescribed. This species was removed from the Welsh Records Panel description list as of 1st January 2000. In the following year there were at least 7 individuals seen in Glamorgan and 2 in Flint.
Breakdown of accepted records:

Gwent	Glamorgan	Carmarthen	Radnor	Meirionnydd	Anglesey	Flint
6	14	2	1	1	1	3

POCHARD *Aythya ferina* *HWYADEN BENGOCH*

(C) Resident breeding species in small numbers in several counties, otherwise a locally common winter visitor, chiefly to coastal counties and passage migrant.

		/1992	92/3	93/4	94/5	95/6	96/7	97/8	98/9	99/00	00/01
Gwent	PW / St. B.		216	346	171		250	160	125	276	
Glamorgan	Rhymney	250	660	200	132	47			105	195	95
	Lisvane	112	40	29	132	47	58	73	88	54	
	Roath Park	97	42	86	180	50		112	84	46	
	Eglwys Nun.		271	280	230	419	267	460	308	125	220
	Cosmeston	101	66	103	80	187	108	59	24		
	Fendrod L.		42	174	205	222	106	25			
	Kenfig	167	74	53	150	177	248	498	200	50	151
Carmarthen	Penclacwydd		155			121	168	105	96	140	82
	Upper Lleidi		128	29		110					
Brecon	Talybont			55	240					50	13
	Llangorse	166	160	128	400	100	101	220	114	125	120
Anglesey	Llyn Alaw	236						204		82	
	Llyn Cefni					380	79				
Flint	Shotwick	169									

* 478 recorded at Kenfig in October 1995.

A widely distributed wintering species but large concentrations are unusual. The most important site is the Severn estuary, where on average 1,831 were recorded over the winters 1985-1990. On the Welsh side of the estuary, Peterstone Wentloog, St. Bride's and Rhymney are the main sites, with birds moving all along the coast between the various sites. During the period of this review, numbers at these sites have been slightly lower than the 500+ in February 1988 (with the exceptions of the winter max. in 1992/3 and 1993/4). Over 100 are recorded from several of the Anglesey Lakes in most winters. Numbers on the Brecon lakes have been on average lower than those quoted in *Birds in Wales*, max. 550 at Talybont in November 1965, however the 1994/5 max. at Llangorse of 400 individuals is the largest count ever recorded at that site. The winter max. November – March, counts at sites that recorded over 100 once are tabulated above.

Large flocks recorded elsewhere: 320 recorded at St. Brides, Gwent on Dec. 28th 1999 and 199 in December 2000; 200 at Cardiff Heliport, Glamorgan in February 1997, 220 in January 1998, 210 on Jan. 10th 1999, 139 on Dec. 16th 1999, 500 on Jan. 16th and 180 in December 2000, 270 at Rumney Great Wharf on Jan. 9th 2000; in Pembroke at Bosherston, 105 in December 1995, 161 in January 1996 and 105 in January 1997; 101 at Usk Res. Brecon November 1994; 145 at Llyn Gwynant, Caernarfon in February 1996.

Pochards are scarce breeding birds in Wales, with a distribution confined to low-lying river valleys and lakes. Anglesey traditionally is the stronghold in Wales, with 8-18 pairs breeding on the island in 1986. During the period of this review, breeding was reported from a total of 13 sites in 6 counties. The numbers at the main sites (i.e. those holding more than 3 pairs in any one year) are tabulated below, but otherwise 1-2 pairs bred in at least one year at Llandegfedd Res., Gwent, Llanishen/Lisvane Res. and Oxwich in Glamorgan, Machynys Pond, Pontarsais, Altwallis and Witchett Pool in Carmarthen, Marloes Mere in Pembroke, Talybont Res. in Brecon, RSPB Conwy in Caernarfon and on Anglesey at Llyn Alaw, Llyn Cefni, and Llyn Maelog. A male bred with a female Tufted Duck at GLWR Goldcliff, Gwent in 2000 fledging 5 young.

Although complete counts in any one year are not available, the Wales totals for 1997, 1998 and 1999 are estimated to be 28, 52 and 31 "pairs" respectively (assuming that only one "pair" is involved at those sites where breeding or summering birds are reported without figures). From the table above it is clear how important the work at Penclacwydd WWT has been for this species.

		1992	1993	1994	1995	1996	1997	1998	1999	2000
Glamorgan	Roath Park Lake	1		3		1	3	1		
Carmarthen	Penclacwydd WWT						5	25	14*	11
Anglesey	Llyn Penrhyn, Treflesg and Dinam			10	12+	8+	16	17	9	
	Other Valley Wetlands complex sites			NC	NC	NC	NC	6	1	
	Malltraeth									3

+ Llyn Dinam not counted in 1995 or 1996   * A minimum figure.

RING-NECKED DUCK *Aythya collaris*
(B) Rare visitor from North America.

HWYADEN DORCHOG

Returning birds clouds the true number of individuals occurring in Wales. *Birds in Wales* quotes 14 records, relating to only 12 individuals. The first a male at Bosherston, Pembroke, in February - March 1967, was closely followed by one that was shot at Llangorse, Brecon, in December of that year. Another male turned up at Bosherston in March - April 1976, returning there for the 1977/8 winter and a female visited Llys y fran Reservoir, Pembroke, in September 1978.

An immature male was at Bosherston, for the 1981/2 winter, with males seen at Talybont, Brecon, February 1980 and Dryslwyn, Carmarthen during the winter 1981/2.

A long staying male in the Welshpool area, Montgomery, from January 1983 - January 1987 was the longest visit to date. A male was recorded at Rhyl, Flint, February - March 1988 with another male at Upper Lliedi Reservoir, Carmarthen January - February 1988 before moving to Old Castle Pond, Llanelli until April. Another male was at Broad Pool, Glamorgan May 1989.

A female visited Skokholm, Pembroke, in October 1986 with another female at Llyn Fanod, Ceredgion January - March 1990 and a male was at Milford Haven, Pembroke in April 1991.

Since then there have been six records of males at: Mynachdy Pond, Anglesey May 1994, Kenfig in April 1995, Holyhead Harbour, Anglesey May 16th 1998, before moving to Llyn Traffwll 18th – 30th, Lisvane Res. Glamorgan Apr. 26th 1999, at Porthmadog, Caernarfon May 25th – 28th 1999 and at Nedern, Gwent May 7th – 21st 2000.

A remarkable influx if 1st year birds into south Wales in October and November 2000. Females at Bosherston, Pembroke Oct. 16th – 17th, at Lisvane/Llanishen, Glamorgan Oct. 17th, at Penberi Res., Pembroke Nov. 11th – Dec. 14th and at Eglwys Nunydd Res., Glamorgan Dec. 31st. The only male at Llyn Pencarreg, Carmarthen Nov. 24th – Dec. 1st.

FERRUGINOUS DUCK *Aythya nyroca*
(B) A rare visitor to Wales.

HWYADEN LYGADWEN

There is a lot of confusion over the true number of individuals of species that have occurred in Wales due to the possibility of birds returning to the same sites for a number of years. *Birds in Wales* quotes 28 dated records of 37 individuals in the twentieth century but at least two of these probably refer to the same individuals.

The first record was of a probable female at Presteigne, Radnor in 1859/60, the next at Churchstoke, Montgomery Two were on Roath Park Lake, Glamorgan in October 1950, a pair at Pontsticill Res. Brecon in December 1956 and a single at Lisvane Res. in November 1957.

Both the 1960's records came from Glamorgan, from Eglwys Nunydd Res. April 1965, Hensol Lake in November 1965. These were followed by a female at Lisvane in January 1970, a male there in September, presumed the same in October 1972.

An adult in Radnor at Hindwell in August 1973 was probably the same as the bird that was seen at Elan Valley the following November, whilst there was an immature seen at Pant yr Eos Res., Gwent in October / November.

One at Kenfig in August 1976, another at Bosherston, Pembroke in November / December 1978, a male at

in 1889/90. In the early 20th century, one was at Pembroke in December 1900, at Builth Wells, Brecon in April 1903 and one out of 7 was shot near Machynlleth, Montgomery, in February 1906.

Males at Afon Wen, Caernarfon in August 1914, at Tonfannau, Meirionnydd Nov. / Dec. 1934, at Orielton decoy, Pembroke in February 1937 and at Lisvane, Glamorgan in December 1939.

Eglwys Nunydd Res. February 1980 – presumed the same in October 1981. Male at Bosherston in November 1982 & 1986 could be the returning Eglwys Nunydd bird or the 1978 individual. Two in December on the Usk at Panty Goitre, Gwent 1982, one at Kenfig October / November 1985 and a pair at Wentwood in March 1987.

Since *Birds in Wales* there has been a returning male in the St. David's area, Pembroke during the winters 1992 – 96 and again in 1999, arriving in late October and sometimes remaining until December.

The only other record was of 2 males at Kenfig, Glamorgan Apr. 28th 1999.

TUFTED DUCK *Aythya fuligula*
(C) Resident breeding species in small numbers in several Welsh counties and a locally common winter visitor.

HWYADEN GOPOG

There are small winter flocks on the majority of low-lying lakes and ponds. The most important wintering area is the Severn estuary, particularly Wentloog and Rhymney but numbers in the period 1992-2000 were lower than the 800 recorded at both sites in the late 1980's. The only other important wintering areas are Kenfig and Eglwys Nunydd Res. The table below gives max. winter, November – March, counts at the main wintering sites, with counts over 100 at least once.

		/1992	92/3	93/4	94/5	95/6	96/7	97/8	98/9	99/00	00/01
Gwent	PW /St.Bride		80	615	365		600	380	380	425	
Glam.	Llan/Lisvane	527	143	76	284	152	193	78	102	63	51
	Rhymney est.	300	200	60	84					110	140
	Kenfig	87	157	62	64	132	88	280	135	89	305
	Eg. Nunydd R.		85	167	259	280	340	297	179	183	407
Brecon	Llangorse				102	150	210	350	230		230
Angle.	Llyn Alaw		121	335	231			211	103		286
	Llyn Cefni					208			252	96	191

Other large counts were of 117 at Ynysfro Res. in September 1993 and 460 at St. Brides on Dec. 28th 1999 both in Gwent; 150 at Cardiff Heliport in January and 140 at Rumney Great Wharf Nov. 15th, both Glamorgan; 220 at Llangorse Lake, Brecon in October 1999.

Large congregations occurred annually in August at Llanishen/Lisvane Res., Glamorgan and Llyn Alaw and Llyn Cefni Anglesey and in August – October at Kenfig and Eglwys Nunydd Res, both Glamorgan.

	1992	1993	1994	1995	1996	1997	1998	1999	2000
Llan/Lisvane	161	189	256	246		177	137	132	205
Kenfig	157	159	180	386	173	43	135	222	270
Eglwys Nunydd				316			200	242	234
Llyn Alaw			700	373	529	488		418 (July)	433
Llyn Cefni							268	279	127

Birds in Wales states that Tufted Duck is absent as a breeding bird from much of Wales, preferring well-vegetated lakes & ponds and sluggish reaches of rivers, rendering most of Wales unsuitable.

Anglesey is the most important area for this species. Lovegrove estimated its breeding population as 40-60 pairs in 1980 but this had risen to 70-80 pairs by 1986. Elsewhere, a handful of pairs in Gwent, Glamorgan, Carmarthen (up to 12 at Witchett), Brecon, Caernarfon and Flint.

Unfortunately the data for the period 1992-2000 is vastly incomplete, from what there is there does not appear to have been any change in the total Welsh population, in the order of 100 pairs. The Carmarthen population is benefiting greatly from the developments at Penclacwydd. A summary of breeding records for each county is shown below, the first figure the number of pairs, the second the total number of young.

	1992	1993	1994	1995	1996	1997	1998	1999	2000
Gwent	2/9		3/			2/		5/	6/37
Glamorgan	2/16	2/	bred	4/18	bred	7/	3/8	5/23+	3/
Carmarthen	3/13+	3/16	6/34	2/		9/	16/	19/	18/
Pembroke	1/5				1/	1/2		1/2	
Ceredigion				1/3					5/22
Brecon			bred		1/5	1/	2/7	6/19+	9/
Radnor			1/8	2/					
Montgomery			6/5	2/12	10-11/	9-10/	9-10/		2-7/
Meirionnydd		bred			2/				1/
Caernarfon							1/		2/
Anglesey	bred	bred	bred	bred	bred	bred		7 sites	24+/
Denbigh					1/5	2/18	1/6	1/	1/5
Flint	11/75		1/4	1/14	2/13		8/	14/	3/14

SCAUP *Aythya marila*   *HWYADEN BENDDU*

(C) Scarce and local winter visitor, also recorded on passage. Has bred once.

The largest concentrations are found around the coasts of north and north west Wales, usually in February and March, with smaller flocks in the sheltered bays off south Wales. The table below gives max. counts at the main wintering areas. The count of 784 in 1995/6 off Denbigh is the highest ever recorded in Wales (*Birds in Wales* states that that area often recorded numbers up to 400 in the late 1980's). The Rhymney / Peterstone Wentloog birds are thought to be the same wintering flock that moves up and down the estuary.

		/1992	92/3	93/4	94/5	95/6	96/7	97/8	98/9	99/00	00/01
Gwent	P.Wentloog						66				
Glam.	Rhymney	18	2	14			40				
	Burry			58	46	93	106	30	30	17	
Cere.	Borth				9	13					
Meiri.	Dwrywd / Glas.	19	6	113	40	80	30				
	Harlech/Black R.					37	80		31		20
Denbigh	Abergele					300+					
	Kinmel Bay	50			120	784	85				
Flint	Dee	46		29		210	36				

Scaup are mainly marine in their preference with few inland records. During the period under review individuals were recorded in winter inland almost annually, mainly singles: at Heathfield Gravel Pits, Pembroke Jan. 2nd and at Trawsfynydd, Meirionnydd Nov. 15th until the end of December 1992, at Llyn Coed y Dinas, Montgomery Dec. 21st 1994, at Pwll Penarth, Montgomery a female Feb. 8th – 25th 1996 and a male there Jan. 9th – 12th 1997, at Llyn Mawr, Montgomery Feb. 5th 1997.

In 1998 there were 2 in Gwent, 18 in Glamorgan, 2 in Pembroke, 2 in Meirionnydd and 3 on Anglesey.

In 1999 there were 5 inland records in Glamorgan in the early months and one at the end of the year, with 9 similarly on Anglesey in the early months and 3 at the end of the year.

There have been no Welsh breeding records since the one at Llyn Traffwll, Anglesey in 1988, although a male bred with a female Tufted Duck on Llangorse Lake, Brecon in 1995, producing hybrid young.

There were 4 other summer records: in 1995 at Ynyshir, Ceredigion on May 13th, at Llyn Alaw, Anglesey on May 7th with probably the same individual at Llyn Traffwll on the 10th – 13th. One summered Foryd Bay, Caernarfon in 1999 and a female was in Red Wharf Bay, Anglesey May 7th 2000.

EIDER *Somateria mollissima*   *HWYADEN FWYTHBLU*
(E) Generally scarce non-breeding resident and winter visitor. Breeding took place for the first time in 1997.

		1992	92/3	93/4	94/5	95/6	96/7	97/8	98/9	99/00	00/01
Glam.	Burry	95	30	70	60	40	95	65	90	118	115
Meiri.	Aberdysynni	14	20	59	107	45	100	110	130		

Rather scarce and nowhere as abundant as Common Scoter around the Welsh coast. *Birds in Wales* describes the rise and fall of the Burry Inlet population from 30 in 1950's to over 200 in December 1988, followed by its rapid decline to the 15-25 individuals present during the early 1990's. Numbers during the period of this review (as shown in the following table) have risen again but as yet as high as the December 1988 figure.

A similar story is described for the Meirionnydd flock and as with the Burry flock, have increased but at a greater level than the 90+ recorded there in the early 1980's. Max. winter, November – March, counts at the two main sites are tabulated above.

The largest flocks recorded elsewhere or at different times of the year were: 55 off Blackpill, Glamorgan Nov. 16th 2000, 38 off Caldey Island, Pembroke Mar. 25th 1998, 50 off Wallog, Ceredigion May 29th 1994. Off Meirionnydd there were 25 off Aberdyfi in April 1994 and off D/Glas 34 in February, 12 in November & 38 in December 1993 (presumably from the Dysynni flock), 102 at Llangelynin (Aberdysynni) on Apr. 24th 1999 and 64 Aberdysynni in July 2000. 44 were in the Menai Straits, Anglesey in November 2000.

Inland records are extremely rare. The only record during 1992-2000 was of a juvenile male at Llangorse Lake, Brecon Oct. 31st 1993.

The first record of successful breeding in Wales took place on Puffin Island, Anglesey in 1997. In 1998 & 1999 a pair attempted to breed again on Puffin Island but failed. In 2000 there were 3 males but no females present. Two nests however were found in Meirionnydd in 1998 and a female with 4 young were seen at Aber Ogwen, Caernarfon in July 2000.

LONG-TAILED DUCK *Clangula hyemalis*   *HWYADEN GYNFFON HIR*

(C) Regular winter visitor and passage migrant, most frequently recorded offshore from counties in north-west Wales. Rare inland.

Formerly a scarce winter visitor to Wale, it is now recorded regularly, mainly from the coast but recently from inland waters. The most important areas are the coasts of Caernarfon and Meirionnydd, particularly Black Rock Sands and the Dwyryd / Glaslyn estuary where numbers wintering reached 48 and 45 respectfully in December 1991. Since then there have been fewer individuals recorded at these and other sites, possibly relating to milder winters. The total number of individuals recorded in each month during 1992-2000 is shown in the following chart.

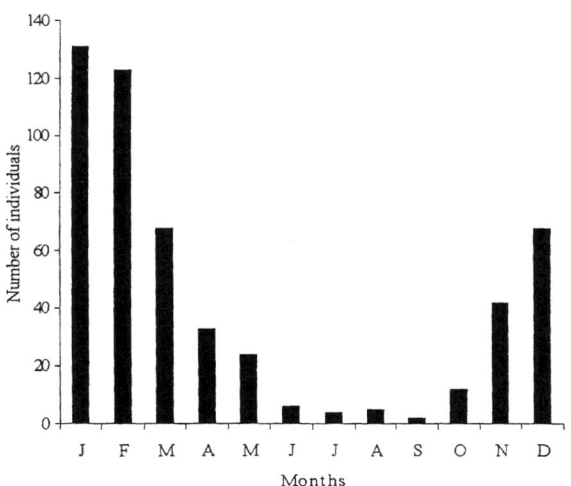

The only records of large flocks were: from Meirionnydd at Dwyryd/Glaslyn estuary, 31 in January, 33 in February and 17 in May 1992, 14-15 January – Apr. 3rd 1999 and 7-8 from Nov. 21st 1999 until the end of the year. 27 at the Artro estuary in February 1993.
From Caernarfon at Black Rock Sands, 7 males + 3 females on Jan. 22nd 1997, 10 on Nov. 29th 1998, 15 there in January 1999 and 20 in January 2000.
The only other count over 10 was of 11 at Pensarn, Denbigh on Jan. 18th 1994.

Inland there were 5 records at: Machynys Pond, Carmarthen Dec. 29th 1995 – Mar. 16th 1996, at Heathfield G.P., Pembroke Jan. 4th 1994; on Anglesey at Llyn Coron Dec. 23rd 1994 until June 10th 1995 and a female at Llyn Alaw July 14th - Aug. 20th 1995, at Llyn Brenig, Denbigh Apr. 15th 1995.
One other summer record of a male at Aberleri, Ceredigion May 29th 2000.

COMMON SCOTER *Melanitta nigra* MOR-HWYADEN DDU

(C) A locally abundant visitor, recorded all year round but mainly during the winter months. The distribution is predominantly coastal, although it is regular in small numbers on inland waters on passage.

	/1992	92/3	93/4	94/5	95/6	96/7	97/8	98/9	99/00	00/01
Carmarthen Bay			5012	17650	10631	4323	6240	17831	21395	11298
Newgale	100	100	250	400	2000	800	600	280	200	200
Dwyryd/Glaslyn		2026	1548	1130	2053	3676		2000		
Mawddach			1023							
Artro Est.		880	1242	613	694					

The most important wintering area for this species in Wales is Carmarthen Bay, where numbers are of international importance. Counts from this area in *Birds in Wales* include c20,000 in early 1973, 25,000 in March 1974 and 16,000 in August 1974. Subsequent counts in the late 1970's and 1980's never produced figures higher than 12,000 (July 1978) and in 1991 the largest count was of only 7,700 birds.

The other main wintering flocks are much smaller. The table above gives max. counts at the main wintering areas, November – March, off Rhosilli, Glamorgan, Carmarthen Bay, Newgale, Pembroke, north Cardigan Bay (Dwyryd/Glaslyn, Mawddach & Arthro estuary, all Meirionnydd). The National Level of importance is 275, while Carmarthen Bay regularly held over the International Level of importance, of 16,000.

The Carmarthen Bay flock is highly mobile and moves between Cefn Sidan, Pendine and Amroth. This flock was heavily affected by the "Sea Empress" disaster in February 1996, when 1818 were picked up dead and 2753 alive. Before the oil spill, the Carmarthen flock was estimated to be less than 4,000, so with so many found oiled or dead one would expect that the flock was totally decimated. Subsequent detailed monthly counts found over 8,000 birds not long after the oil spill, increasing to 10,000 by March.

This provides evidence that Carmarthen Bay is not only a nationally important wintering ground for this species but also extremely important as a stop over for migrating birds. One interesting ringing recovery as a result of the Sea Empress, of an individual picked up oiled, cleaned then released off Dorset in March 1996. It was subsequently shot at Udorskiy, Komi Assr, Russia on July 1st 1998. This provides useful evidence to the origin, Fenno-Scandinavia, of the Carmarthen Bay flock.

In north Wales 3,022 were counted in a co-ordinated count from Traeth Lafan, Caernarfon to Mostyn, Flint in January 1998, of which 848 were off Penmaenmawr, 966 off Llandulas and 800 off Abergele. Other large counts in this area were 1,000 off Llandulas Jan. 26th 1992, 5-6,000 off Conwy Mar. 27th 1996, 4,700 off Rhyl Dec. 14th 1997 and 2-3,000 off Penmaenmawr Jan – Feb. 1998. In 1999 there were 800 in Conwy Bay in January and 1200 off Pensarn on Jan. 31st. 300 were off Traeth Lafan in January and 600 in April, 500 off Llandulas on Feb. 2nd presumably were the same birds.

Large counts elsewhere were: 700 at Broughton Bay, Glamorgan November 1993 and 800 in Red Wharf Bay, Anglesey Feb. 11th 1996.

Co-ordinated counts were carried out of north Cardigan Bay. The results of which are tabulated below. Within this area, large congregations collected off Harlech, Meirionnydd, 2,000 there Dec. 13th 1997 and 1,338 in January 1998, off Black Rock 1,313 and 846 off Shell Island also in January 1998, 700 off Morfa Bychan Dec. 1st 2000. The counts from Dwyryd / Glaslyn, Mawddach and Artro estuaries are given in the first table.

	1991/92	92/93	93/94	94/95	95/96	96/97	97/98
October		2060	3063		2705		
November	6421		2373	3606	4544	3999	
December		4172				5144	
January	4075					4318	5220
February	2968	5085	4872	4755	6720		

Year	Number	Dates	Year	Number	Dates
1992	2793	June 6th – Nov. 15th	1997	2549	June 21st – Dec. 31st
1993	1492	June 12th – December	1998	2474	June 13th – Dec. 28th
1994	2384	July 3rd – Dec. 31st	1999	4392	June 2nd – Dec. 12th
1995	2596	May 30th – Dec. 2nd	2000	6093	June 15th – Dec. 30th
1996	2150	June - December			

Autumn passage noted annually off Strumble Head, Pembroke, with peak movements occurring in June and July and thought to consist largely of birds heading into Carmarthen Bay to moult. The annual total and passage periods for the years 1992-2000 are tabulated above.

Passage was also noted off Port Eynon, Glamorgan 752 flew west on June 26th 1994, 500 off Moelfre, Anglesey June 13th 1995 and 308 passed Point Lynas June 24th – Nov. 10th 2000.

Birds in Wales states that inland records are scarce. Data from the period under review suggest that Common Scoters are annual visitors to inland waters with 48 recorded in Gwent, 4 in Glamorgan, 71 in Brecon, 21 in Radnor, 6 in Ceredigion, 6 in Montgomery, 3 in Meirionnydd and 13 in Flint. The largest flocks were: 14 at Llandegfedd Res. Gwent July 17th 1993, 13 at Llangorse, Brecon July 17th 1995, 15 at Talybont, Brecon June 25th 1994, 19 at Claerwen Res. Radnor Nov. 4th 1994 and 12 at Shotwick Lake, Flint June 15th 1998.

There were two accepted records of Black Scoter *M. n. americana*. The first for Wales was at Newgale, Pembroke Dec. 25th 1991 until Feb. 8th 1992, although bird may have been present from Nov. 24th. The second at Llanfairfechan, Caernarfon Mar. 10th – May 8th 1999.

SURF SCOTER M*elanitta. perspicillata*   *MOR-HWYADEN YR EWYN*
(B) A rare trans-Atlantic visitor to Wales.

The true picture of this species occurrence in Wales is clouded by the possibility of returning individuals. Analysis of data suggests 22 individuals recorded in Wales up to 1991 and 9 since.

Breakdown of records by county / site:

Glamorgan	imm. female, Mumbles, March – April 1976
	Imm. male Eglwys Nunydd Res. and Kenfig, October – November 1981
	male, Worms Head, February 1984
	2 males, Rhossili, March 1991.
Carmarthen	dead male, found Ginst Point, January 1971
	imm. male, Burry Port, October 1984
	male, Marros / Telpyn / Amroth (Pemb), October 1990 – January 1991
	male, Ragwen Point, January - February 1993
	3 males, Telpyn / Amroth (Pemb), November 1994 – January 1995
	male, Telpyn / Amroth (Pemb), February – March 1997, presumed the same returning individual there March 1998, September 1998 – April 1999 and November 1999 – March 2000.
Pembroke	male, Druidston, October – November 1979
	Imm. Male, passed Strumble Head, November 1982
	4 males, passed Strumble Head, November 1987, 2 males presumed from same party off Nolton November – March 1988
	male, Skokholm 1990
	plus the above records from Carmarthen Bay.
Meirionnydd	2 males & a female, Harlech, December 1988 – January 1989, presumed returning male off Shell island in December 1989
	male, Morfa Harlech, November 1997.
Caernarfon	male, Penychain, April 1972
	male, Llanfairfechan, Dec. 1983 – Apr. 1984, presumed same returning bird at Abergele (Denbigh) December 1985 – January 1986, Llanfairfechan February 1986 & February 1987 and off Penmaenmawr in January 1988
	male, Bardsey Oct. 24th 1986
	male, Porthmadog, February – March 1987.
Denbigh	above Abergele record 1985/6 plus male, Llandulas, April 1993.
	male, Llandulas, October 1997.
	Male, Llandulas, Jan. 23rd – Mar. 25th 2000.

VELVET SCOTER *Melanitta fusca* MOR-HWYADEN Y GOGLEDD
(C) A scarce and local winter visitor, regular in small numbers to some coastal counties. Occasionally recorded on passage. Rare inland.

Birds in Wales states that the only sites to regularly hold over 15 wintering individuals are north Cardigan Bay, particularly Black Rock Sands (Caernarfon), Aberdysynni (Meirionnydd) and Llanfairfechan (Caernarfon). Smaller numbers, usually up to 10 or so are seen at the mouth of the Dyfi (Ceredigion) and less than 5 in Carmarthen Bay (Carmarthen / Pembroke) and St. Bride's Bay (Pembroke).
Monthly distribution shown in the chart.

During 1992-2000 small groups were reported all around the coast but mainly from the sites mentioned earlier. The max. counts at these sites are tabulated below.

The total number of individuals recorded in each month is shown in the graph opposite. Passage peaks were in November and March. 3 individuals summered at Llanfairfechan, Caernarfon in 1992.

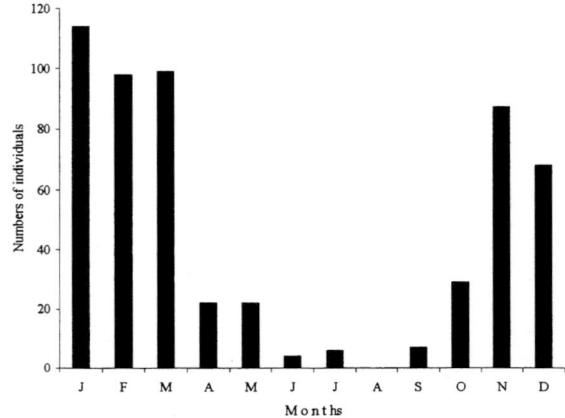

	/1992	92/3	93/4	94/5	95/6	96/7	97/8	98/9	99/00	00/01
Carmarthen Bay		11	7	7	5	4	1	5	1	5
St. Bride's Bay	3			1	1	3	4	2		
Borth/Aberdyfi			8		4	3				
D/Glas/M. Harlech	9		19		2	7	10	15	14	10
Penmon/Llandulas		15	17				1	4		

Spring passage was noted at four sites, 14 at Amroth on Mar. 27th 1996 and 6 there Mar. 15th – 22nd 1998, a single at Ramsey, Pembroke June 22nd 1995, 24 at Morfa Harlech, Meirionnydd on Mar. 30th 1997 and 20 at Pensarn, Caernarfon May 7th 1994.

Autumn passage in 2000 was noted in October at Strumble Head, Pembroke 4 on the 15th, 11 passed Bardsey, Caernarfon on the 24th and 2 off Point Lynas, Anglesey on the 12th.

GOLDENEYE *Bucephala clangula* HWYADEN LYGAD-AUR
(C) Locally common winter visitor and passage migrant, with an increasing number of individuals summering on inland lakes.

Small wintering flocks are widely scattered on lakes, reservoirs and rivers throughout Wales. The largest concentrations are usually found on the north Wales coast around Llanfairfechan (Traeth Lafan) but the numbers there in 1994/5 winter were far higher than ever recorded before. In south Wales the Cleddau, Llangorse Lake, Ogmore estuary and Eglwys Nunydd Res. are important sites with 30-45 at each. The following table gives the max. counts at the main wintering areas, November – March.

		/1992	92/3	93/4	94/5	95/6	96/7	97/8	98/9	99/00	00/01
Gwent	Llandegfedd R.		15	12			25				
Glam.	Ogmore	45	39	39	34	29	74	32	47	41	
	Kenfig	33	15	14	19	10	17	10	20	10	
	Eglwys Nunydd		30	60	42	41	41	47	40	44	66
Pemb.	Cleddau	51	34	45	23	38	62	41	59	56	27
Cere.	Dyfi	30	47	27	30	30	35	3	12	30	11
Brecon	Talybont	23	15	20	13			24	20	21	5
	Llangorse				23	22	24	74	45	30	26
Meiri.	Dwyryd/Glas	39	29	57	39	16		23	25	30	
Caern.	Traeth Lafan	269	73	204	465	250	120	86	138	132	157
Angle.	Llyn Cefni		26								
	Llyn Maelog		25		34						
	Llyn Alaw		12	20	20		26	35	30*	33	

* in April.

Co-ordinated counts in Cardigan Bay found 40 in February, 40 in November and 43 in December 1996, 74 in January 1997 and 52 in January 1998. Co-ordinated count of north Wales, coast, Traeth Lafan – Mostyn, Caernarfon - Flint January 1998, found 179 individuals, of which 77 were off Traeth Lafan and 38 off Penmaenmawr. Other large counts were: 61 at Borth y Gest / Pwllheli, Caernarfon Feb. 2nd 1995, 100 Aber Ogwen, Caernarfon during the early months of 1999, 54 on the Inland Sea, Anglesey Feb. 8th, 36 Trearddur Bay, Anglesey Feb. 15th and 60 Point of Air, Flint Jan. 27th all 1998.

Still no breeding record in Wales, although individuals have summered: 2 at Nantymoch Res., Ceredigion in June 1992, Montgomery, Denbigh and on Anglesey at Llyn Traffwll in 1994 & 9 at Llyn Coron in May 1999. In 1997 a pair was recorded along a suitable stretch of river in Meirionnydd on May 13th – 14th.

SMEW *Mergus albellus*   *LLEIAN WEN*
(E) Scarce winter visitor, not recorded annually in any county except in Glamorgan, though subject to cold weather influxes from the Continent.

In Wales this species is very scarce in most winters, with most records relating to red-heads (adult females or immatures). Influxes are associated with severe weather to the East, often with a smaller influx in the subsequent winter. The following table summarises the number of individuals recorded in each winter month.

The only area to which birds regularly returned was in Glamorgan where birds often moved between the three sites of Kenfig, Eglwys Nunydd Res., and Baglan Pool. The chart below summarises the number of wintering individuals in this area.

In October 1999 a male arrived at Inner Marsh Farm, Flint on Oct. 7th and remained there until Mar. 3rd 2000.

	January	February	March	April	October	November	December
1992	1	1	2			1	3
1993	2	1	1			2	5
1994	7	4	4	1		5	8-9
1995	1						9
1996	13	14	7				12
1997	44	24	12	1		4	15
1998	13	13	11			5	13
1999	10	7	6	1	1	1	5
2000	5	4	2				3

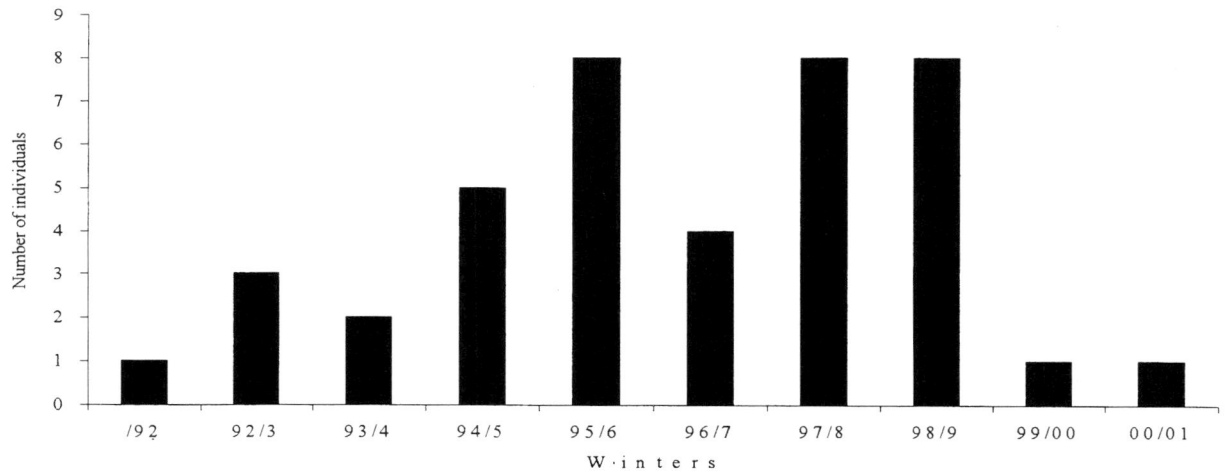

Number of wintering Smew at Kenfig / Eglwys Nunydd Res. / Baglan pool

HOODED MERGANSER *Mergus cucullatus*   *HWYADEN BENWEN*
Formerly one Welsh record but this species has now been deleted from the British List following a review: "all records are considered no longer acceptable and likely to relate to captive birds".

RED-BREASTED MERGANSER *Mergus serrator*   HWYADEN FRONGOCH
(C) Locally common winter visitor, particularly to the north Wales coast. Small numbers breed, chiefly around the coast - line from Ceredigion to Anglesey.

The most important sites are the Dyfi and Dwyryd / Glaslyn estuaries. The table below gives the max. winter, November – March, counts at the main sites.

		/1992	92/3	93/4	94/5	95/6	96/7	97/8	98/9	99/00	00/01
Glam.	Burry S.	25	25	25	20	35	25	25	35	35	20
Carm.	T/T/G	37	31	35	18	12	21	13	15	3	
Pemb.	Amroth		40	17	47	17		20	14	11	
	Cleddau	9	23	20	11	14	12				
Cere.	Dyfi	65	64	80	62	63	112	60	56	51	39
Meiri.	D/Glas	28	93	109	42	62	25	120	26	45	7
	Mawddach	34	60	52	46	48	50	22	34		30
	Dysynni	17	12	52	60	27	26	75	15		

Co-ordinated counts of north Cardigan Bay produced the following results:

	/1992	92/93	93/4	94/95	95/96	96/97	97/98	1998
October		64	84	92				
November			177	48		130		
December		121				186		
January	45						283	294
February	56	208		68		268		

Within this area, 150 were off Ynyslas, Ceredigion and a further 67 off Criccieth, Caernarfon Feb. 2nd 1995, 85 off Pwllheli, Caernarfon Jan. 25th 1997, 75 off Aberdysynni, Meirionnydd and 11 off Criccieth during the January 1998 count.

The only other co-ordinated count was of the north Wales coast, Traeth Lafan, Caernarfon to Mostyn, Flint on Jan. 29th 1998, when 209 were counted of which 132 were off the Great / Little Orme. Other large winter flocks in this area were: 40 Llandulas Jan. 8th 1994 and 75 off Colwyn Bay Jan. 11th 1994, Denbigh.

Unlike the Goosander, the Red-breasted Marganser is predominantly a coastal breeder in Wales. *Birds in Wales* states that the first confirmed breeding in Wales took place in 1953 on Anglesey. There was a slow increase in numbers to around 50 pairs by the mid 1970's. In 1981 an RSPB survey estimated 27-43 pairs breeding inland (9-20 of which were on the Dyfi and 7-9 on the Dee). The survey was repeated in 1985 when 24-41 pairs were estimated, with a similar distribution. By 1991 the Welsh population was estimated at 150 pairs.

Little breeding expansion was noted during the period under review and breeding was restricted to a handful of areas. Available data is very sparse and non-existent for coastal areas. There is no reason to assume any change in the total Welsh population to that published in *Birds in Wales*.

The main breeding area remains the Dyfi (Ceredigion / Meirionnydd and Montgomery), other sites in Ceredigion include the Afon Ystwyth and Afon Rheidol (prospecting was observed on the Teifi in 1995 and 1996); in Brecon at Talybont and Newbridge-on-Wye; in Montgomery at Lake Vyrnwy, at Cwm Llinau; in Caernarfon at Aber Ogwen and on the Afon Glaslyn in 2000; on Anglesey at Cemlyn. Single breeding records: in Montgomery at Afon Banwy; in Caernarfon at Afon Conwy and Llyn Dinas; in Meirionnydd at Afon Tryweryn, Llyn Trawsfynydd, Tal y Llyn, Llyn Tegid and Dysynni. In Pembroke the first ever breeding was recorded in 1995 at the Gann, when a female and two ducklings were seen.

A large moulting flock congregates annually off Traeth Lafan / Aber Ogwen during late July and August. Counts from there were:

1992	1993	1994	1995	1996	1997	1998	1999	2000
220	380	553	288	330	234	453	255	317

GOOSANDER *Mergus merganser*   HWYADEN DDANHEDDOG
(C) Breeding resident on most Welsh rivers in mid and north Wales, the breeding population slowly expanding. In winter, numbers are augmented by birds dispersing from farther north.

In winter this species frequents the larger lakes and reservoirs of Wales, with few wintering around the coasts. The largest numbers are found on the lakes and waters of upland mid Wales. Although an annual winter visitor to many sites, few have large numbers most years. Those that did are tabulated below (no counts from 2000/01).

		/1992	92/3	93/4	94/5	95/6	96/7	97/8	98/9	99/00
Gwent	Llandegfedd	12	45	42	31	52		49	24	
	Bryn Bach		18	25	18		22			
Brecon	Talybont	30	45	106*		32	38	34	27	60**
	Glasbury		14	15			23	14		
	Usk Screthiog		25	25	20			13		

* 106 at Talybont in 1993/4 is the largest count recorded in Wales.
** this total probably included the 44 at Llangorse on Dec. 24th.

Counts of between 20 and 30 recorded at many sites: in Gwent at Garnlydan Res.; in Ceredigion at Rheidol GP & Rheidol Res.; in Brecon at Llangorse; in Radnor at Llanbwchllyn and Llandrindod Lake; in Montgomery at Dolydd Hafren and Llyn Du Meifod; in Flint at Hanmer Mere. In Carmarthen, 40 at Dinefwr Ponds January 1994, 60 on the River Tywi, Llandeilo in September 1995 and 31 at Llyn Pencarreg, November 1993.
90 were counted on Feb. 4th 1996 on the River Wye between Newbridge and Hay. 86 were counted at the William Condry reserve, Montgomery in August 2000 and 46 were on Talybont, Brecon in September 2000.

The fast flowing unpolluted rivers and streams of upland Wales, with their plentiful supply of fish are an ideal habitat for breeding Goosanders. According to *Birds in Wales* breeding in Wales was not proved until 1968 on the Dyfi, since then many rivers have been colonised. In 1985 an estimated 100 pairs bred on 14 Welsh rivers, with the largest numbers on the Vyrnwy and the Wye. The Welsh breeding population was estimated at 150 pairs in 1991. Data from the period under review is far from complete but what there is does not indicate any sort of change from the levels already discussed.

During the period 1992-2000 there were breeding records from the following river systems:
Gwent: Wye.
Glamorgan: Neath & Tawe present Pontypridd and Abercynon.
Carmarthen: present on all rivers in Carmarthen except the Teifi below Newcastle Emlyn.
Ceredigion: Ystwyth and Dyfi.
Brecon: Usk, Irfon, Usk Res., Llyn Brianne, Caban Coch,
Radnor: Elan Valley (6 pairs in 1999).
Montgomery: Vyrnwy, River Clywedog, Dolwen, Dolydd Hafren, and Llanerfyl.
Meirionnydd: Trawscoed, Llanwchllyn.
Caernarfon: Ogwen, Conwy, Rivers Llynfwy, Seiont, Foryd Bay and on Llyn Padarn.
Denbigh: Dee, Clwyd.
The following table summarises available data. The first number refers to the number of pairs, the second the number of young.

		1992	1993	1994	1995	1996	1997	1998	1999	2000
Glamorgan	River Neath			3/26	1/2	2/17	1/3	3-4/	3/	2/8
	River Tawe					1/1	2/7	1/	2/	1/6
Carmarthen	Tywi	8/9	7/8		5/	2/		5/		
	Afon Cothi	1/		2/	1/		2/			
	Afon Teifi						1/10	2/11		
Brecon	Usk		3/16		9/	7/	4/	7/	6/	7/
	Wye		6		9/	2/25	4/31	3	5/	4/
	Irfon		1					2/		
Montgomery	Morfa Dyfi						3-4/29			
	Dolydd Hafren						6/26			
	Lake Vyrnwy		5-6/	5/		6/21	3/4	3/14	7/31	6/
Denbigh	Clwyd			1/9	5/35			2/4	1/12	
	Dee			1/10				2/	1/4	

RUDDY DUCK *Oxyura jamaicensis*   *HWYADEN GOCH*
(C) Scarce, but increasing breeding resident in several counties. Locally common on Anglesey, but elsewhere a scarce winter visitor.

The first breeding record in Wales occurred in 1976, since which this species has become a well-established breeder on Anglesey and on a handful of other sites. The Welsh population was estimated at approximately 70 pairs in 1990, with over half of them breeding on Anglesey. Large flocks have only ever been recorded on the Anglesey Lakes, with no more than 10 on any other water. With recent conservation action to reduce numbers on Anglesey, the apparent decline since 1998 is probably genuine, indeed there were reduced numbers there in winter 2000/01. The following table gives max. winter, November – March, counts at the main sites, all Anglesey.

	/1992	92/3	93/4	94/5	95/6	96/7	97/8	98/9	99/00	00/01
Llyn Alaw	85	85	82	91		184	225	221	159	
Llyn Traffwll		81	90		80	156	122	147		
Llyn Coron		10					210			
Llyn Penrhyn		22			56			52		

Largest groups elsewhere were 7 at Kenfig, Glamorgan in both winter periods of 1994 (6 there on Nov. 21st 1999), 7 at Llangorse, Brecon Jan. 3rd – Apr. 25th 1999, 5 at Llyn Coed y Dinas, Montgomery at the end of 1995 and from Flint in 1997, 21 at Hanmer Mere Feb. 15th and 21 on Erddig Pond, Denbigh Sept. 21st.

Breeding records restricted to only a dozen or so sites, mainly in north Wales: in Radnor at Llyn Heilyn; in Montgomery at Gungrog, Lymore Pool and Llyn Coed y Dinas; in Caernarfon at RSPB Conwy; on Anglesey at Llyn Alaw, Llyn Cefni, Llyn Traffwll, Llyn Rhos Du, Llyn Cors Erddreinog, Llyn Penrhyn and Malltraeth; in Flint at Llyn Helyg and at a Deeside industrial site. A pair was seen displaying at Llangorse, Brecon May 5th 1999.

	1992	1993	1994	1995	1996	1997	1998	1999	2000
Llyn Heilyn	1 pr.						1 pr., 5y		
Gungrog	3 prs.	2 prs.		1-2 prs.	1 pr.			male	
Lymore Pool		1 pr.			1 pr.				
Llyn Coed y Dinas								1 pr., 1y	2 prs.
Conwy RSPB						2 prs, 9y		6 prs.	4 prs., 10y
Malltraeth				1 pr., 4y		5 prs.			
Llyn Dinam, Penrhyn and Treflesg			34 prs.	22+ prs.	26+ prs.	11 prs.	23 prs.	12 prs.	
Other Valley Wetlands			NC	NC	NC	NC	10 prs.	6 prs.	
Llyn Alaw					bred		bred		bred
Llyn Cefni					bred		bred		bred
Llyn Traffwll					bred		bred		
Llyn Rhos Du							bred		
Cors Erddreinog					bred			2 prs.	
Llyn Helyg		prob. 1pr.	1 pr, 3y						
Deeside Industrial site			1pr., 2y	2 broods 7y	2 broods 8y	bred			
Inner Marsh Farm						3 broods		*	

* at least 6 males + 4 females present all year but no evidence of breeding.

HONEY BUZZARD *Pernis apivorus* — *BOD Y MEL*

(E) First confirmed breeding in 1992, now breeds in small numbers. Formerly a rare summer visitor, numbers have increased throughout the 1980's and its is now recorded annually.

The first breeding in Wales was recorded in 1992 and this species has done so every year since, with numbers slowly increasing to 8 pairs in 2000. Passage is recorded almost annually with records during 1992-1999 from Gwent (4), Glamorgan (2), Carmarthen (1), Pembroke (8), Brecon (1), Caernarfon (7), Anglesey (1), Flint 4).

A significant number of records in autumn 2000, all tying in with a large movement over the UK. Records were: a juvenile at Kenfig, Glamorgan Oct. 1st; in Pembroke at Skomer on Sept. 16th and at Whitchurch on Sept. 30th; north over Plas Gogerddan, Ceredigion on Oct. 2nd; in Brecon at Mynydd Epynt on Sept. 4th and 3 over Honddu valley near Upper Chapel on Sept. 29th; at Elan Valley, Radnor Sept. 30th; Caernarfon: a juv. at Rhiwlas on Oct. 7th, at Moel y Ci Nov. 21st – 22nd and on Bardsey in September on the 19th, 2 on the 25th and one on the 30th; south over Rhosneigr, Anglesey on Sept. 30th; south over Ruabon Moors, Denbigh on Sept. 30th.

BLACK KITE *Milvus migrans* — *BARCUD DU*

(B) A vagrant.

Two records, singles at Aberdaron, Caernarfon June 4th 1995 and Skomer, Pembroke May 20th 1998. These were the 7th and 8th Welsh records, the others coming from Llangorse, Brecon in October 1976, Overton, Glamorgan in April 1979, Cardiff, Glamorgan in May 1979, Lake Vyrnwy and Dyfi forest, Montgomery & Meirionnydd in April 1985, Llansadwrn, Anglesey in April 1987 and Skokholm, Pembroke in May 1990.

RED KITE *Milvus milvus*   *BARCUD COCH*

(E) Formerly widespread, but almost eradicated between the late 18th and early 20th centuries. Now an increasing resident breeder in eight counties of central Wales; rather a scarce and erratic visitor elsewhere.

The Welsh kite population increased quite steeply through the 1990s, following a considerable improvement in productivity during the late 80s, probably related to the gradual reoccupation of lower and more fertile country. The numbers counted by the annual census, and given in the table below, show an increase from 101 pairs (84 definitely breeding) in 1992 to 232 pairs (184 definitely breeding) in 1999. These figures need qualification, however. It is generally agreed that up to the early 1990s the annual survey missed about 10% of the population. Coverage certainly deteriorated through the 90s, with decreasing enthusiasm, less financial support, and other demands on the watchers' time. An 'informed guesstimate' is that by 1999 we were failing to locate perhaps 20 to 25% of territorial birds, and failing to obtain proof of breeding for an increasing proportion of the known pairs. The proportion of 'other pairs' in the record increased from around 10% up to 1990, to about 25% by 1999; an improbable situation in real life. Many of them must in fact have attempted to breed.

On this admittedly hypothetical basis, the Welsh territorial population in spring 1999 is likely to have been around 270-280 pairs, with perhaps 240-250 attempting to breed. In addition there would be some 200 unmated individuals, mostly yearlings, giving a total spring population of around 750 birds.

Monitoring changed permanently in 2000, when it was decided that the attempt to locate all pairs was no longer viable, and to concentrate instead on a tetrad-based survey, funded by RSPB. It was hoped that this could be replicated periodically, perhaps every ten years, to give estimates of the total territorial population and to indicate changes. The figures given for 2000 in the table are therefore less indicative than before, and no attempt was made to count unmated birds. The 2000 survey of 1300 tetrads produced an estimated breeding population of 259 pairs (with very wide 95% confidence limits of 200-318 pairs) and a total estimate of territorial pairs (including non-breeders) of 337 pairs (268-406). The central figure for breeding pairs (259) is likely to be on the low side, while that for non-breeding pairs (78) seems highly improbable. Clearly the methodology will need refining before a long-term solution to the problem of estimating kite populations is found.

With the increasing population came a gradual expansion of the kite's range in Wales, though this was less dramatic than the increase in numbers. Most kites settle to breed very near their birthplace, and they infill rather than pioneer new ground. Being sociable birds, and only weakly territorial, they are prepared to nest at quite high densities and in close proximity. Those that set up outlying territories often fail to persist. There were however some notable gains during the 90s, probably favoured by the availability of new food-sources on the periphery of the earlier range, such as the feeding-stations at Rhayader and Trefilan, the large abattoir at Llanybydder, and another slaughterhouse near Sennybridge.

These must have habituated large numbers of immature kites to areas formerly little used, leading to colonisation by breeding pairs. These new resources also replaced to some extent the now abandoned land-fill refuse tips, where kites had previously scavenged. In 1996 Glamorgan and Caernarfon were added to the list of old counties with breeding pairs, making eight in all; and there were spring birds in two or three more. By the late 1990s Welsh kites were being regularly seen over the border in Shropshire and Herefordshire, and pairs had bred within a few miles of the border, if not beyond it.

RED KITE
NORTH WALES
MARCH 2001

Expansion is probably still restricted by persecution, particularly to the north-east and east. A farmer shot at a pair near Old Radnor in 1998, killing at least one of them; and the disappearance of the only three established pairs around the Berwyn mountains during the late 90s must surely indicate human intervention. Egg collecting ceased to be a major problem, though a few nests were still robbed. The major cause of nest failure continued undoubtedly to be the Welsh weather, especially rain around hatching time.

Birds in Wales 1992-2000

The widespread use of illegal poison baits, together with other legally used poisons, remained the greatest single cause of full-grown kite mortality. The discontinued warble-fly treatment Fenthion kept its position as the most commonly-used chemical on poison baits affecting kites in Wales, but several birds died from alphachlorolose poisoning, and the first death by Carbofuran (now widely misused by keepers in Scotland and N.England) was recorded in Wales in 1999. Accidental or secondary poisoning occurred too, associated with Warfarin and more recent third-generation rat poisons such as Brodifacoum, and also with various sheep-dip chemicals. These intractable problems apparently showed little improvement during the period under review.

Despite a large increase in the numbers of young kites ringed, the proportion of Welsh kites reported outside Wales was considerably lower in the 1990s than it had been between the 1960s and 1980s. This could perhaps be a consequence of the establishment of feeding-stations, attracting and holding large numbers of young birds. A 1993 chick stayed at Tatton Park in N.Cheshire from October 1993 until April 1995, except for a brief excursion to Stafford in April 1994. Two young of 1997 were seen (and one later died) in Dorset in early 1998. Much more remarkable was the recovery of a 1995 chick in Co.Wexford in October 1999. Not only was it the first Welsh-ringed kite in Ireland, but it was also the first distant record of an 'adult' bird. All previous long-distance movements had occurred during the first year or so of life, and most four-year-old kites have settled down to breed, and become quite sedentary. Possibly this individual had been in Ireland for several years before it died, but there seems to be no evidence to support this contention.

Visitors to Wales from the new introduced populations in England and Scotland occurred annually during the 1990s. Most were first winter birds, but some made repeat visits in their second winter, after returning to the release areas for their first summer. We also had visits from two young reared by a naturally-occurring 'wild' pair in Suffolk. No older birds were recorded up to 1999, but in 2000 a third-year bird of Spanish origin from the East Midlands release area was found dead on a nest near Llanidloes. There were no eggs in the nest, so the record falls short of actual proof of breeding, but this was the first recorded nesting attempt in Wales by one of the introduced kites.

Red Kite Population Counts in Wales 1991-2000

	Breeding pairs	Other pairs	Total	Successful pairs	Young reared	Unmated Birds	Total birds	
							April	August
1991	76	16	92	41	62	-	260	318
1992	84	17	101	60	96	92	294	386
1993	104	11	115	61	82	88	318	397
1994	111	27	138	70	99	94	370	469
1995	127	19	146	79	117	106	396	514
1996	130	31	161	90	119	122	444	556
1997	152	28	180	99	129	135	495	618
1998	167	33	200	112	174	132	536	700
1999	184	48	232	120	167	155	618	774
2000	201	48	249	103	141	-	-	-

WHITE-TAILED EAGLE *Haliaeetus albicilla*   *ERYR Y MOR*

(B) A very rare visitor.

Two records both of immature birds, the first at Skomer, Pembroke Nov. 10th & 11th 1993 was the first in Wales since 1910. The other record was of a first year bird at Cors Caron, Ceredigion Nov. 12th 1997. *Birds in Wales* quote 17 records 1818 – 1910, mostly of shot specimens, the last was at Abersoch, Caernarfon in November 1910.

MARSH HARRIER *Circus aeruginosus*
(C) Scarce but increasing passage migrant and winter visitor.

BOD Y GWERNI

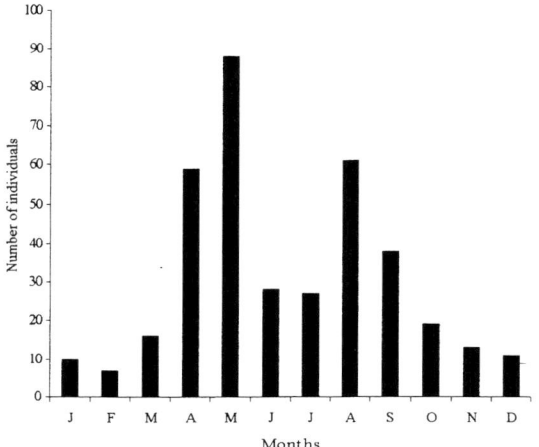

The revival of the British population since 1972 has led to an increase in the number of individuals seen in Wales. The total number of individuals recorded in each county during the period 1992-2000 is shown on the map above. Pembroke accounts for more records than any other county.

The total number of individuals recorded in each month is shown in the chart. This species appears to be becoming commoner, it is now an annual wintering species in Wales, mainly on Anglesey with other wintering records in Pembroke 1994/5, Carmarthen 1996/7, Meirionnydd on the Dyfi 1993/4, two birds in Montgomery 1997/8.

Passage peaks in April / May and in autumn during August / September.

Breeding not recorded in the period although summering noted in Ceredigion at Cors Fochno in 1993, 1995 & 1996 with two there in 1994; a male resident on Anglesey all of 1995, display noted at two sites that year (a nest was built but no breeding suspected), male seen displaying at three sites in 1998 and 1999. In 2000 a male and female were present, display and nest building were observed but there was no evidence of successful breeding.

HEN HARRIER *Circus cyaneus*

BOD TINWEN

(C) Small numbers breed on the uplands of north and mid-Wales, otherwise a winter visitor and passage migrant in small numbers, chiefly to coastal counties.

Birds in Wales quote six main roost sites in Wales, with a total roosting population of around 14 individuals. During the period under review the main roosts were at Llanrhidian, Glamorgan; Dowrog, Castle Martin, Skomer and Marloes Mere all in Pembroke; Cors Caron & Cors Fochno in Ceredigion. The max. number of birds using these roosts in each winter during the period 1992-2000 is tabulated below.

	/1992	92/3	93/4	94/5	95/6	96/7	97/8	98/9	99/00	00/01
Llanrhidian			4	6	4					
Dowrog	8	6	3	8	5	3	1	5	4	3
Castle Martin		4	1		1	1		1	1	
Skomer / Marloes Mere	2	1	5	3	5	3	4	3	3	1
Cors Caron	2	2	3+	4	6	4	3	4	5	5
Cors Fochno	2	2	4	2-4	4-5	4	4	3	3	2-3

Birds in Wales 1992-2000

Wintering individuals were recorded from every Welsh county. The total number of wintering individuals in each county during 1993-2000 is shown on the map opposite. With the largest figures coming from those counties with regular roosts.

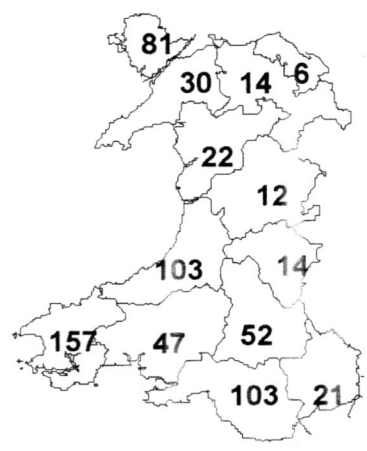

Data on the sexes of the wintering birds are not complete, however from those that were sexed, the following breakdown of records is possible:

1992	20-40 in each winter, 40% males, 60% ring-tails
1993	c80 birds in total, 43% males, 57% ring-tails
1994	c80 birds in total
1995	100 birds in total
1996	100+ birds in total, 40 % males, 60% ring-tails
1997	75 birds in total
1998	100+ birds in total
1999	16 males + 20 ringtials in the 1st winter period, 14 males + 24 ring-tials in the 2nd
2000	21 males + 13 ring-tails in the 1st winter period, 12 males + 26 ring-tilas in the 2nd.

Birds that had been wing-tagged, originating in Scotland or north Wales were reported almost annually.

Year	Site, County	Date	Origin
1992	Tumble, Carmarthen	Oct. 12th	Pitlochry, Scotland
1993	Skomer / Ramsey, Pembroke	October / November	Islay
1994	Penclacwydd, Carmarthen	May	north Wales
1995	Cors Fochno, Ceredigion	March	north Wales
	Cors Caron, Ceredigion	November	Scotland
1997	Cors Fochno, Ceredigion	November / December	Langholm, Dumfries, Scotland, 1st W male
1999	Stackpole, Pembroke	Oct. 27th	Forest of Bowland, Lancs. Ringed as nestling in June.

Birds in Wales states that this species has ceased to breed regularly in Wales by 1910 but recolonisation started in the late 1950's. The total number of breeding pairs remained around 5 pairs until 1977 when 27-30 pairs were recorded. The Welsh population has remained fairly stable since, with a maximum of 28 pairs rearing a total of 55 young in 1992. Any inferred changes indicated by the following figures are likely to be due to changes in observer coverage not a change in population.

	number of pairs laying eggs	other territorial pairs	number of successful nests	min. number of young fledged
1992	21	7	14	55
1993	16	11	10	37
1994	18	9	7	23
1995	14	9	7	19
1996	16	9	8	26
1997	22	2	10	31-32
1998	20	8	-	-
1999	c20	-	-	-
2000	c20	-	-	-

MONTAGU'S HARRIER *Circus pygargus*   *BOD MONTAGU*
(B) Rare summer visitor and passage migrant which formerly bred in small numbers.

The only accepted records of this species since 1991 were of a displaying male at Mynydd Hiraethog, Denbigh in 1993, a male at Fairwood Common, Glamorgan May 22nd 1994, immature male at Cors Fochno, Ceredigion May – July 1994, on Skomer, Pembroke Apr. 30th 1995, a ringtail at Porthmynawyd, Pembroke May 26th 1998, a juv. at Skomer, Pembroke Aug. $27^{th} – 28^{th}$ 1998 and a male at Cors Fochno, Ceredigion May $1^{st} – 22^{nd}$ 2000 (also observed at the William Condry reserve, Montgomery).

The *Welsh Bird Report* for 1992 included three other records from Cors Caron, Ceredigion that were never described and have therefore been removed from the Welsh record.

GOSHAWK *Accipiter gentilis*   *GWALCH MARTH*

(E) An increasing breeding resident in all Welsh counties.
Birds in Wales states that the number of records increased during the 1950's & 1960's and the first confirmed breeding record this century was in 1969 in Carmarthen. During the next two decades the numbers of breeding Goshawks increased rapidly as the species colonised the mature coniferous woodland of upland Wales and the mixed lowland woods. The total Welsh breeding population was estimated at 150 pairs by 1991.
During the period of this review, this expansion in range and abundancy continued and breeding was recorded in every county. Using data supplied by county recorders and the Welsh Raptor Study Group the population is now estimated at 200-250 breeding pairs in Wales. The main strongholds are east and mid Wales, with few recorded in Anglesey and Pembroke.
In 1995, 105 nests were monitored, 84 of which (79%) were successful. The average clutch size of 54 nests was 3.2, brood size of 2.31 in successful nests.

SPARROWHAWK *Accipiter nisus*   *GWALCH GLAS*

(E) Common and widespread breeding resident throughout Wales, some evidence that it is declining where Goshawks are widespread.
Using data from the National Atlas, the Welsh population is estimated at 3,100 pairs.

BUZZARD *Buteo buteo*   *BWNCATH*

(C) Common and widespread breeding resident throughout all of Wales.
Using data from the National Atlas, the Welsh population is estimated at 3,800 pairs. There has been a marked increase in 4 10-km squares of north Brecon, during 1983 – 1999, of +115%, paper on which was published in *Welsh Birds* (2:251-256).
Large congregations: 14 in a re-seeded ley, Brecon 1993, 38 near Radnor in December 1993, 36 in a re-seeded ley Llanwenog, Ceredigion Oct. 28th 1994, 55 feeding on slaughter-house slurry near Llanybydder, Ceredigion Oct. 10th 1995, 25 worming near Brecon Nov. 29th. 1998 and 30 worming near Llanfilio, Brecon Jan. 10th 1999. Over 40 mid October 1999 – January 2000 near Newcastle Emlyn, Ceredigion, max. 57 on Dec. 15th (*Welsh Birds* 2:292).

Birds in Wales 1992-2000

ROUGH-LEGGED BUZZARD *Buteo lagopus*   BOD BACSIOG
(B) A scarce visitor in autumn and winter.

Since 1991 there have been 5 accepted records in Wales, relating to probably only 4 individuals.
A rather mobile bird, first seen Kerry, Newtown Mar. 4th, then Rhyader Apr. 11th and Lake Vyrnwy Apr. 23rd Montgomery / Radnor 1995. Two records of what was probably the same bird in Gwent at Garnlydan Res. on Oct. 9th and at Peterstone Wentloog on Nov. 15th 1998. In 1999 there were two records, at Bardsey, Caernarfon Oct. 18th – 19th and an immature at Trichrug, Ceredigion on Dec. 14th.

OSPREY *Pandion haliaetus*   GWALCH Y PYSGOD
(E) An annual spring and autumn passage migrant which has increased in parallel with the Scottish breeding population. Single birds over summered in 1986, 1992, 1994 and 1997. Several have stayed for long periods.

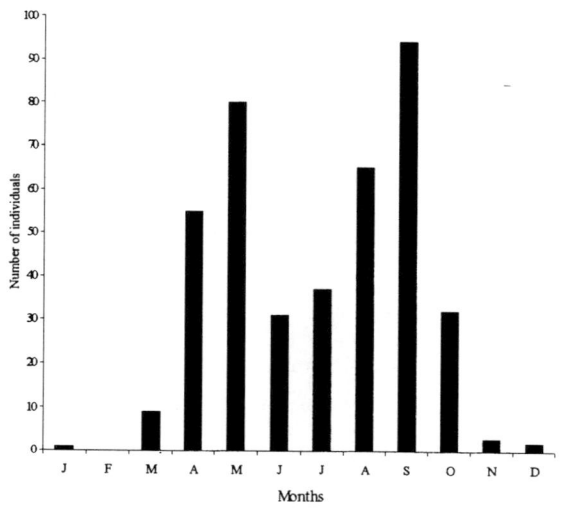
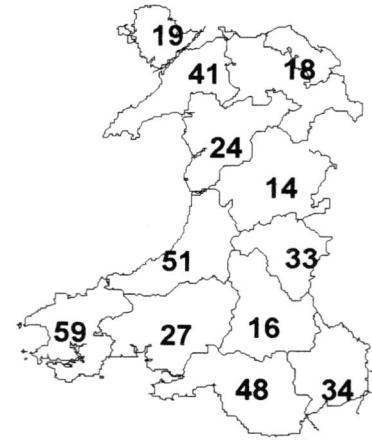

The number of Welsh records has increased in parallel with the increase in the Scottish population and from 1980 onwards Ospreys have become annual spring and autumn visitors, with a few summering. The total number of individuals recorded in each month, during 1992-2000 shown in the chart and the total number of individuals recorded in each county is shown on the map.
Passage starts late March, peaking in May. Larger numbers recorded during autumn passage, especially in September.
In 1994 one was observed at Llandegfedd Res., Gwent on Dec. 3rd and probably the same bird was subsequently seen at Llanrhidian Marsh, Glamorgan Jan. 19th 1995.

An individual was seen on Dec. 18th 2000 on the navigation buoy in the Teifi estuary, Ceredigion / Pembroke.

Another species that has yet to breed in Wales, but evidence suggests an increase in spring and summering records. The most intriguing records are from Montgomery, where following two being present in summer in 1997, a single individual was present 30th June – 26th August in 1998 and the same bird (identified by a dark green colour ring on its left leg) was again present 8th June – 28th July 1999. 2 were also recorded carrying fish in the summer of 1996 in Meirionnydd.

LESSER KESTREL *Falco naumanni*   *CUDYLL COCH LLEIAF*
The record of an individual in the Vale of Neath, Glamorgan Nov. 7th 1973 has now been rejected by BBRC. There has therefore not been an acceptable Welsh record of this species.

KESTREL *Falco tinnunculus*   *CUDYLL COCH*
(C) Rather scarce breeding resident throughout Wales. Partial and passage migrant in small numbers to some counties.

Considering the extent of upland rough grazing in Wales, this species scarcity is rather surprising. *Birds in Wales* suggested a breeding population of 800 – 1,000 pairs for the early 1990s. It has certainly declined since, particularly in improved farmland in Gwent, Brecon, Carmarthen, Pembroke, south Ceredigion, Anglesey, Denbigh and Flint. The main reason for its increasing scarcity is almost certainly the decline of vole-rich grasslands with rising stocking rates of sheep. The bald pastures and degraded hedges so typical of much Welsh farmland today are totally unsuitable habitat for voles or, indeed, anything else. At present the main areas for Kestrels in Wales are the cliff coasts of Pembroke and Ceredigion, and no doubt north to Caernarfon although we lack records, and some of the hill ranges, particularly the Black Mountains in Brecon / Gwent and Ruabon / Berwyn in Denbigh. Kestrels are also fairly widely distributed through the lowland farmland of the Borders but become progressively scarcer as one moves west.

The Welsh Ornithological Society has been surveying Kestrels since 1997 and the table below summarises the records so far. The overall breeding performance from 53 nests in the period was 3.03 young fledged per nest; for 177 known breeding results together, mainly records of broods fledged, it was 2.19 young per pair. This overestimates breeding performance because pairs which fail early are likely to be missed. Of 523 records of occupied sites, 50% did not relate to proven breeding pairs. Although often casual records, some of these were almost certainly pairs which failed early in the breeding cycle. The occupation rate of 83% of sites checked is also high. Where areas have been regularly monitored, occupation rates have not exceeded 55%. The data are insufficient at this stage to allow a good estimate of total population.

Small numbers of Kestrels from outside Wales winter, these birds mainly leaving by March. There is a larger movement through the Country in autumn, mainly in late August and September and often of birds of the year. Numbers are not well known but there have been recent records of up to 30 – 40 on Mynydd Epynt in September for example but none at all in 2000.

Breeding Kestrels in Wales 1997 - 2000

	Sites checked	Sites occupied	Occupied territory	Single birds	Pair	Pairs bred	Young Reared	Young by pairs	Pairs failed
1997	129	101	13	14	9	65	112	46	8
1998	209	181	41	22	27	91	139	52	7
1999	230	203	59	25	39	82	127	57	8
2000	183	140	40	8	27	65	88	39	5

Occupied territory denotes sites where birds known to be present but no more.
Single birds denotes sites where it was known that only one bird was present.
Pairs denotes sites where a pair was known to be present but no more.
Pairs bred includes 60 pairs for which the actual result was not recorded.

RED-FOOTED FALCON *Falco vespertinus*   *CUDYLL TROEDGOCH*
(B) A rare visitor.

An additional record to those published in *Birds in Wales* of a male at Ramsey, Pembroke on May 24th 1959, this was only recently submitted to and accepted by BBRC. This makes a total of 13 records in Wales up to the end of 1991, 5 in Glamorgan and Pembroke, one in Ceredigion, Anglesey and Denbigh.

The only records since then were in 1992 both of females at: Rhandirmwyn, Carmarthen May 29th and at Bosherston, Pembroke June 2nd.

MERLIN *Falco columbarius*   *CUDYLL BACH*
(C) Scarce breeding resident, passage migrant and winter visitor.

Birds in Wales states that Merlins have suffered a long, steady decline throughout Wales since the turn of the century with a slight recovery in the 1980s, when pairs were noted in old corvid nests on the edge of upland conifer plantations. This increase continued into the early 1990s.

The most recent all Wales survey in 1993 (*Welsh Birds* 1(1): 14-20) checked 126 territories, of which 84 were occupied and 73 held pairs. Coverage was incomplete and the total population was estimated as 90-100 pairs, an apparent increase since the early 1990's when the total Welsh population was believed to be at least 70 pairs (*Birds in Wales*).

In 1993 nest sites were located in heather moor (44%), moorland trees (4%) and conifer plantations (52%), with the first predominating in north Wales and the last in south. Overall nesting performance was 2.0 young fledged per pair from ground nests and 1.45 from tree nests. Overall density was 1.87 territories per occupied 10-km square. None were found nesting in Anglesey, Carmarthen, Flint or Glamorgan but single pairs have bred in Carmarthen and Glamorgan since the survey. Records since have been incomplete but give little indication of much change.

Widely distributed all round the coast in winter, with records for every month and not uncommonly noted up to mid May and from early August. Numbers are not well understood but in most years 40-60 birds have been involved in each winter period. Also winters in inland farmland in mild winters, these perhaps being local breeding adults from nearby hill areas.

An individual of the Icelandic race *subaesolon* trapped at Afonwen, Caernarfon Nov. 15th 1996.

HOBBY *Falco subbuteo* — HEBOG YR EHEDYDD

(E) Rare breeding summer visitor in at least four counties with a total breeding population of approximately 30 pairs. Elsewhere a rare passage migrant, although a few summer on occasion.

Hobbys have been sporadic breeders in south east Wales and have increased slightly since the early 1960s so that by 2000 pairs breed in Gwent, Brecon, Radnor, Montgomery, Denbigh, Flint and probably in 2 other counties. No comprehensive survey has been undertaken in Wales. Some local populations are studied in detail, but records are incomplete even at a county level. Published reports suggest a Welsh population of 20-25 pairs. Williams (2000) considered the population to be in excess of 30 pairs, which seems a reasonable assessment, based on incomplete coverage and the difficulties in surveying this species.

The map opposite summarises all non-breeding records by county.
An individual at Good Hope, Pembroke Nov. 16th 1996 is the latest recorded in Wales.

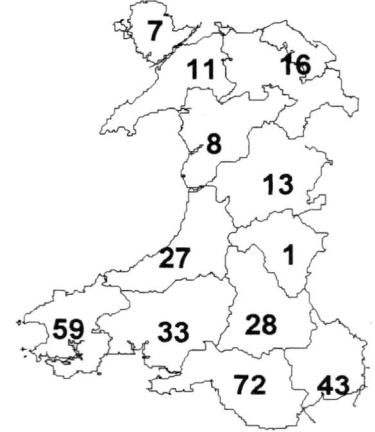

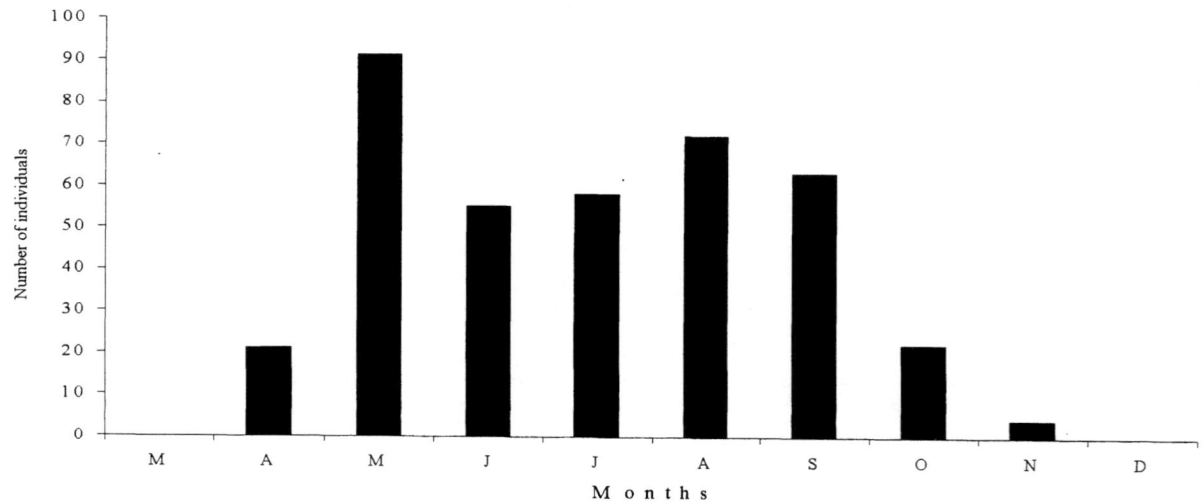

Monthly distribution of non-breeding Hobbys 1992-2000

PEREGRINE *Falco peregrinus*   *HEBOG TRAMOR*
(C) Resident breeding species in all counties, increasing in numbers to a current breeding population (1991) of 280 pairs.

	1993	1994	1995	1996	1997	1998	1999	2000
Gwent	6	8	10		9	5	13	9
Glamorgan	10	14+	19 *	15	21	27	29+	18
Carmarthen			5		9	8	6	
Pembroke		33-34	35-37	38	43	43	43	48
Ceredigion	14	15+	14				8 inland	
Brecon	15	16	14	12	13	13	17	12
Radnor	14		8		3		18	
Montgomery	25+	13	20		11	9	20	43
Meirionnydd	28							
Caernarfon							4	
Anglesey						2	2	4
Denbigh			19	21	8	8		7-8
Flint								

* East Glamorgan only.

Birds in Wales states that this species had fully recovered from the population crash in the 1950s & 1960s, due to contamination by persistent organochloride pesticides, by 1981. Indeed by 1991 the population of inland and coastal areas had reached unprecedented levels. Although the table above, of the number of breeding pairs in each county during the period 1992-2000 is incomplete, that data there is suggests that this increase has continued. Several of these figures are very far from complete surveys. Where there is a good run of counts, it appears that the population in Brecon is fairly stable, while it is still increasing in Pembroke and Glamorgan. The Montgomery population has increased considerably over the period. The total Welsh population was estimated at 300 prs. in 1992 (50 prs. failed that year as a result of human persecution of one sort or another). The total Welsh population by 2000 is probably of the order of at least 350 pairs.

RED GROUSE *Lagopus lagopus* *GRUGIAR*

(C) Breeding resident on heather uplands, declining in most areas.

The Welsh population has declined rapidly since the Second World War due to afforestation, agricultural improvement and overgrazing. *Birds in Wales* states that the bulk of the Welsh population by 1991 was confined to half a dozen or so keepered moors, although birds were widely but very thinly distributed throughout the uplands.

Total Welsh population in 1991 was believed to be less than 1,000 pairs. Strongholds include Berwyn, Mynydd Hiraethog and Migneint.

Breeding data from the period 1992-2000 is incomplete, what records there are are broken down into their counties:

Gwent: 22 shot at Coity Mountain in 1996, recorded at 9 sites in 2000.

Glamorgan: 3 north of Merthyr Tydfil on Dec. 31st 1992 was the first record for the county since 1953. Single at Cefn Drum, near Pontardulais on Mar. 21st 1998 and 8 at Waun Wen Nov. 9th 2000.

Carmarthen: 9 birds at 3 sites in 1992 and fresh pluckings found at Garreg Lwyd in 2000.

Ceredigion: recorded at one site in 1999. In 2000 there was one unconfirmed report of breeding in the county and birds were seen at 2 other sites.

Brecon: 10 prs. at 7 sites in 1995 (away from Elenydd). None found by RSPB survey of 15,000 ha of Mynydd Du (Brecon / Carmarthen) and only 4 non-breeding birds on 4,000 ha of Cnewr Estate in 1996.

Thought to be declining.

Radnor: survey of Elenydd in 1995 found 23 prs. in 150 km^2 (9 of which in Brecon, 6 in Ceredigion & 8 in Radnor). Survey in 1997 reported the species as being thinly scattered in 19 1-km squares, mainly singles. 120 counted post breeding season on Ireland Moor in 2000.

Montgomery: at Lake Vyrnwy: 107 prs. in 1995, 92 prs. in 1995, with a density of up to 4 prs./km^2, 59 prs. in 1997, 108 prs. in 1998, 122 prs. in 1999 and 124 prs. in 2000.

Meirionnydd: recorded at 9 sites in 1999 and at 4 in 2000.

Caernarfon: individuals at 6 sites in 2000.

Denbigh: on Ruabon Moor, autumn transects found 33 birds in 1994, 81 in 1995, 59 in 1996, 62 in 1997 and only 6 in 1998 – the lowest total in over 20 years.

Desk survey by CCW estimated that there were no more than 5,000 birds in Wales – similar numbers to a RSPB survey during 1990/1.

BLACK GROUSE *Tetrao tetrix* *GRUGIAR DDU*

(C) Scarce breeding resident confined to the moorlands of north and mid-Wales, declining over most of its range.

The habitat requirements of black grouse are complex. They are a sedentary species that require a continuous mosaic of upland habitats such as wet heath, blanket bog, and open forest for feeding, displaying, breeding, roosting and moulting. In Wales, during the 1950s to 1980s these semi-natural, transitional habitats have slowly disappeared because of changes in land-use policy. The majority of black grouse in Wales (over 90%) are now associated with upland commercial conifer plantations that have a mosaic of vegetation communities along the forest edge with a well-developed ground layer of young heather and bilberry.

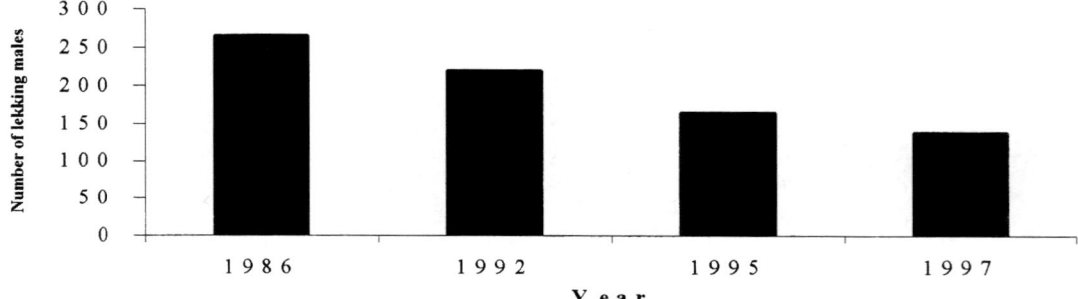

The number of displaying male black grouse during 1986 to 1997

Changes in the number of displaying males during 1986-1997

County	Number of displaying males				1986-1997
	1986	1992	1995	1997	
Caenarfon	7	10	3	2	-71%
Carmarthen	4	1	0	0	-100%
Ceredigion	13	13	7	7	-46%
Denbigh	96	65	59	64	-33%
Meirionnydd	126	113	77	41	-67%
Montgomery	18	17	18	17	stable
TOTALS	264	219	164	131	50%

Similar to other areas within the UK, the Welsh black grouse population has showed a chronic decline from the 1980's through to the late 1990's. Because black grouse are secretive and elusive, the only practical survey method to measure their populations is by counting displaying males on their lek sites in spring (Gillian et al., 1998). The first systematic Welsh survey of black grouse undertaken in 1986 recorded 264 males at 91 leks (Grove et al., 1986). With evidence indicating range contraction and a decrease in lekking males, three further surveys were conducted in 1992, 1995 and 1996. These surveys (using the same methodology) gave an estimate of the number of displaying males. Comparisons between years and sites provided an indication of population change and site persistence.

The 1997 Welsh census recorded 131 males on 66 leks and suggests the population had dramatically dwindled by 50% since the 1986 survey (Williams et al., 1997). The pace of decline is illustrated in the figure above.

In the early 1990's, black grouse disappeared from Carmarthen with small isolated populations remaining in Caernarfon and north Ceredigion. During this period, there was also a rapid decline in Meirionnydd that held 48% of the Welsh population in 1986 but only 32% in 1997. The table above shows the pattern and intensity of decline is not uniform throughout the species range in Wales. Nevertheless, black grouse are now restricted to core populations in parts of mid and north Wales. The number of leks and the average lek size has declined during 1986 to 1997. The tabulated data shows a loss of 25 (27%) active leks and an increase in the proportion of solitary displaying males over a 10 year period. It is widely thought that single displaying males if they are widely dispersed and in isolation from core populations represent a declining population.

Various causes of decline in black grouse have been identified, ranging from factors that may cause direct loss of habitat (Baines, 1994; Baines, 1996), to other, more subtle factors which may reduce breeding success and/or adult survival (Picozzi & Hepburn, 1986; Cayford & Hope-Jones, 1989). However, opinions agree that in Wales the decline is strongly associated with complete habitat loss, habitat degradation or fragmentation. Whilst major land-use policies have undoubtedly had a considerable effect on habitat quality and availability, climate change may have attributed to poor breeding success.

Numbers of displaying black grouse at leks in Wales between 1986 & 1997

Year		1	2	3	4	5	6	7	8	9+	No. of leks
					Number of displaying males						
1986	No.	43	18	14	4	2	3	0	4	3	91
	%	47	20	15	4	2	3	0	4	3	
1992	No.	70	20	10	9	1	0	0	1	2	113
	%	62	18	9	8	1	0	0	1	2	
1995	No.	47	23	4	3	2	0	1	1	1	82
	%	57	28	5	4	2	0	1	1	1	
1997	No.	35	17	6	2	2	3	0	1	0	66
	%	53	26	9	3	3	5	0	2	0	

Due to the historical decrease in both range and numbers and the severity of the current decline, the RSPB and other conservation organisations believed that a large-scale recovery project was necessary to prevent black grouse becoming extinct in Wales by 2015. Following funding by the European Union, the National Assembly of Wales and the RSPB, the Welsh Black Grouse Recovery Project (WBGRP) was launched in 1999. Continuing work initiated in 1997, between 1999-2001, the project integrated a comprehensive suite of habitat prescriptions at 6 black grouse key areas in mid and north Wales (Llandegla and Ruabon Moors, Clocaenog, North Berwyn, Pale, Llanbrynmair and Migneint/Dduallt). Collectively these sites are referred as the project core area and comprised 80% of the remaining male population in Wales in 1997 (Williams et al., 1997).

Persistence of black grouse populations is likely to be closely related to the number of cocks and hens in a population and the extent and suitability of favourable habitat. An integral part of the WBGRP was monitoring the response of black grouse numbers to habitat management within the 6 key areas. Displaying males at all lek sites within the 6 key areas were counted and late summer brood counts were undertaken within a 1.5km radius of 18 focal leks (3 for each key area).

The table below shows the annual lek counts within the WBGRP key areas between 1997 and 2000. The lek counts conducted in 2000 recorded 171 displaying males, an overall increase of 63 (%) males from the 1997 lek counts. Though this comparison should be treated with caution, due to incomplete lek coverage in 1997, the 2000 lek counts suggest the number of displaying males on North Berwyn, Clocaenog, Ruabon/Llandegla and Pale/North Vyrnwy have showed signs of recovery and increase. However, on both Llanbrynmair and Migneint/Dduallt, males have continued to decline. The reasons for these different population trends are not clear but may be attributable to predator control and habitat availability and quality.

It is highly probable that when accounting for other remaining black grouse populations in Wales coupled with those counted within the WBGRP key areas the total Welsh population may be approximately 210 lekking males. This would represent a 60% increase from the 1997 survey.

Number of lekking males counted on the six key areas (1997-2000).

Core Area	Lek counts				
	1997	1998	1999	2000	change 1997-2000
North Berwyn	23	31	44	39	16
Ruabon/Llandegla	21	38	54	73	52
Llanbrynmair	13	16	10	5	-8
Clocaenog	18	16	16	24	6
Pale/North Vyrnwy	19	34	24	21	2
Migneint/Dduallt	14	8	3	9	-5
TOTAL	108	143	151	171	63

From 1997 to 2001 black grouse hens and their broods were surveyed within the focal lek radii within all key areas (Pale Moor was surveyed by the Game Conservancy Trust), using trained dogs. From the late summer brood counts the number of chicks per hen for the 3 focal leks for each key area were pooled to estimate breeding success (Figures below).

Number of hens   *Number of chicks*

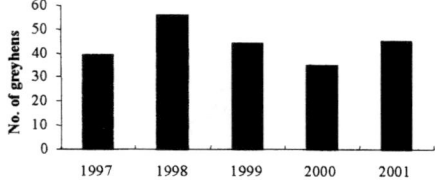
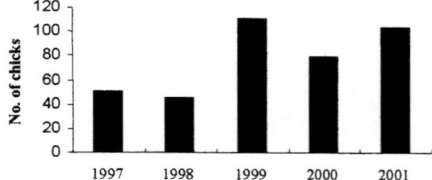

Number of broods *Number of chicks per hen*

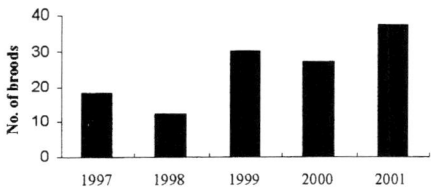
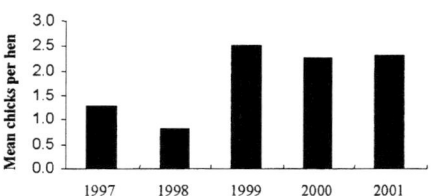

[Note: In 1998, North Berwyn was added to the other 5 key areas due to its importance for black grouse.]

In 1997, 39 hens were located (from 18 focal leks), 18 with broods (45 chicks) representing 1.3 chicks per hen. In comparison, during 2001, 45 hens were located (from 10 focal leks), 37 with broods (104 chicks) giving 2.3 chicks per hen. Baines (1993) calculated a yearly adult mortality of 30% and estimated that 1.5 juveniles per hen per year are needed to maintain a population. However, there are a number of pitfalls in applying this data to Wales. First, the figure is derived from data from a wide range of sites that do not include any in Wales. Second, the differences in adult and juvenile mortality rates between sites could cause major differences in the number of juveniles required to maintain stability. Used with caution, the productivity data for Wales from 1999-2001 suggest that the majority of key areas are rearing sufficient chicks to maintain their populations.

The results of a radio-tracking study in the North Pennines suggest first-year and adult males do not move greater than 2 km from natal areas. In contrast, first-year hens may disperse up to 19km during either late autumn or early spring (Baines, in press). If these findings (non-dispersal of males) apply to Wales and suitable habitat for black grouse is heavily fragmented into comparatively small areas, the chance of local extinction increases with increasing isolation of a lek and degrading habitats. The degree of such isolation will also have implications for recovery projects, specifically re-colonisation of formerly occupied sites.

RED-LEGGED PARTRIDGE *Alectoris rufa* *PETRISEN GOESGOCH*
(E) Almost extinct in the majority of Wales except for a scattering of records. The only county where there are more than this is on Anglesey. In most places the population is maintained by releases and clouded by the occurrence of hybrids with A. chukar.

By 1970 this species was confined almost exclusively to the Welsh border counties. Since then numbers and distribution has increased due to widespread shooting releases. The only place in Wales where this species is described as numerous is at City Dulas, Anglesey where 5,000 were put down for shooting in 1999.
Releases were also noted at; Treleddyn Farm, St. David's in Pembroke, December 1998, Pennant, Ceredigion in 1995, Llanhamlach and Llanwrtyd Wells, Brecon in 1997 and Ruabon Moor, Denbigh in 1997 & 1998.
The only large coveys were: 9 at Skirrid Fawr, Gwent Dec. 12th 1995, 18 at Llanfyllin, Montgomery Nov. 22nd 1995, 12 at Caldicot, Gwent Oct. 26th 2000 and 15 at Carmel Head, Anglesey Jan. 23rd 2000.

GREY PARTRIDGE *Perdix perdix* *PETRISEN*
(E) Almost extinct as a breeding species in much of Wales. The population is maintained by releases in some places.

Birds in Wales states that this species has gradually declined throughout Wales, probably since the 1940s. This decline has been attributed to a combination of factors, including agricultural intensification, over grazing, use of herbicides and a change in weather patterns to cold springs and wet summers.
During 1992-2000 this decline continued with very few records being reported. By 2000 it is highly likely that this species had become extinct as a breeding species in most Welsh counties and that populations were only maintained by releases for shooting.
In Gwent by 1995 it had all but disappeared from the usual sites around Abergavenny and was mainly found on the Gwent Levels. In neighbouring Glamorgan its distribution was described as almost exclusively coastal in 1994.

Birds in Wales 1992-2000

In Pembroke the only Partridges were relics of shooting releases in the Angle area, while releases were noted at Treleddyn Farm, St. David's in December 1998. 500 were put down for shooting at Dulas, Anglesey in 1999.

In Denbigh none were found in 1997 on Ruabon Moor / Llangollen or Alyn Waters C.P., where this species was formerly resident. In 1997 an RSPB survey of 73 1km squares on Anglesey, found Grey Partridges in only three, compared to 13 in 1986 - a 77% decline.

QUAIL *Coturnix coturnix*   *SOFLIAR*
(C) A scarce and irregular breeding species, irruptive in some years; scarce passage migrant.
Quail are unpredictable summer visitors to Wales, occurring sparingly or not at all in many years. The following table summarises the number of calling individuals recorded in each county during 1992-2000.

	1992	1993	1994	1995	1996	1997	1998	1999	2000
Gwent		1	4		1				
Glamorgan	2-3		5	7	4+	3	1	5	
Carmarthen	1		2			5			
Pembroke	5-6		6	2	4	8	4-5	3	
Ceredigion	3		1			7	7-8	1	1
Brecon	5		3	1	2	2	2		
Radnor						6		1	
Montgomery	3	2							
Meirionnydd			1		1		3		
Caernarfon		1	1	1		1	1	2	1
Anglesey		2	5	1		5		3	1
Denbigh			4-5	2				2	1
Flint	2			1		1			
Total	21-23	6	32-33	15	18	32	17-19	17	4

The only confirmed breeding was of a brood seen at Walton, Radnor in 1996.

Unseasonal records included one on the Cardiff shore Apr. 12th 1992, one picked up at Barmouth, Meirionnydd and taken into care on Dec. 6th 1994 and one at Llyn Helyg, Flint Apr. 13th 1992.

PHEASANT *Phasianus colchicus*   *FFESANT*
(D) A locally common breeding resident, widespread throughout Wales.

Birds in Wales 1992-2000

GOLDEN PHEASANT *Chrysolophus pictus*   *FFESANT EURAID*
(E) Feral population appears to be declining, if not extinct.

The information on this species is incomplete, but the latest reports are that they are now extinct at one of the Anglesey sites and that there have been no recent records from the other. Williams (2000) considered the current population to be 30-35 individuals, but it is possible that the numbers are much lower than this and this species may already be extinct or on the verge of extinction as a breeding species in Wales.

LADY AMHERST'S PHEASANT *Chrysolophus amherstiae*   *FFESANT AMHERST*
(C) A small and declining feral population in Flint.

The current population at Halkyn, Flint is thought to be no more than 10 individuals by Williams (2000), but there is no other information to support this figure.

WATER RAIL *Rallus aquaticus*   *RHEGEN Y DWR*
(C) A scarce breeding resident, more commonly recorded as a passage migrant and winter visitor to all counties.

Water Rails are common winter visitors to Wales. Max. winter counts during 1992-2000 were: 10 at Kenfig, Glamorgan December 1992, 15 there Dec. 20th 1998 and 14 in early 1999, 14 at Penclacwydd, Carmarthen in January 2000 and 11 in December, 30 Teifi Marshes, Ceredigion / Pembroke both winter periods 1998 & 1999 and 15 in December 2000, 14 Llangorse December 1992, 10 there Nov. 28th 1998, 10 RSPB Conwy, Caernarfon 1998. At least 18 different individuals recorded on Bardsey, Caernarfon Sept. 11th – end of season 1999.

Water Rails are a scarce breeding species in Wales, absent from most of north Wales, the Cambrian Mountains and the south Wales uplands where suitable habitat does not exist. A summary of breeding sites, max. number of territorial males indicated in brackets, is given below.

Although the data are incomplete, what there is suggests a total Welsh population of 100 - 200 pairs, no change from that quoted in *Birds in Wales*.

Gwent	Uskmouth (5)
Glamorgan	Kenfig (9), Oxwich (2), Hensol Lake, Nelson Bog, Neath Abbey (4), Square Pool Neath, Crymlyn Bog (6), Crymlyn Burrows (2), Margam Burrows, Pant y Sais Fen (4)
Carmarthen	Ffairfach (2), Witchett Pool (3), Machynys, Ashpit Ponds (4). 18 prs. located in a survey of the SE of Carmarthen in 1996
Pembroke	Skomer (2), Skokholm, Bosherston, Teifi Marshes (2), Dinas Cross
Ceredigion	Ynyshir (3), Cors Fochno, Aberleri (2), Teifi Marshes (5)
Brecon	Llangorse (2), Mynydd Illtyd, Gors Llwyn, Cwm Berwyn, Caer Beris
Radnor	Pentrosfa, Rhosgoch, prob.
Montgomery	Newbridge Dolydd Hafren
Caernarfon	Spinnies, Conwy RSPB
Anglesey	Malltraeth (4), Cors Erddreiniog (4), Llyn Alaw, Cors Goch, Rhos Goch, Valley lakes (2)

Figures for the Teifi Marshes were split into those in Ceredigion and those in Pembroke.

SPOTTED CRAKE *Porzana porzana* *RHEGEN FRAITH*
(B) Rare vagrant and occasional breeder.

An almost annual visitor to Wales although no breeding has been proved since the 1980s and very few sites hold calling males on a regular basis. Breeding was suspected in 1993 when six males were heard calling at three sites (one each in Brecon, Caernarfon and Flint), single males calling in Caernarfon & Cors Erddreiniog, Anglesey in 1998 and at Cors Caron, Ceredigion & Gwenfro Isaf, Anglesey in 1999.

Otherwise records of spring passage birds: Ffairfach, Carmarthen June 7th & 11th 1992, Uskmouth, Gwent Apr. 24th & 25th, Ffrwd Fen, Carmarthen Apr. 15th and Goodwick, Pembroke Apr. 14th 1993, calling at Marloes Mere, Pembroke May 13th - 15th 1995.

Autumn passage recorded at Llangorse Aug. 23rd 1992, when a juvenile was trapped, a juv. at Peterstone Wentloog, Gwent Sept. 2nd - 24th. 1994, juv. Llyn Coed y Dinas, Montgomery Sept. 24th - Oct. 3rd 1995, juv. Penclacwydd, Carmarthen Sept. 8th 1996, adult at Kenfig Sept. 19th & 20th and a juv. at Teifi Marshes, Ceredigion / Pembroke Aug. 24th 1997.

CORNCRAKE *Crex crex* *RHEGEN YR YD*
(B) Formerly a numerous and ubiquitous summer visitor to all counties. Now extinct as a breeding species.

The demise of the Corncrake in Wales is well described in *Birds in Wales* and by 1980 this species was little more than a very rare passage migrant, mainly in autumn.

Of the 33 individuals recorded since 1991, there have been only 5-6 breeding attempts, 2 pairs at one site and 2/3 at another on Anglesey in 1992, one on Anglesey in 1994.

It is interesting to note that *Birds in Wales* states that in the 1980s there were more passage records in autumn than in spring. This pattern is different to that of the 1990s, when there were 17 spring records compared to just 6 autumn records. This may just reflect the ease of finding calling birds rather than any change in migration routes.

Spring passage: eleven records in Pembroke, in the St. David's area, 3 in 1992, around May 25th, one Apr. 28th onwards for a week 1993, 2 in April / May 1994, one Apr. 24th 1995. Elsewhere in Pembroke, singles at Camrose May 18th and Marloes Mere June 3rd 1993, at Nolton June 27th 1995 and Skokholm May 26th 1999.

In Ceredigion, one in late spring at Llandewi Brefi 1993, near Lampeter July 19th - Aug. 2nd 1993; at Painscastle, Radnor Apr. 30th 1996; at Dolydd Hafren, Montgomery June 19th 1996; in Caernarfon a male calling at Llanfairfechan in 1996 and a male at Morfa Madryn on June 19th 1998; at Amlwch, Anglesey May 6th 1993.

Autumn passage: at Gorslas, Carmarthen Sept. 22nd, killed by a cat; in Pembroke, Llangloffen Fen Aug. 13th 1997 and a juv. on Skokholm Sept. 14th – 17th 1999; one picked up dead at Boduan, Caernarfon Oct. 3rd 1998; one Braint Est. Anglesey Sept. 13th 1999.

MOORHEN *Gallinula chloropus* *IAR DDWR*
(C) A locally common breeding resident and winter visitor, widespread thoughout Wales.

The only large counts were from Penclacwydd, Carmarthen and Bosherston, Pembroke.

Penclacwydd	1995	64 in the early months	135 in December
	1996	121 January / February	215 in November
	1997	220 in January	281 in November
	1998	194 in February	198 in November / December
	1999	195 in February	190 in November / December
	2000	210 in February	175 in November
Bosherston	1995		115 on Nov. 21st

The breeding biology was studied at Penclacwydd in 1997, when there were 37 monogamous pairs, 11 polyandrous (f + 2m) and 3 polygamous pairs (M + 2f).

Using data from the National Atlas, the total Welsh population is estimated at 13,440 pairs. BBS data 1994 - 1999 however suggests a population change of 118%, producing a total Welsh population of 16,000 pairs.

COOT *Fulica atra*   CWTIAR
(C) *A locally common breeding resident and winter visitor, widespread throughout Wales.*

		/92	92/3	93/4	94/5	95/6	96/7	97/8	98/9	99/00	00/01
Gwent	Llandegfedd R.		154	244			100	35	114	30	
	Ynysfro Res.		193	230	218		320	94	93		
Glam.	Llan/Lisvane		192	252	172	292				66	108
	Kenfig	272	238	285	262	278	116	179	361	471	405
	Eglwys Nunydd					59	232	44			148
Carm.	Penclacwydd				73	112	210	132	82	95	110
Brecon	Llangorse	320	270	168	350	520	800	650	830	520	700
Meiri.	Llyn Mwyngil	260		120	196	136					
Angle.	Llyn Alaw					253		678			
	Llyn Cefni				215	414	382	96			
	Cemlyn					121	403	479			

Birds in Wales states that in Wales, large eutrophic lakes are restricted to lowland areas and Coots are absent from most of the upland waters. Anglesey is a stronghold for this species and the islands many lakes and open waters occasionally support high breeding densities. In winter these lakes often support large numbers, over 200 individuals, e.g. 380 at Llyn Alaw in 1984/5, 543 at Cemlyn in 1986/7 and 520 at Llyn Llywenen in January 1989. The table above gives the winter max. at the major wintering sites that were counted regularly and contained over 200 birds on at least one occasion.

Sites which have held over 100 birds in winter include: Ashpit Ponds, Machynys and Sandy Water Park Carmarthen; Bicton Res., Llys y fran Res. and Bosherston in Pembroke; Malltraeth, Llyn Rhos Ddu and Llyn Maelog on Anglesey; Llyn Helyg, Inner Marsh Farm, Shotwick and Flint Industrial site in Flint.

Large autumn counts on Anglesey in 1999 include: 322 at Llyn Cefni in July, 1046 at Llyn Alaw in September, 225 at Cemlyn in September. In 2000 there were 300 on Llyn Alaw in August, 273 on Llyn Cefni in July and 359 on Llyn Llygerian in September 2000.

Large counts at Sandy Water Park, Carmarthen were of 210 on Sept. 19th 1999 and 309 in September 2000.

The only annual survey of breeding birds during the period under review was at Llandegfedd Res. in Gwent and in Brecon from Llangorse & other sites. Data tabulated below, as number of successful pairs / number of young fledged.

A decline was reported in Ceredigion at Ynyshir in 1994, due to Mink *Mustela vison*. The population of Carmarthen was estimated at 60+ pairs in 1995. The only other breeding data were from 1998 but these may not have been complete surveys: 16 pairs in Eastern Glamorgan, 10 in Gower at 4 sites, one in Pembroke, 6-8 pairs in Montgomery at 2 sites, one in Caernarfon, bred at 5 sites on Anglesey and 3 pairs in Denbigh.

The total Welsh breeding population was estimated at 1,500 – 2,000 pairs in 1991. Using data from the National Atlas, the total Welsh population is estimated at 2,530 pairs. BBS data 1994 - 1999 however, suggests a population change of 133%, producing a total Welsh population of 3,500 pairs.

	1992	1993	1994	1995	1996	1997	1998	1999	2000
Llandegfedd Res.		38/	34/85						
Llangorse	30/19	31/11	31/22	30/70	30/50		47/14	51/17*	55**
Other sites, Brecon		21/45	28/42	15/22	22/31		12/33	14/27	11/20

The 1992 figure at Llangorse represented a 75% decrease since 1979. * 40 pairs failed completely. ** 35 prs. failed.

CRANE *Grus grus*   GARAN
(B) *A rare visitor to Wales.*

Birds in Wales quote a total of 18 Welsh records, 16 of which were in the 20th century. Since 1991 there have been a further 6 individuals in Wales. Singles: at Marloes Mere, Pembroke Oct. 20th – 22nd 1994, one over Bardsey, Caernarfon Apr. 27th 1995 was subsequently seen the next day at Rhoscolyn, one at Llantrisant, Anglesey Sept. 28th – Oct. 11th 1997 and at Hensol, Glamorgan Mar. 25th – 26th 1999.

Two near St. David's, Pembroke Dec. 15th – 31st 1999 proceeded to take a tour of south Wales before arriving in Avon, being seen at Cardiff Bay, Glamorgan and St. Brides, Gwent on Jan. 2nd 2000. Singles at Lisvane Res., Glamorgan Feb. 24th and at Cors Fochno, Ceredigion Apr. 7th 2000.

Breakdown of Welsh records of Crane, graphed by month, by county shown on the map.

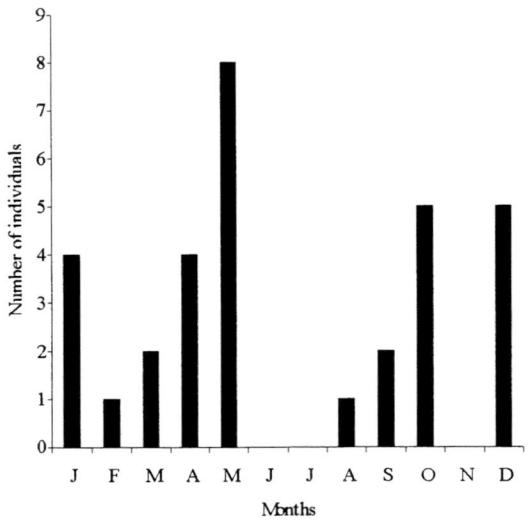
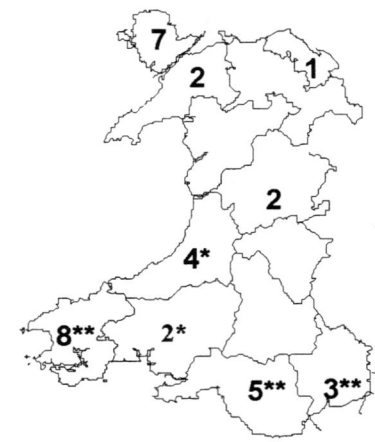

* 2 at Llandewi Brefi, Ceredigion, then moved to Pembrey, Carmarthen
** 2 at St. David's, Pembroke then moved through Glamorgan and Gwent

OYSTERCATCHER *Haematopus ostralegus*   PIODEN Y MOR

(C) A common, chiefly coastal, breeding species and abundant winter visitor to all coastal areas of Wales.

		1992	1993	1994	1995	1996	1997	1998	1999	2000
Glamorgan	Blackpill	2300	2200	2230	2650	3470	2910	2600	3106	3943
	Burry S.	11175	18000	9075	12780	12915	12850	18260	25310	16135
Carmarthen	Burry N.			2021	1531	4750	3549	3238	5540	4990
	TTG	11135	4894	5090	3119	2524	2044	1462	6284	3803
Flint	Point of Air	8000	14060	20920	16060	8590	13200	6200	8340	6150
	CQ'Oak	5000	3500	1700	8400	6200	3000	2600	6000	6167

2,000 roosting on the west side of the Great Orme, Caernarfon Oct. 16th.

Non-breeding flocks are present throughout the year on intertidal areas, with Wales holding more than 6% of the East Atlantic Flyway wintering population and over 20% of the UK total wintering population, totalling c70,000. The table above gives peak autumn counts, July – October, at the main sites, with over 2,500 at least once. The peak winter counts, November – March, at the main sites, with over 2,500 at least once are tabulated below. The National Level of Importance is 3,600 and the International Level is 9,000.

		/1992	92/3	93/4	94/5	95/6	96/7	97/8	98/9	99/00	00/01
Glam.	Blackpill	2980	2500	2556	2750	3327	3170	3636	2800	3320	2880
	Burry S.	10385	13800	10897	12080	19900	16850	17750	22355	23300	16891
Carm.	Burry N.				1868	1994	7045	4795	9464	7929	5860
	TTG	2779	5138	5224	5741	4464	2799	3250	2881	3495	2249
Caern.	Traeth Lafan	4068	2820	5055	5935	4520	5780	3163	3262	5781	6897
Flint	Point of Air	10000	5300	14420	17160	8340	7720	6437	5200	6030	5040
	CQ'Oak	3500	1400			1030				533	3500

Interesting ringing return of an individual caught and colour ringed at Sker, Glamorgan in May 1992 was subsequently seen at Plouharnel Bay, France in November 1996 & February 1997.

Birds in Wales quote figures for Average Winter Peak Counts for the five main estuaries for the period 1986-90. During the period under review there is no data available for the Dee as a whole. The Average Winter Peak Counts for the other 4 estuaries, 1996-2000, are higher than 1986-90, particularly on the Burry Inlet. Data tabulated below.

	1986-90	1996-2000
Burry Inlet	16,934	25,657
Swansea Bay (Blackpill)	2,491	3,250
Carmarthen Bay (T/T/G)	2,602	3,378
Traeth Lafan	3,623	4,501

The Oystercatcher is mainly a coastal bird in Wales, well distributed as a breeding species along rocky, pebbly and sandy shores, salt-marshes and the well-vegetated top of islands. *Birds in Wales* states that inland it is a scarce breeder but increasing along the major river systems.

During the period under review very little data was collected on Oystercatchers breeding in Wales. Apart from a survey on Anglesey in 1997, when 73 1-km squares were surveyed, with Oystercatchers found breeding in only 10 squares, compared to 23 in 1986, a 57% decline, there has not been any comparative fieldwork. Assuming that this 57% decline is representative of the change in the island's population as a whole, then a revised estimate of Anglesey's population is c150 pairs. There is no other reason to assume that there has been any other significant change in the Welsh breeding population. This generates a modified figure of 600 – 800 pairs breeding on the Welsh coast.

As for inland, *Birds in Wales* quoted 25-30 prs. nesting on shoals in 1991. During 1992-2000, there was breeding in Gwent (2-4 pairs), Brecon / Radnor (1-3 pairs at Glasbury), Montgomery (up to 10 pairs, Dolydd Hafren, Afon Vyrnwy, River Severn, Llyn Coed y Dinas and Leighton) and Denbigh (1 pair Llyn Brenig). The lack of records reflects the lack of observations and not a decline in habitats.

Of interest were records of pairs nesting on roofs at Saltney, Flint in 1992, fledging 2 young.

The table below summarises the number of breeding pairs on the Welsh islands, where compared to data in *Birds in Wales* there has been little change.

		1992	1993	1994	1995	1996	1997	1998	1999	2000
Pembroke	Skokholm	43	47	52	69	59	43	41	38	41
	Skomer	93	94	108	107	113	103	115	94	77
	Ramsey	25	20	25-27	27-30	24-26	22-26	26		24
Caernarfon	Bardsey		71	78	73	71	70	63	76	83

BLACK WINGED STILT *Himantopus himantopus* *HIRGOES*
(B) A vagrant which has probably bred once in Wales.

Three were at Cemlyn, Anglesey Apr. 10th – 21st 1993 and a single was at Cors Geirch, Pwllheli, Caernarfon Mar. 28th – Apr. 1st 1997. These represent the 10th and 11th Welsh records. Apart from the possible breeding in Gwent in the 1950's, of the 12 dated individuals recorded in Wales, all have been in spring March - May, except for singles in August 1965 and July 1967 and of these there has been a total of 4 individuals in Pembroke, Caernarfon and Anglesey.

AVOCET *Recurvirostra avosetta* *CAMBIG*
(B) A scarce visitor to Wales, recorded chiefly on spring passage but occasionally wintering.

A total of 83 individuals were recorded in the nine years, of which over half were from Gwent. The total number of individuals recorded during 1992-2000 is shown by month in the following chart and by county on the map.

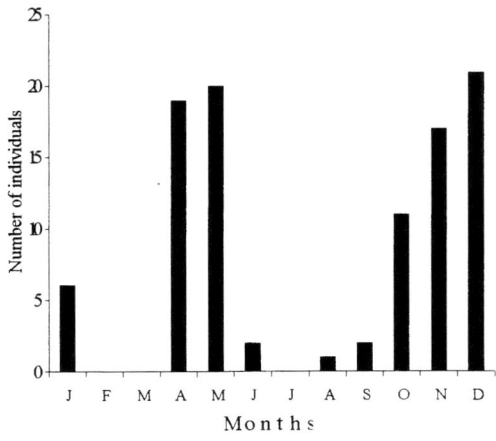
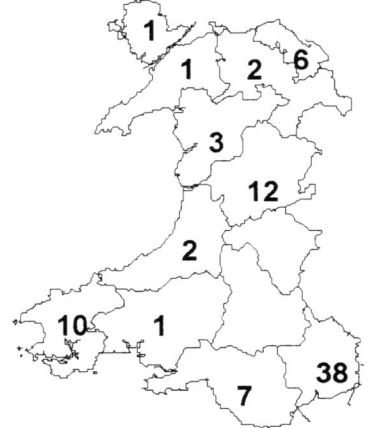

The vast majority were spring records, April – May, usually of singles or small groups but there was a small influx in April 1995 with a total of 16 birds turning up in three different counties in a three day period. Of the winter individuals, 18 were from Gwent, 4 in Glamorgan, singles in Carmarthen & Pembroke and 2 in Denbigh. Most records referred to ones or twos but records involving 4 birds or more included: in Gwent, at Goldcliff 7 on May 7th Wentloog Nov. 30th – Dec. 14th 1994, 4 were at the Nevern Estuary, Pembroke May 10th – 12th 1993 and a remarkable 1994 and 9 on Oct. 7th 1998, up to 13 were at Peterstone record of 12 at Lake Vyrnwy, Montgomery Apr. 22nd 1995 was the first county record.

STONE CURLEW *Burhinus oedicnemus*   *RHEDWR Y MOELYDD*
(B) A rare visitor.
An additional record to 26 published in *Birds in Wales*, of a single at Llanasa, Flint June 5th 1991.
Since then six individuals were recorded: in Gwent at Gobion Apr. 14th 1995, Offa's Dyke Apr. 9th 1997 and at Redwick Aug. 8th 1998; at Teifi Estuary, Ceredigion May 13th 1997; at Bardsey, Caernarfon May 20th 1992; at Penmon, Anglesey June 21st 1997.

Breakdown of all dated Welsh individuals by month, shown in the chart and all records by county on the map:

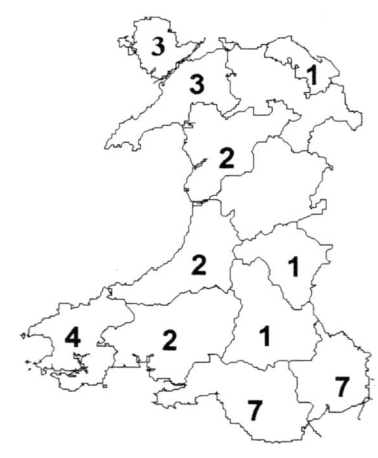

LITTLE RINGED PLOVER *Charadrius dubius*   *CWTIAD TORCHOG BACH*
(C) A summer visitor, breeding in small numbers on the gravel shoals of rivers in south and east Wales and on industrial sites in north-east Wales. Elsewhere it is an uncommon passage migrant.

The first record of this species in Wales was in Glamorgan in May 1949 and the first breeding record was in Flint in 1970. Since then this species' colonisation has continued.
The map opposite shows the maximum number of pairs recorded breeding in each county. Although incomplete, the data suggest a total Welsh population of 100 pairs, compared to the 60 pairs quoted in *Birds in Wales*.
The main rivers are the Usk in Gwent & Brecon, Wye in Brecon & Radnor and in Carmarthen the Tywi, Cothi, Sawdde and Bran. Other important sites are Dolydd Hafren, Llandinam Gravel Pits and Llyn Coed y Dinas in Montgomery, RSPB Conwy in Caernarfon and Inner Marsh Farm in Flint.
The most important river system in the UK is the Tywi and its tributaries, which holds 4-5% of the UK population, with an average of 1 pair per 1.2 km and up to 5 pairs per km in places. In 1999, a pair on the Tywi for the first time in 10 years & 400 clutches fledged 4 young from 4 eggs. There was no complete survey of the Cothi & Tywi in 2000 but pairs were located at several new sites suggesting range expansion (including onto man-made pools along Carmarthen bypass), county total 45-50 pairs.

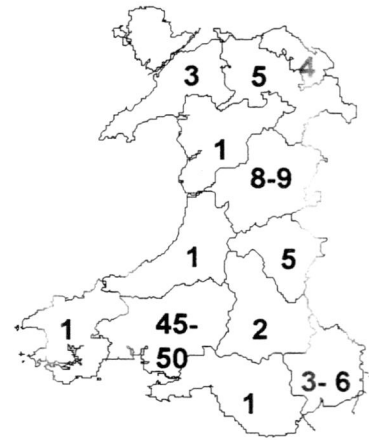

RINGED PLOVER *Charadrius hiaticula*   *CWTIAD TORCHOG*

(C) A scarce breeder, concentrated principally in the north-west (Llyn peninsula and Anglesey) and the mid-section of the south coast. It is also a common passage migrant and winter visitor to all coastal areas.

Present throughout the year in most coastal areas in Wales but there are marked annual peaks in numbers in May and August. The Dee is the principle site, with *Birds in Wales* quoting an average August count of 695 for the period 1986-90. During the period under review a large spring passage was recorded annually in May at Ynyslas, Ceredigion, data tabulated below.

1992	1993	1994	1995	1996	1997	1998	1999	2000
140	200	240	141	256	154	155	193	296

275 were also recorded at Rhymney Great Wharf on May 12th 1993 and 230 at Point of Air, Flint in May 2000.

Maximum autumn counts, July – October, at the main sites, where over 150 recorded at least once, tabulated:

		1992	1993	1994	1995	1996	1997	1998	1999	2000
Gwent	St. Brides/Usk			255			150			157
Glamorgan	Rhymney	365								
	Blackpill	246	263	306	304	307	550	138	303	285
	Burry S.	326			500	120	160	215	480	390
	Jersey Marine							198	87	62
	Port Talbot		159		200	142	89	132	100	
Carmarthen	Burry N	200	141						55	
	TTG	252	58	106	134	171	180			
Ceredigion	Ynyslas	136	186		130	208	242	323	307	305
Flint	Point of Air	500	220	330	201	395	250	310	415	432

A large autumn passage was also recorded at Nevern Estuary, Pembroke, with 300 on Aug. 31st 1997.

Winter counts are traditionally lower than autumn figures, with small numbers dispersed all around the Welsh coast. *Birds in Wales* estimated that the total number of wintering birds was approximately 800, c2% of the East Atlantic Flyway wintering population. Maximum winter counts, November – March, at the main sites, where over 150 recorded at least once, tabulated. The National Level of Importance is 290.

		/92	92/3	93/4	94/5	95/6	96/7	97/8	98/9	99/00	00/01
Glam.	Blackpill	206	206	326	267	272	298	257	258	300	356
	Jersey Marine				133	200	110	200	153	100	80
	Burry S.	112	176				90	100			
Carm.	TTG	131	70	198	236	51		59			

160 at Beddmanarch Bay, Anglesey in January 2000.

Birds in Wales quote figures for Average Winter Peak Counts. During the period under review the only comparative figure is for Swansea Bay, with an average of 277 in the period 1996-2000, compared to 254 for 1986-2000, a slight increase.

Birds in Wales states that at the beginning of the 20th century, this species was a widespread and common breeding species on all low parts of the Welsh coast, most numerous around the northern part of Cardigan Bay, the west coast of Anglesey and Glamorgan. Industrialisation and tourist development have caused a gradual and substantial reduction in suitable breeding areas, with a total Welsh population of 182 pairs at the time of the last full survey in 1973. An incomplete survey in 1984 calculated the total to be 224 pairs breeding on the coast (Denbigh and Flint were not surveyed, an estimate of 6 pairs was used [from the 1973 survey] while 15 prs. bred there in 1989). The table below summarises the total number of pairs recorded breeding in each county for the period 1992-2000. Although incomplete, figures appear fairly stable and there is no reason to assume that the total Welsh population has changed from the c250 pairs at the end of 1980s. The increase in Carmarthen is due to better coverage and not due to a real increase.

	1992	1993	1994	1995	1996	1997	1998	1999	2000
Gwent		none	1*					2+	1
Glam.	5	11-12	14	3	8+	18-19+	10+	16-19	11-13
Carm.			20-21		5				3
Pemb.	4	1		3		2	4-5	1-2	1
Cere.		2-3	3**	2***	6-7			6	
Cere./Meiri					1			1	
Meiri.		6 sites			2 sites			1	2
Caern.			6-8	2	2	2	10-12	1	5-8
Anglesey	1			Cemlyn	5+ Ynys Feurig		3 Cemlyn	6	1
Flint		9-11		8-10	20-22	15-17	2 sites	11-12	6

In Gwent *1st breeding in since 1984. ** 1st breeding at Tan-y-bwlch in 20 years, *** 1st breeding at Aberaeron since 1975.

KILLDEER *Charadrius vociferus*   *CWTIAD TORCHOG MAWR*
(B) A trans-Atlantic vagrant.
The third Welsh record came from Holyhead, Anglesey Dec. 30th 1993 – Jan. 2nd 1994. The other records were from Pwll, Llanelli, Carmarthen in February- March 1978 and from Bardsey, Caernarfon March 1982.

KENTISH PLOVER *Charadrius alexandrinus*   *CWTIAD CAINT*
(B) A rare visitor.
23 records listed in *Birds in Wales* involving 28-32 individuals. Since then there has been a further 7 records, all singles: from Gwent at Goldcliff Sept. 19th 1993 & Apr. 9th 2000 and Cold Harbour Pill Apr. 18th 1998; in Pembroke at Nevern Estuary May 13th – 14th 1998 and at Newgale Apr. 25th 2000; Ynyshir, Ceredigion Apr. 12th 1993; Rhosneigr, Anglesey May 13th 1994.

Breakdown of all Welsh records by month in the chart, and by county on the map:

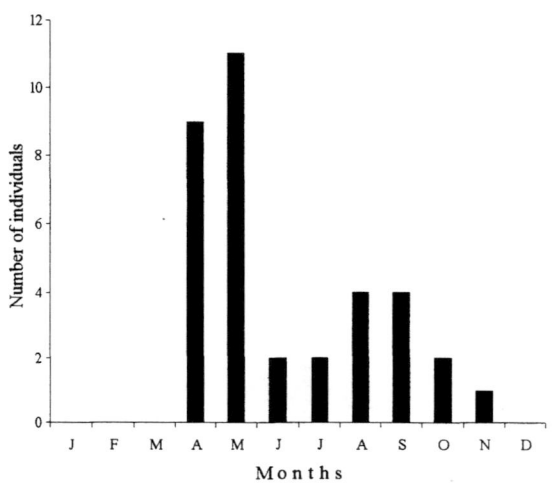
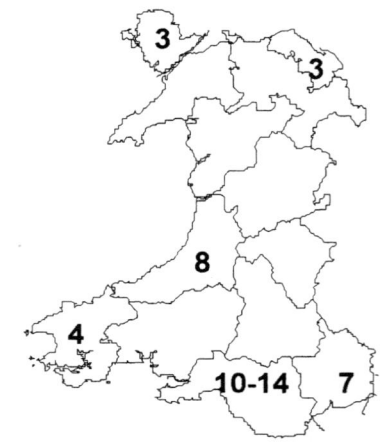

DOTTEREL *Charadrius morinellus*   *HUTAN Y MYNYDD*
(B) An occasional breeder, otherwise a regular but uncommon migrant in spring and autumn.

	1992	1993	1994	1995	1996	1997	1998	1999	2000
Glamorgan	8 & 0				1 & 0		1 & 0		1 & 0
Carmarthen	8 & 0	9 & 0	4 & 1	18 & 0	45 & 7	17 & 22	21+ & 7	7 & 3	11 & 2
Pembroke	1 & 0	4 & 0		7 & 4	7 & 3	2 & 0	0 & 1**	0 & 4	7 & 4
Ceredigion	4 & 0		3 & 0	19 & 0	3 & 1				13 & 0
Brecon	1 & 0	15 & 0	0 & 1	8 & 0	0 & 1	10 & 0		3 & 0	6 & 0
Radnor								10 & 0	
Montgomery					4 & 0		2 & 0		
Meirionnydd		3 & 0	7 & 2					3 & 0	
Caernarfon		42 & 0		0 & 1	24 & 1	15 & 3	30 & 5	54+ & 43+	17 & 0
Anglesey		2 & 0	2 & 0		5 & 1	5 & 0			

** one found with a wintering flock of Golden Plovers at Castle Martin, Pembroke Dec. 13th 1998.

Spring passage birds visit Wales in April – June and in autumn, August – October. The table above summarises the total number of individuals in each county, split into spring & autumn.
The main sites for spring passage were in Carmarthen at Tair Carn Isaf, Garreg Lwyd, Pen Rhiw-ddu, Tair Carn Uchaf, Foel Fraith, Garreg Las and Mynydd Ddu; Pumlumon in Ceredigion; Foel Fras and Carnedd Llywelyn in Caernarfon.
The large autumn total in 1999 was due to a flock of 35 at Nantlle Ridge, Caernarfon on Oct. 16th. Outside these periods there were 2 records both of singles at Ynyslas, Ceredigion Nov. 13th – 20th 1996 and at Castle Martin, Pembroke Dec. 13th 1998.

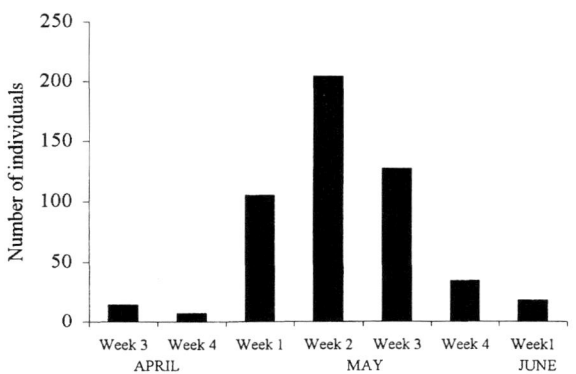

Weekly Spring Passage of Dotterels 1992-2000

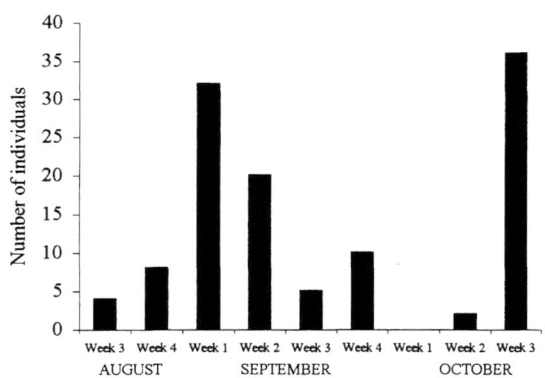

Weekly Autumn Passage of Dotterels 1992-2000

GOLDEN PLOVER *Pluvialis apricaria*   CWTIAD AUR
(C) *A scarce and declining breeding species, found chiefly on the upland plateaux of Denbigh, Radnor, Ceredigion and Brecon.*

		/1992	92/3	93/4	94/5	95/6	96/7	97/8	98/9	99/00	00/01
Glam.	Sker Point	645	640	260	1600					1000	
	Burry S.	2230	1800	2000	3080	3110	1500	1700	420	1060	600
Carm.	Ginst	2150	7000	7000	9000	10000	10000	10550	8955	12267	7307
	Gwendraeth	430	1000		733	3500	1300	300	500	300	1000
	Upper Loughor					1400	720				
Pemb.	Castle Martin					800	2000	550	1000		
	Roch		1000		1200				1500		
	Angle			1000			1000	1000			
Cere.	Ynyslas		785	300	900	1040	850		490	850	
Caern.	Dinas Dinlle				600	200			1000	1040	301

Birds in Wales stated that wintering flocks are scattered thinly throughout most areas of lowland Wales but are particularly concentrated in coastal areas. The Welsh wintering population thought to be in decline, a marked contrast to the rest of Britain and Ireland. The table above gives peak wintering counts at the main sites, where over 1,000 recorded at least once. The National Level of Importance is 2,500.

The numbers on the Burry Inlet appear to be in decline, the average peak winter counts were 2450 during 1981-1985, 1410 during 1986-90 but slightly higher, at 1,799 for 1996-2000. Numbers at Ginst however are increasing.

Large flocks, over 1,000 recorded elsewhere; in Glamorgan: 1,000 at Scurlage & Cilliborn in January 1995 and at Llandimore on Feb. 26th 1996; in Pembroke 1,000 at Marloes Dec. 19th 1996, 1,500 at Panteg Dec. 2nd – 19th 1998 and 1,000 at Lleithyr Jan. 19th 1999 and 1,000 at Hook on Jan. 26th, 1,000 at Lawrenny Jan. 31st and 1080 at Tegryn Dec. 26th 2000; in Ceredigion 1,000 at Penparc in February 1993, at Post Bach Nov. 9th 1996 and Blaenporth Jan. 19th 1999; 700 – 1,000 on the west side of M. Epynt,

Brecon on Nov. 11th 2000; in Caernarfon 1,120 at Penrhos Pwllheli Feb. 2nd 1995; in Anglesey 1,000 at Inland Sea Dec. 16th 1998.

The Welsh breeding population has declined considerably in range and numbers during this century. *Birds in Wales* quotes 215-224 pairs found by the RSPB survey in 1975-78 but not more than 100 pairs by 1990. Although the data for the last 9 years are not complete, what there is does suggest that this decline is continuing. By 2000 the total Welsh population is probably less than 80 pairs.

The Elenydd, which lies partly in 3 counties, Brecon / Radnor / Ceredigion, is now the main Welsh breeding site. RSPB / CCW survey of 150 km^2 of the Elenydd in 1995 found 43 prs. / territories, plus one other outside the SSSI. The actual breeding population is thought to be a little larger but even so a decline of 40 –60% in the last 20 years. RSPB / CCW survey found 12 prs. in 11 1-km squares of the Elenydd in 1997, little change compared to the 1995 survey.

Breeding data elsewhere is incomplete:
Carmarthen less than 5 pairs.
Ceredigion: 12 pairs with young found on Pumlumon in 1994, one pair seen there in 1997.
Brecon: 6-9 pairs at Abergwesyn Common (part of the Elenydd) in 1992, 8-9 prs. there in 1998 but only 2 prs. in 1999 and 3 prs. in 2000. 7 territories on Drygarn Fawr (part of Abergwesyn Common) in 1994. Extinct on Mynydd Ddu, Brecon / Carmarthen in 1996 (cf 2 prs. in 1978). Within the Black Mountains, a pair at Waun Fach 1995, a total of 5 prs. Black Mountains in 1999 and 3 prs. in 2000, 9 adults with 5 young seen in June. Recorded at 2 other sites in 1997, including one where a male mated with 2 females.
Radnor: Single pairs at 2 sites Glascwm Hill 1997 & 1998. One pair Aberedw Hill in 1996 & 1998. 15-18 pairs at 4 sites in Elan Valley in 1999.
Meirionnydd: population thought to be stable, present at 2 sites in 1992, 10-15 prs. at 2 sites in 1993, small numbers at one site in 1995. 25 – 30 at Cadair Idris on Apr. 30th.
Denbigh: one pair Ruabon Moor in 1992, a male seen displaying there in 1997 and a pair there in 1998. Survey of Mynydd Hiraethog in 1994 found a 83% decline, only 8 prs. (cf. 46 prs. in 1977).

PACIFIC GOLDEN PLOVER *Pluvialis fulva* *CORGWTIAD Y MOR TAWEL*
(A) A vagrant.
The first accepted Welsh record was at Oakenholt, Flint on Aug. 2nd – 4th 1990. It is interesting to note that *Birds in Wales* quotes the accepted record of a American / Pacific Golden Plover at the same site the year before, on July 23rd.

GREY PLOVER *Pluvialis squatarola* — CWTIAD LLWYD

(C) A widespread and common winter visitor to Welsh estuaries and spring & autumn migrant.

		/1992	92/3	93/4	94/5	95/6	96/7	97/8	98/9	99/00	00/01
Gwent	Peter/Went.			100	61		300	60	100	100	
	St. Brides						110	32	104	80	
	Severn shore	235	148	240							
Glam.	Blackpill	97	112	82	181	135	99	56	74	33	
	Burry S.	420	830	600	605	630	590	495	740	565	105
Carm.	Burry N.					203	92	18	311	89	
	TTG	82	42	81	125	73	58	40			
Pemb.	Cleddau	36	10	68	125	181	211	131	77	221	43
	Frainslake					141	93				
Flint	Point of Air	350	98	128	100	120	92	143	106	100	61

Well distributed around the Welsh coast. Peak winter counts from the main sites, over 100 recorded at least once are tabulated above. The National Level of Importance is 430. *Birds in Wales* quote figures for Average Winter Peak Counts for the five main estuaries for the period 1986-90. During the period under review there is no data available for the Dee or Severn as a whole. The Average Winter Peak Counts for the other 3 estuaries, tabulated below show a slight increase on the Cleddau, a decrease in Swansea Bay but no change on the Burry.

	1986-90	96-2000
Burry Inlet	743	749
Swansea Bay (Blackpill)	130	80
Cleddau	128	164

Other large counts over 100 were: 100 at Undy, Gwent in January 1994 & October 1999 and 112 at Jersey Marine in September 1998.

Inland records are rare, with *Birds in Wales* quoting 7 in Brecon but none from Radnor or Montgomery.
During the period 1992-2000 there were 7 records: from Brecon singles at Llangorse Apr. 10th – 14th 1994, Jan. 7th & May 14th – 19th 1995 and Apr. 14th – 17th 2000, at Glasbury on May 14th 1994 with 4 there in early November 1996; single at Dolydd Hafren, Montgomery Dec. 12th 1996 – first for the county.

LAPWING *Vanellus vanellus* — CORNCHWIGLEN

(E) Rapidly declining breeder – almost extinct as a breeder in many counties. It is also a winter visitor, principally in coastal areas and especially when there is hard weather to the east.

		/1992	92/3	93/4	94/5	95/6	96/7	97/8	98/9	99/00	00/01
Glam.	Burry S.	3123	2755	4389	4885	2620	1815	4930	3220	2775	300
Carm.	Burry N.		1152	1301	1320	7133	2639	2665	1255	1473	788
	TTG	7150	5120	4985	12096	5870	1819	3400	709	4062	2974
Pemb.	Cleddau	1045	1403	680	5405	2205	1958	2434	2287	3099	1217
Cere.	Dyfi	607	1351	1236	2055	1098	1026			2180	439
Flint	C 'Holt	3000	2000	2000	2100	3435	1550	1210	700	540	750
	Inner Marsh				2500	1500		4600	2600	3000	1800

More numerous in winter particularly on coastal pastures and marshes. The principle Welsh wintering sites are the Dee, Burry Inlet, Carmarthen Bay and Cleddau Estuary. The table above gives the peak winter counts from the main sites, over 2000 recorded at least once. The Average Winter Peak Counts, during the period 1996-2000 were: 5,595 for the Burry Inlet (using whole site counts), 3,172 for Carmarthen Bay (T/T/G) and 2,397 for the Cleddau.
Flocks of over 2,000 were also recorded at: 3,000 at Llyn Alaw, Anglesey in January 1992, 2,000 at Tre-reddol, Ceredigion Dec. 1st 1994, an influx onto the Teifi estuary, Pembroke/Ceredigion, 3,000 there on Dec. 15th 1997 with 2,000 still present in February 1998, 3,800 at Shotwick Jan. 20th 1998, in 1999 2,000 at Lleithyr on the 28th & 3,000 in the Angle-Castle Martin area in December – January 2000, both Pembroke, 2,000 at Llyn Coron, Anglesey in December 1999 and 2,000 Traeth Lafan, Caernarfon in February 2000.

Birds in Wales 1992-2000

Birds in Wales states that the Lapwing has been declining as a breeding bird in Wales since the 1920s. The total Welsh population was estimated in 1987 at 7,500 pairs, mostly in Gwent, Glamorgan, Anglesey, Denbigh and Flint. The loss of its preferred nesting habitat, lowland wet grassland is thought to be a main factor. By 1992 the population had further reduced to an estimated 1,000 pairs. This decline appears to have slowed so that by 2000 the revised Welsh population estimate is 800-850 pairs. The map opposite shows the number of pairs reported breeding in each county in 2000 (* represents 1999 data). The 114 in Ceredigion include 110 prs. On the Dyfi NR, which is in Ceredigion, Meirionnydd and Montgomery.

In 1998 the BTO repeated its 1987 survey of the breeding population in England and Wales, visiting the same tetrads as covered in 1987. In Wales 48 pairs were recorded in 152 tetrads, giving an estimate for the whole of Wales of 1700 pairs (95% confidence limits of 814-2782), compared to an estimate of 7448 pairs (95% confidence limits of 4274-11451) in 1987. Although the wide confidence limits on both estimates suggest they should be regarded a little cautiously, the surveys were strictly comparable and show that this once common farmland bird has lost 77% of its Welsh population in a decade. The main distribution shown by the survey was in Gwent, Glamorgan and the border counties and, in 75 tetrads surveyed in Carmarthen, Pembroke, Ceredigion, Brecon, Meirionnydd, Caernarfon and Anglesey, only one was occupied, by a single pair. The survey was, of course, a sampling survey so it did not

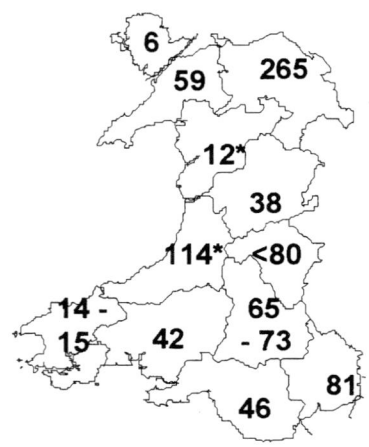

mean that these counties lack breeding Lapwing, for example the important Dyfi SSSI population was missed completely. Nevertheless it is impossible to doubt the picture of decline revealed.

The species is now the subject of a major recovery project by RSPB and part of this is to monitor all sites known to have held 10 pairs or more recently, known as Key Sites. Counts for 1998 & 1999 are shown in the table below.
Summary of breeding pairs, at Key Sites, by county:

County & site	1998	1999
GWENT		
Rhyd y Blew	16	0
Waun y Pound	16	0
Cefn Manmoel	12	0
Newton Farm		6
GLAMORGAN		
Fochriw	16	17
Parc Slip	16	8
Tyle Coch		8
Baglan Moors	9	
CARMARTHEN		
Penclacwydd	22	28
PEMBROKE		
Castlemartin	10	4
Ramsey	13	13
CEREDIGION/MEIRIONNYDD		
Dyfi Est. NNR	91	110
BRECON		
Mynydd Llangatwg	8+	
Mynydd Illtyd	12	10
Ty'r Ash	14	
Llangasty		7
RADNOR		
Pant y Dwr	10-12	
Dyfnant	8	1
Begwns/Ireland Hill	13	
Maelienydd	8	

County & site	1998	1999
CAERNARFON		
Dinas Dinlle	28	28
Madryn	21	21
Tremorfa	15	14
Conwy RSPB	12	9
Penrhyn Bay		9
ANGLESEY		
Bodorgan Estate	12+	3
DENBIGH		
Kilford Farm	12	28
Holt	13	20
Sarphle Farm	10	8
Banhadla Farm		10
Pulford/Trevalyn	9	27
Mwdwl Eithiny		9
FLINT		
Saughall Sealand		27
Shotton Tip	3	0
Sealand Ranges	12	22
Beeches Farm	50-55	50
Burton Meadows		20
Broughton	8	
Hawarden A/F	13	
Shotwick fields	21	15
Willow Farm		10
Maes Gwyn Farm	14	
Inner Marsh Farm	29	34

MEIRIONNYDD		
Morfa Harlech	26	12
Glaslyn Marsh	7	7

A substantial proportion of the Key Sites are in the Dee flood plain, where a detailed survey found 248 pairs (and 60 more in Cheshire). Most of these birds breed in maize where fledging success has proved to be very poor at 0.14 young/pair. Elsewhere in the Key Sites productivity was 0.49 young/pair and overall in Wales, excluding the Dee flood plain, it was 0.61, which may be just sufficient to maintain the population.

Habitat use is of interest. In 1987 23% of pairs were on tilled land, 77% on agricultural grassland and 3% on non-farmland. In the repeat survey in 1998 33% were on tillage, although the area had declined, 31% on agricultural grassland and 31% on non-farmed habitats, a statistically significant difference. The Key Sites survey has produced similar results with 36% of pairs on spring tillage, 39% on agricultural grassland, extensively managed, 16% on industrial sites and 9% on heath & salt-marsh (A. Pickup pers. Comm.). The interesting concentration on the tiny area of tillage in Wales (spring tillage occupied less than 3.5% of land) and the emerging importance of industrial sites point directly at the management of agricultural grassland as a cause of decline. This was once the Lapwing's primary habitat.

KNOT *Calidris canutus* *PIBYDD YR ABER*

(C) An abundant winter visitor to some of the major estuaries, notably the Dee, Burry, Traeth Lafan and Severn. Relatively scarce on other estuaries. It is also an uncommon passage migrant in spring & autumn and it is rare inland.

		/1992	92/3	93/4	94/5	95/6	96/7	97/8	98/9	99/00	00/01
Gwent	Rhymney	300	45	600	5000						
	Peter./Went.		950	2000	40		800	250		250	
	St. Brides		1,100	6000							
Glam.	Taff / Ely			1000	3000						
	Burry S.	3630	1270	5980	3360	8090	8200	3800	5270	2280	2500
Carm.	Burry N.			3000	3164	1410	800	971	3500	1	
	TTG	800	16		434	646	240				
Angle.	Traeth Abermenai			1000	750	500			1020	1400	520
Flint	Point of Air	300	1000	2730	5800	3000	1230	500	1090	460	490
	CQ'Holt	200	1500	4000	800	550					

Wintering flocks of this gregarious species are concentrated in a very few areas, particularly the Dee, the Severn and the Burry Inlet. *Birds in Wales* quotes a decline in the Welsh wintering population since the Birds of the Estuary Enquiry in the 1960s, with Wales holding little more than 7% of the East Atlantic Flyway population. The table above gives peak winter counts at the main sites, with over 500 at least once. The National Level of Importance is 2,900, while counts on the Burry go over the International Level of 5,000.

Birds in Wales quote figures for Average Winter Peak Counts for the five main estuaries, Dee, Severn, Burry Inlet, Carmarthen Bay and Traeth Lafan for the period 1986-90. During the period under review the only site with comparable data is the Burry Inlet, where an average of 4,243 during 1986-90, compared to 5,848 for 1996-2000 – a substantial increase.

Large flocks, over 500 recorded elsewhere at Swanbridge, Glamorgan, 500 on Nov. 5th 1993, 700 at Cefn Sidan, Carmarthen Feb. 7th 1992 with 1,000 there Jan. 12th 1998 and 600 in November 1999, 450 Traeth Lafan, Caernarfon Dec. 26th 1999 and 800 at Flint Sands on the Dee, Flint in February and 1121 in November 2000 . 578 logged passing Strumble Head, Pembroke from Aug. 21st – Nov. 6th, max. 142 on Sept. 11th and 160 on Aug. 25th.

Inland there were 7 records: singles at Llandegfedd, Gwent Aug. 29th 1994, at Cors Caron, Ceredigion Apr. 8th 2000, 6 flew south over the Epynt, Brecon Aug. 21st 1996 (the first county record since 1980), at Llangorse Apr. 6th 2000 and in Anglesey 10 at Llyn Alaw Sept. 8th 1995, with 2 the following day and a single there Sept. 10th 1998.

SANDERLING *Calidris alba*   PIBYDD Y TYWOD

(C) A winter visitor, locally common on sandy beaches in Flint, Carmarthen and Glamorgan. A spring & autumn migrant on the coast. Scarce inland.

		/1992	92/3	93/4	94/5	95/6	96/7	97/8	98/9	99/00	00/01
Glam.	Kenfig	200	140	7							
	Blackpill	454	171	163	420	344	360	250	407	320	550
	Port Talbot		108	156	130	340	214	220	151	189	31
	Jersey Marine			143	250	384	300	400	318	230	101
Carm.	Cefn Sidan		450	400	374	266	384	470	398	108	130
Flint	Point of Air				165	160	80	117	450	159	
	Gronant						220	58	70	300	

Birds in Wales states that the Sanderling's distribution in Wales emphasises its dependence on long sandy shores and is concentrated on 3 areas; the Dee / Clwyd estuaries and the coast towards Llandulas, Cefn Sidan (the second most important site in the UK) and Swansea Bay. The table above gives the peak winter counts, November – February, at the main sites, with over 150 at least once. Large flocks were also noted in Glamorgan at Aberavon beach, 179 in January & 86 on Nov. 9th and 550 at Weobley Feb. 5th 1999.

Birds in Wales quote figures for Average Winter Peak Counts for the main sites (but not Cefn Sidan) for the period 1986-90. During the period under review there are no comparable figures available for the Dee / Clwyd estuary and the coast towards Llandulas. The Average Winter Peak Counts for 1996-2000 for the other sites are tabulated below.

	1986-90	1996-2000
Swansea Bay (Blackpill)	262	336
Cefn Sidan	-	400

Large numbers are often recorded on passage at these main sites. The following table summarises peak passage counts at the main sites, spring March - May, autumn July - October:

		1992	1993	1994	1995	1996	1997	1998	1999	2000
Kenfig	Spring	150	120	148	219	114			520	-
	Autumn	128	130	110	300	255			248	72
Port Talbot	Spring			1	150	140	397	42	209	320
	Autumn			172	138	240	90	104	184	160
Cefn Sidan	Spring	400		398	283	323	661	640	-	592
	Autumn	528		396	306	305	7	258	250	290

Spring passage also noted at Ynyslas, Ceredigion, peaks tabulated:

1992	1993	1994	1995	1996	1997	1998	1999	2000
85	235	400	35	66	95	39	107	45

Large spring counts also recorded in Glamorgan at 120 Margam Sands May 10th 1996; at Frainslake, Pembroke, 40 on May 18th & 57 on July 21st 1996; 80 Kinmel Bay, Denbigh Oct. 28th 1999.

Inland records are rare and during the period under review there were singles in 2000 at Llangorse Apr. 14th – 17th and at Glasbury May 13th, both Brecon.

LITTLE STINT *Calidris minuta*   PIBYDD BACH

(C) An uncommon passage migrant in small numbers to coastal areas, principally in autumn and has occasionally wintered. It is rare inland.

The number of records of this species continues to increase. The total number of individuals recorded in each month in Wales during the period 1992-2000 is shown in the following chart.

Winter records included birds at: 3 at Penclawdd, Glamorgan Dec. 19th 1992, one Ogmore, Glamorgan Dec. 25th 1996, one at Lawrenny, Pembroke Jan. 30th 1993, an over-wintering bird on the Dyfi, Ceredigion 1993/4, an over-wintering bird at Llyn Alaw, Anglesey 1995/6, and in Flint one Inner Marsh Farm November – December 1992 & 2 in January, one in February 2000 and 3 on the Flint coast in December 1998.

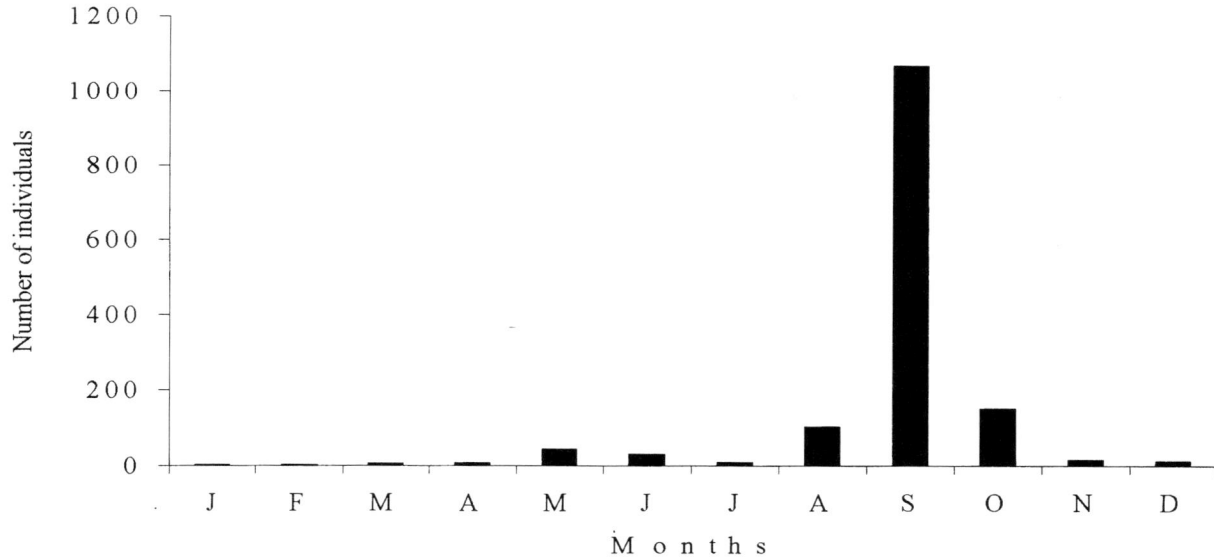

Large influxes included 220+ individuals in autumn 1993 including 53 at the Dyfi Est., Ceredigion on Sept. 21st, but unprecedented numbers in autumn 1996, with over 600 birds between July 21st and December (the largest quoted in *Birds in Wales* was in 1960, when 350 were at Shotton, Flint in September). The main influx was in September, peaking on the 21st – 25th. Large congregations were: 36 at Llandegfedd Res. Gwent on the 23rd, 37 at Ogmore, Glamorgan on the 21st, in Pembroke 34 at Angle Bay on the 25th, 24 at Nevern estuary on the 20th, 29 on the Teifi estuary (Pemb. / Cere.) on the 25th, on Anglesey 25 at Llyn Alaw on the 21st, 43 at Malltraeth on the 19th, and at Gronant, Flint 50 on the 21st, 57 on the morning of the 22nd, increasing to 75 by evening, 102 on the 23rd, 98 still there on the 25th.

Inland records are very rare, *Birds in Wales* quotes 10 in Brecon, 2 in Radnor and only one in Montgomery up to 1991. Since then there were: 2 at Dynefwr Ponds, Carmarthen May 12th 1992; at Llangorse, Brecon, one on Sept. 10th 1992, 6 on Sept. 10th 1993, singles Oct. 1st & 3rd 1993, 10 on Sept. 21st 1996 (4 other records in the county that year) and Sept. 20th 2000; 2 at Llyn Heilyn, Radnor Aug. 24th 1995; in Montgomery one at Dolydd Hafren Apr. 8th, 2 at Llyn Coed y Dinas Sept. 15th both 1995, 7 at Dolydd Hafren Sept. 12th, presumed same 6 at Llyn Coed y Dinas on the 14th 1998.

TEMMINCK'S STINT *Calidris temminckii*

(B) A rare visitor.

PIBYDD TEMMINCK

26 individuals recorded in *Birds in Wales*, since then a further 11 recorded in Wales. Singles at Goldcliff, Gwent Sept. 10th – 15th 1999; at Gann, Pembroke Sept. 10th 1995, in Ceredigion at Teifi Marshes, May 12th – 13th 1993 (first county record) and at Ynyslas Sept. 12th 1998; at Dolydd Hafren, Montgomery May 12th – 15th 1993 (first county record) and in Caernarfon at Bardsey, the first county record on Sept. 20th 1996 and a juv. at Conwy RSPB on Aug. 27th – 28th 2000. In 1992 there were at least 4 birds in May at Inner Marsh Farm, Flint, single on the 15th, 2 on the 25th and another 28th – 29th.

Breakdown of Welsh records by month in the chart and by county on the map:

Birds in Wales 1992-2000

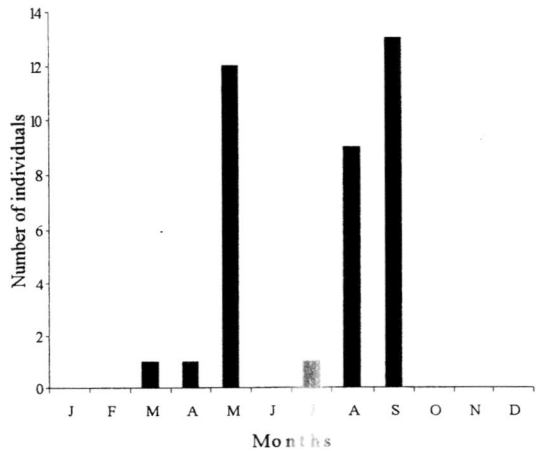
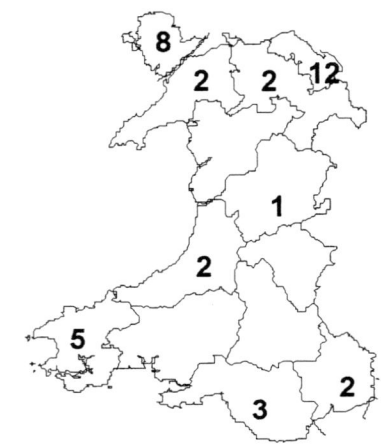

WHITE-RUMPED SANDPIPER *Calidris fuscicollis*
(B) A vagrant.
Additional record to the 6 published in *Birds in Wales*, of an individual at the Teifi Estuary Ceredigion / Pembroke Aug. 5th 1991. Since then only two further records both at Goldcliff, Gwent Aug. 11th – 17th 1995 (then at Severn Beach 20th – 31st) and Sept. 14th – 21st 1999. The previous 6 records were from Glamorgan (2) both at Blackpill / Swansea bay, Carmarthen (2) both at Pembrey / Cidweli, singles at the Gann, Pembroke and at Shotton, Flint.

PIBYDD TINWEN

BAIRD'S SANDPIPER *Calidris bairdii*
(B) A vagrant.
Four records all of juveniles, at Llandegfedd Res., Gwent Sept. 26th – Oct. 4th 1997, at GLWR Goldcliff, Gwent Oct. 3rd – 13th 2000, at Tan y Bwlch, Ceredigion Sept. 19th and at Conwy RSPB Sept. 24th both 1998. Nine previous records in Wales, all August – November, 3 from Pembroke, 2 each from Glamorgan, Carmarthen and Flint.

PIBYDD BAIRD

PECTORAL SANDPIPER *Calidris melanotos*
(B) A scarce visitor.
Amendment to *Birds in Wales* following the review of Glamorgan records in 1995 (Landsdown and Hurford) only 10 acceptable records in that county and a Welsh total of 65 records concerning 69 individuals. Seventeen records since: Goldcliff, Gwent Sept. 9th – 21st 1999, an adult June 4th and a juv. Sept. 23rd- 24th 2000; Penclacwydd, Carmarthen July 25th – 27th 1996; Skokholm Aug. 23rd 1994 and at Herbrandston, Pembroke Oct. 3rd 1998; Dyfi, Ceredigion Aug. 27th – 28th & Sept. 9th 1997, juv. Cors Caron Sept. 17th & 22nd 2000; Shell Island, Meirionnydd May 18th (1st county record) 1997; Caernarfon at Conwy RSPB Aug. 9th 1996, Sept. 8th – 15th with a second individual Oct. 2nd 1999 and an adult May 16th – 19th 2000; on Anglesey at Malltraeth Aug. 22nd – Sept. 7th 1994 & Sept. 3rd 1999, Rhosneigr Aug. 8th – 9th 1997 and Cemlyn May 9th – 15th 1999.

Breakdown of all Welsh individuals, by month in the chart, and by county on the map:

PIBYDD CAIN

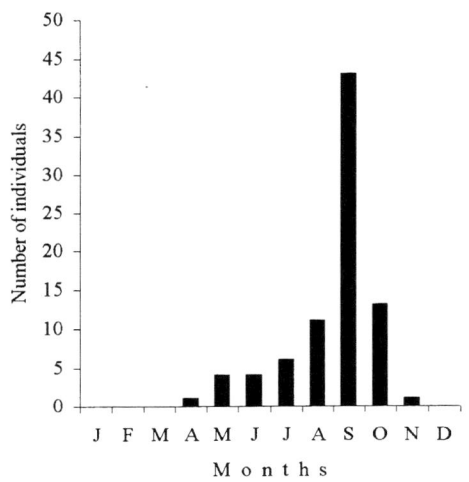
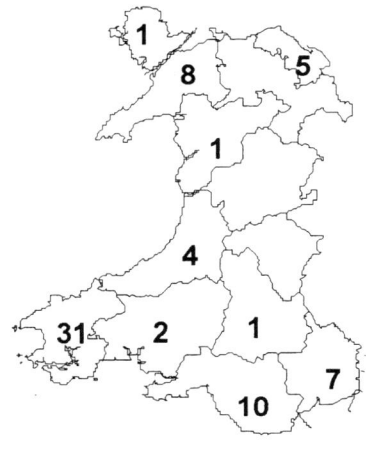

SHARP-TAILED SANDPIPER *Calidris acuminata*   PIBYDD GYNFFONFAIN
(B) A vagrant.
The third Welsh record was of an adult at Dinas Dinlle & Foryd Bay, Caernarfon Aug. 25th – 28th 1996. Both the other two were in October 1973, at Shotwick, Flint and at Morfa Harlech, Meirionnydd.

CURLEW SANDPIPER *Calidris ferruginea*   PIBYDD CAMBIG
(C) An uncommon passage migrant to coastal areas, recorded in every month but principally in autumn, relatively numerous in some years. Rare inland.

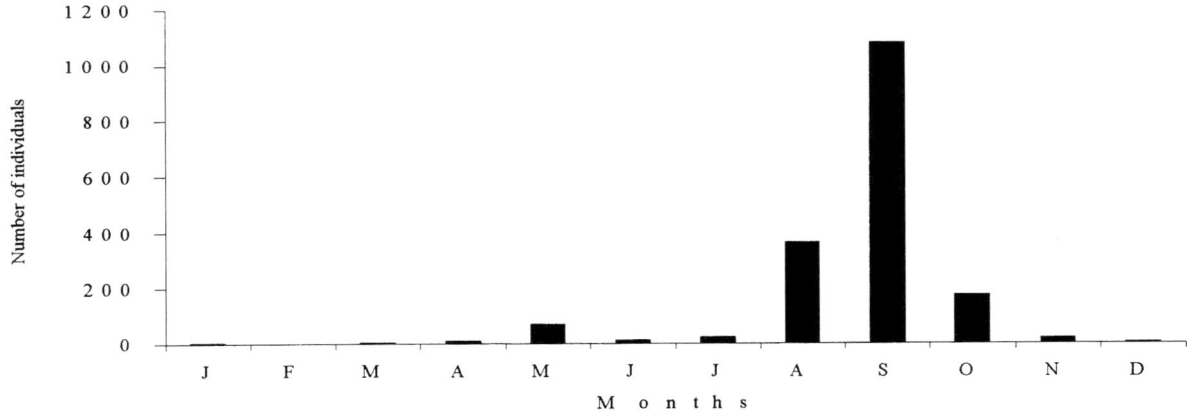

The pattern of occurrence of this species has changed little in the last 100 years according to *Birds in Wales* but the number of records has increased considerably, partly as a result of increased coverage. The total number of individuals recorded in each month during the period 1992-2000 is shown in the following chart.

Early / late records included individuals at Point of Air, Flint Mar. 6th 1992, at Beaumaris, Anglesey Jan. 12th 1998 and at Rhymney, Glamorgan Dec. 23rd 1998. There were no inland records of this species in the period.

A large spring count was of 13 at Ynyshir, Ceredigion on May 18th 1992. There were large influxes during the autumns of 1996 and 1998. The 1996 influx started around the 22nd of September when 44 were at Malltraeth, Anglesey, on the 29th in Pembroke there were 26 at the Nevern Estuary and 29 in Angle Bay, with 79 recorded at Penclacwydd, Carmarthen the following day. The 1998 influx was not as dramatic, with a total of 220 recorded in autumn, from July 22nd – Nov. 27th, max. 58 at Penrhyngwyn, Carmarthen on Sept. 12th.

A large flock of 100 on the Dyfi, Ceredigion Aug. 28th 1999, accounted for more than half of that month's records.

PURPLE SANDPIPER *Calidris maritima*   PIBYDD DU
(C) A fairly common winter visitor in small numbers to coastal areas with rocky shorelines.

Purple Sandpipers have always been very local in their distribution in Wales owing to their feeding preferences for the tidal edge of rocky shores. *Birds in Wales* quote several traditional sites, Sker Point and Rhossili in Glamorgan, Aberystwyth Harbour in Ceredigion, Trearddur Bay on Anglesey, Rhos Point in Denbigh and the islands of Skokholm, Small and Grassholm in Pembroke, Bardsey in Caernarfon and Puffin off Anglesey. During 1968-74, an average maxima. of 229 individuals were wintering in Wales, of which 50 on Anglesey, 70 in Caernarfon, 10 in Ceredigion, 20 in Denbigh, 2 in Flint, 67 in Glamorgan and 10 in Pembroke. In the 1980s numbers increased on Anglesey but decreased in Glamorgan and the Winter Shorebird Count of December 1984 – January 1985 gave a figure of 162 in Wales.

During the 1990s the number wintering on Anglesey was thought to be decreasing, with an estimated 75-125 wintering there in 1995. Numbers at Skomer and Aberystwyth, although higher than the 1968-74 figures, were also declining. In light of these changes the total Welsh wintering population is in the range 150-200 individuals.

Data from the period under review is incomplete, with few sites counted on a regular basis. The main wintering sites, with counts over 20, are tabulated below.

Large counts include 22 at Crabart Rhossili in January & February 1999, 21 at Port Eynon in December 2000, both Glamorgan and 100 at Bardsey in May 1999 and 90 on Apr. 17th 2000. 6 on Skokholm, Pembroke on May 26th 2000 were late individuals.

		/1992	92/3	93/4	94/5	95/6	96/7	97/8	98/9	99/00	00/01
Pemb.	Skomer	29-33	22	22		18	24	7	14		
Cere.	Aberystwyth	22	18	17	20	17	17	14	14	11	
Caern.	Bardsey		46	36	40	40	32	20	35		2
Angle.	Trearddur					15	5		15	24	

DUNLIN *Calidris alpina*   PIBYDD Y MAWN
(C) A scarce and local breeder, restricted to a few poorly drained upland moors with scattered pools. An abundant passage migrant and winter visitor.

A common visitor to all coastal areas, inland it is a regular spring and autumn passage migrant in small numbers and occasionally in winter. The table below gives the max. winter counts, November – March, at the main sites, where over 3,000 recorded at least once. The National Level of Importance is 5,000, while the count on the Burry in 1996/7 just goes over the International Level of 14,000.

		/1992	92/3	93/4	94/5	95/6	96/7	97/8	98/9	99/00	00/01
Gwent	Peter. / Went.		13000	7000	8000		10000	5500	2800	5000	55
	Goldcliff / Redwick		10000	10000	10000		5000	10000	5000	3000	1000
	St. Brides / Usk			1200	6000		2000	2000	4470	5000	260
	Undy			6000	200		500	1000	3000	3000	1500
Glam.	Rhymney	8950	4500	11110	3200	4000			1050	3000	
	Blackpill								3790	2520	3250
	Taff / Ely	4000	4500	2000			3000	3000	4050	70	
	Burry S.	4488	9300	12950	10905	7090	14460	8700	9500	9795	6800
Carm.	Burry N.			2560	1647	2050	2104	3130	3515	1350	375
	T/T/G								3093	3299	360
Pemb.	Cleddau	3366	3062	4160	4197	4426	8561	5318	5973	4884	1433
Caern.	Traeth Lavan	3000				2206	4074	1752	1000	4032	1663
Flint	Point of Air	3700	1500	2143	6200	4340	2900	1600	2080	5850	4650
	CQ' Holt / IMF			5000	4000	2000	10000	3400	1500	2300	250

The degree of overlap in Gwent is unknown.

Other large counts were 10,000 on the Cardiff foreshore, Glamorgan Dec. 7th 1994, 3,000 at Gronant, Flint in January / February 1997. Count of the whole of the Dee, Flint / Cheshire found 25,487 in January and 31,619 in December 1998.
Large passage was noted at Ynyshir, Ceredigion: 500N in 3 hrs. May 16th with 855 on the 17th 1993, 1500 on May 17th 1994. Influx in May 2000 with a grand total of 6,000 on the 6th.

	1986-90	1996-2000
Burry Inlet	7,731	11,408
Swansea Bay (Blackpill)	2,131	-
Carmarthen Bay (T/T/G)	1,382	-
Cleddau	3,743	5,832
Traeth Lafan	4,718	2,613

Birds in Wales quote figures for Average Winter Peak Counts for the main sites for the period 1986-90. During the period under review there are no comparable figures available for the Dee and the Severn estuaries while the Taff / Ely estuary has been lost due to the Cardiff Barrage. The Average Winter Peak Counts for 1996-2000 for the other sites are tabulated above. A large increase in the Burry, more in the Cleddau but a lot fewer at Traeth Lafan.

Birds in Wales states that this species was probably never a common breeding species in Wales and the total breeding population numbered 50-70 pairs. Although there is a lack of evidence over the past 8 years, what there is, suggests that there has not been any change. Below, organised by counties is a summary of breeding information during the period 1992-2000.

Ceredigion: nested on the Elenydd in 1998.
Brecon: 3 prs. Pwll Gwy-rhoc in 1993 & 1994, 2 on territory at Abergwesyn Common in 1998, 3 prs. on the Black Mountains in 2000.
Radnor: 43 prs. found during the 1995 RSPB / CCW survey of 150 km^2 of Elenydd and one other within the area of the SSSI – no suggestion that numbers have changed. 5 territories in 3 1-km squares of Elenydd in 1997. This is the stronghold of this species in Wales, its population was estimated at 15 prs. in 1976/7 (Elan / Claenwen catchment only), 28 prs. in 1990 and 37-40 prs. in 1991.
Meirionnydd: 3-5 prs at one site in 1993.
Caernarfon: one displaying at Morfa Madryn on June 17th 2000 and a pair at Llyn Conwy in 2000.

BROAD-BILLED SANDPIPER *Limicola falcinellus*   *PIBYDD LLYDANBIG*
(B) A vagrant.
The 5th Welsh record (and the 6th individual) was at Conwy RSPB, Caernarfon May 24th 1999. The previous 5 individuals were at Shotton, Flint 2 individuals in September 1960, Peterstone Wentloog in May 1979 & Sluice Farm May 1988 both Gwent and Malltraeth, Anglesey in June 1984.

BUFF-BREASTED SANDPIPER *Tryngites subruficollis*   *PIBYDD BRONLLWYD*
(B) A rare visitor.
Birds in Wales quote 26 records involving 31 individuals, since then there has been a further 8: all were of singles at GLWR Goldcliff, Gwent a juv. Sept. 12th – 30th 2000, at Sker Farm, Glamorgan Sept. 14th 1992, in Pembroke, juveniles at Dale Airfield Sept. 5th – 13th 1994 & Sept. 6th – 21st 1996, at Ramsey Sept. 23rd 1997 and at Castle Martin Sept. 23rd 2000, in Flint at Point of Air Sept. 9th 1994 and an adult at Inner Marsh Farm Aug. 4th 1999.
Breakdown of all Welsh individuals by month, shown in the chart and by county on the map;

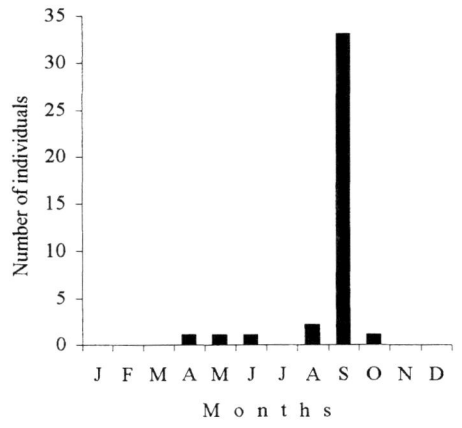

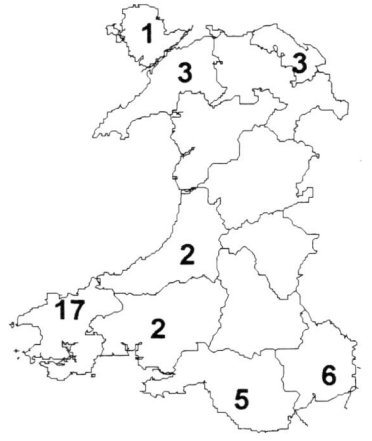

RUFF *Philomachus pugnax*   PIBYDD TORCHOG
(C) A fairly common passage migrant in small numbers, principally in autumn. Occasionally winters.

Most records are in the period August – September, with smaller numbers on spring passage. Approximate breakdown of totals for each county are shown in the following map and the total number of individuals recorded each month during 1992-2000 is shown in the chart.

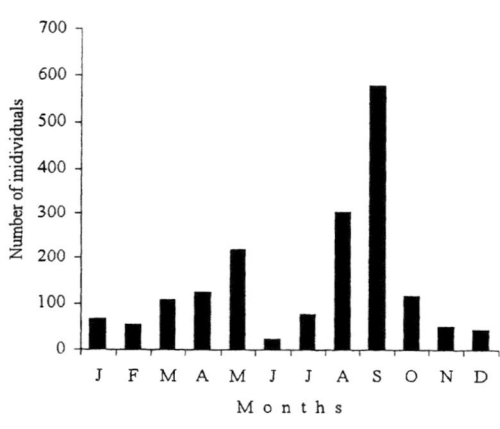

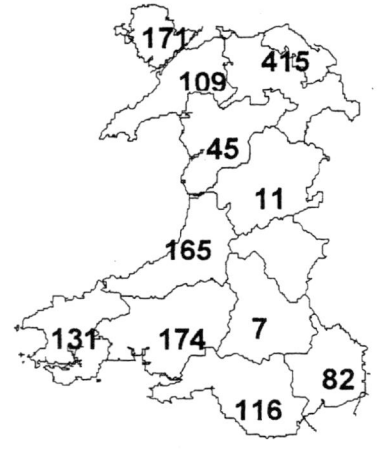

Inner Marsh Farm, Flint / Cheshire was the only site to regularly hold large numbers of this species, in 1992 there were 11 in March and 15 in September, monthly figures for 1995 were:

Jan.	Feb.	Mar.	Apr.	May	June	July	Aug.	Sept.	Oct.	Nov.	Dec.
18	12	14	15	12	4	10	7	10	7	14	16

Birds in Wales states that wintering records were not recorded in Wales until 1960-1965, since then numbers have increased. During the period 1992-2000 Ruff were a regular wintering species in Wales but numbers in this period were no where near the average of 26 individuals recorded in the winters 1966-1971.

Eleven wintered in 1997/8, 7 in 1998/9, 8 in 1999/2000 (including 5 at Inner Marsh Farm, Flint), 11 in 2000/01 (10 of which were at Inner Marsh Farm). Large counts were: 15 at Wernffrwd, Glamorgan May 17th 1992 and 22 in the Clwyd Est., Denbigh Aug. 24th 1994. A count of 85+ at Cors Fochno / Dyfi estuary on May 1st 1994 is the highest ever recorded in Wales.

Inland there were records of singles at Llangorse Lake, Brecon Sept. 1st 1996, Apr. 15th, 3 on Aug. 18th and a single 23rd & Sept. 1st 2000, a single at Brechfa Pool Aug. 15th 2000 and in Montgomery at Llyn Coed y Dinas Aug. 30th 1995 and at Dolydd Hafren Apr. 8th 1997.

JACK SNIPE *Lymnocryptes minimus*   GIACH FACH
(C) A fairly common passage migrant and winter visitor.

Vastly under recorded throughout the whole of Wales. Small numbers reported from every county but few records of more than 5 birds together. Multiple records, of 5 or more were: 5 at Goytre, Gwent Nov. 27th 1993; in Glamorgan: 5-6 at Baglan Jan. 26th – 27th 1998, 25 on the Burry Inlet in February 1999, 10 at Forest Farm in the early months and 14 at the end of the year 1999 and 7 at Trowbridge on Jan. 7th 2000; 6 at Penclacwydd, Carmarthen Nov. 27th 2000; in Brecon 4-5 at Llangorse Nov. 2nd 1997 and 7 there Nov. 7th 1999; 9 at Foryd Bay, Caernarfon Feb. 6th 1998; on Anglesey: 8 at Red Wharf Bay, Jan. 1st & 7 there Feb. 12th 1994, 6 at Traeth Dulas Nov. 19th 1995, 8 at Penrhyn Glas, Feb. 7th 1998 and 6 at Mynydd Bodafan Nov. 27th 2000; in Flint 7 at Gronant Feb. 5th & 10 on Mar. 1st 1995 with 30 there Dec. 5th 1999.

Birds in Wales quoted a record of 50+ at Gelli Aur, in 1969 and 40 at Tumble in 1977, both Carmarthen, both of these are now thought to be unlikely.

SNIPE *Gallinago gallinago* *GIACH GYFFREDIN*

(C) A widespread breeding bird, nesting in all counties but in small and declining numbers. Substantial numbers in winter.

		/1992	92/3	93/4	94/5	95/6	96/7	97/8	98/9	99/00	00/01
Gwent	Llanwern								125	182	164
Glam.	Clyne Com.				55	44	38	102		42	
Carm.	Burry N.	70	40	77		50	178		60	21	
	Penclacwydd			70	165	54	58	116		73	28
Pemb.	Cleddau	141	126	166	203	226	114	133	189	154	215
Flint	Point of Air			72	177					90	51

A numerous winter visitor, with influxes in October. *Birds in Wales* states that groups of 80-200 were not unusual but during the 1990s there were few such large counts. The table above summarises the max. winter counts at selected sites, where over 100 recorded at least once.
Other counts of over 100: in Carmarthen 132 at Pwll Jan. 21st 1994, 200 at Llannon Feb. 8th 1995, 150 at Ffairfach Dec. 2nd 1996, 100 at Pwll Feb. 26th 1998 and 100 at Johnstown Dec. 4th 2000; 105 at Oxwich, Glamorgan Jan. 2nd 1995; in Pembroke 100 at the Dowrog Feb. 12th and at Newgale Nov. 27th 1994, 106 at Herbranston Nov. 12th 1997, 100 – 150 at Newgale Marsh Dec. 21st 1997 (100 there Jan. 1st 2000), 300+ Castle Martin Cors Nov. 24th 1998; 141 Teifi Marshes, Pembroke/Ceredigion Dec. 4th 1998; 140 at Foryd Bay, Caernarfon, November 1998; 100+ at Malltraeth Nov. 28th 1998 and 100 Cors Crigyll Dec. 15th 2000, both Anglesey; 187 at Gronant, Flint March 1995.

Birds in Wales states that from being described as a common breeding species in several parts of Wales in the late 19th and early 20th centuries, the Snipe is now greatly reduced in numbers and range. The current breeding distribution is very patchy and confined to a relatively few favoured hill bogs and some lower marshy ground.
The Elenydd, which lies partly in 3 counties, Brecon / Radnor / Ceredigion, is now the main Welsh breeding site. Breeding records during 1992-2000 are incomplete but the map opposite gives approximate county totals from data exists.

Birds in Wales quotes a total Welsh population at 300 – 500 pairs. It is hard to detect any sort of pattern from such scant information but this figure seems to still stand although this species appears to be declining in some parts, e.g. no confirmed breeding in Pembroke, where at least 10 prs. in 1978-88.

The results of the main surveys were:
Carmarthen: 42 territories found by the RSPB on 15,000 ha of Mynydd Du in 1996.
Ceredigion: 8 pairs Dyfi Marshes in 1997, a 50 % decline but numbers bounced back to 17 pairs in 1999. 42 drummers on Cors Caron in 2000.
Brecon: 11 – 31 at Epynt, 5-11 on Mynydd Illtyd, 8 territories on 4,000 ha of Cnewr Estate 1996.
Radnor: 150 km^2 of Elenydd surveyed by RSPB / CCW in 1995 found 113 territories, with a further 18 in the rest of the 240 km^2 SSSI.
Montgomery: 6-15 pairs at Morfa Dyfi (part of the Dyfi SSSI) and 4-6 pairs at Lake Vyrnwy.
Caernarfon: RSPB survey of 19 key wader sites found 64-72 pairs in 1993. 17 sites on the Llyn surveyed in 1993 15-16 pairs found (cf. 28 in 1986/7).

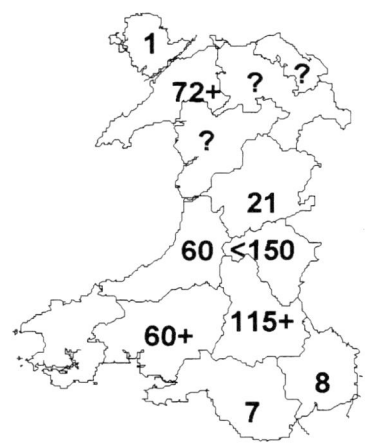

LONG-BILLED DOWITCHER *Limnodromus scolopaceus* *GIACH GYLFIN-HIR*
(B) A vagrant.
The first records for Meirionnydd, Brecon and Carmarthen with singles at Glaslyn Marshes Oct. 17th – 26th 1993, Llangorse Oct. 18th – 21st 1997 and intermittently at Penclacwydd Nov. 30th – Dec. 10th 1997. These are the 6th – 8th Welsh records.

The previous 5 records came from Nefyn, Caernarfon October 1963, Llyn Alaw, Anglesey 1978, Peterstone Wentloog, Gwent September 1985, Gann, Pembroke December 1987 - January 1988 and Sluice Farm & Rhymney, Gwent & Glamorgan in March – April 1989.

WOODCOCK *Scolopax rusticola*   *CYFFYLOG*

(C) A regular breeder in small numbers in all counties except Pembroke & Anglesey. It is widespread and common / abundant in winter.

During the period of this review there were few breeding records.

Gwent: Roding heard in 6 areas in 1993, 5 in 1994, 4 sites in 1997, 5 at 3 sites in 1999 plus 3 birds seen in a territorial dispute in Broad Mead in 1999. Recorded at 3 sites in 2000.

Glamorgan: Bred at 2 sites in 1997 & 1998 and one in 1999 & 2000.

Carmarthen: 3-4 prs. at Dinas/Gwenffrwdd in 1992, roding heard in the Caio & Crychan Forests and breeding reported at Cynghordy in 1993.

Ceredigion: Nest with 4 eggs found at Glandyfi in 1992 was the first proof of breeding for many years, also heard roding at Ynyshir 1992 & 1994.

Brecon: 4-5 prs. in 1992 but incomplete survey. 3 prs. in north of the county in 1993 & 1994, 6 prs. at 10 sites in forests in the north of the county, 3 prs. bred in Wye / Elan and roding heard at 2 sites in the south 1995. Only one pair in 1997, none found in the Crychan, a regular site. None found in 1998. 2-3 at Llanwrthwl in 1999 & 2000.

Montgomery: Present at Lake Vyrnwy, a nest with 4 eggs found at Glantwymyn in 1993 and one near Nelson in a larch wood in 1997.

Caernarfon: Bred at Llyn Bodgynydd in 1999 & 2000 and at Gwaun Gynfi in 2000.

Denbigh: 4 at Pen-y-ffridd in 1997, present in 1998 but none found at World's End or Horse Shoe Pass. A nest containing hatched egg shells found at Bonwm Forest in 2000.

Flint: Roding heard at Llyn Helyg in 1997, 4 there in 1998. 4 Penmynydd and 2 at Hawarden in 1999. 4 at Nercwys in 1997, present in 1998 & 2000 and 6-8 at in 1999.

In addition reported as being widespread in Clwyd, roding at 8 areas in 1992, at one in 1993 & 1994.

Although data are incomplete, it does suggest a total Welsh population of less than 100 pairs.

BLACK-TAILED GODWIT *Limosa limosa*   *RHOSTOG GYNFFONDDU*
(C) A common spring and autumn passage migrant; it is common to abundant and increasing in winter on the Dee.

		/1992	92/3	93/4	94/5	95/6	96/7	97/8	98/9	99/00	00/01
Glam.	Burry S.	20	75	3	30	25	105	60	85	55	
Carm.	Burry N. *	101	72	85	100	88	227	271	98	99	17
Flint	O' Holt	1500	2000	1100	1400	960	1010	1600	1450	290	1800
	Inner Marsh Fm.						330	270	235	600	202
	Bagillt Bank **							1100	850	914	1590

* includes figures for Penclacwydd.
** Bagillt Bank is a low tide count and figures may overlap those from the other 2 Flint sites.

The number of records of Black-tailed Godwits has been increasing in Wales since the 1930s and by the 1960s large autumn flocks were recorded in the Burry Inlet and Carmarthen Bay. Numbers on the Dee rose to c1600 in the winter of 1989/90, approximately one third of the British Wintering Population. The table above gives the max. winter counts, November – March, at the main sites, with over 100 at least once. The National Level of Importance is 70 and the International Level is 700. Counts from the whole of the Dee in 1998, found 1,607 in February and 1,602 in November.

Large numbers were recorded on passage in spring, the max., March – June, counts at the main sites are tabulated below.

	1992	1993	1994	1995	1996	1997	1998	1999	2000
Burry S.	20	10			25	210	1	10	5
Burry N.						245	249	76	61
O' Holt						1100	900		1000
Inner Marsh Farm						1350	1400	1360	400

Similarly autumn passage brought large numbers to these sites, max., August – October, counts given below.

	1992	1993	1994	1995	1996	1997	1998	1999	2000
Penclacwydd	134	76	66	84	289	327	281		
O' Holt	2000	1200	1250	1600	1800	1400	1200		2500

Other large counts were of: 134 at Burry N. September 1992 with 75 in the Burry S. in August. 70 Peterstone Wentloog, Gwent September 1997 and 515 at Inner Marsh Farm in September 1999.

Inland records are scarce:
Singles at Dinefwr Ponds, Carmarthen Aug. 9th 1999 & Apr. 9th 2000 and Cors Caron, Ceredigion Nov. 1st 1998; in Brecon: 3 at Llangorse Mar. 1st – 3rd 1995, with one remaining until the 8th, single Mar. 15th, 7 on Aug. 16th, one on Nov. 22nd 1998, single Apr. 6th with 5 there Aug. 5th 1999 and 2 Dec. 10th 2000, at Brechfa Pool singles June 29th 1998 and Apr. 8th 1999; 8 at Glasbury, Radnor Apr. 23rd 1996; 2 at Dolydd Hafren, Montgomery Mar. 28th and a single July 22nd – 23rd 2000; 3 Llyn Traffwll Apr. 26th and 8 Llyn Alaw May 25th, both Anglesey 1997; singles on River Clwyd Mar. 28th and at Ruabon Nov. 3rd, both Denbigh 1997.

BAR-TAILED GODWIT *Limosa lapponica*   *RHOSTOG GYNFFONFRITH*
(C) A common passage migrant and winter visitor, more numerous in autumn than spring. It is rare away from the coasts but has been recorded in all inland counties.

More widely distributed in Wales than the Black-tailed Godwit, found on both sandy and muddy shores, whereas the latter is mainly on muddy estuaries. The Welsh wintering population is estimated at 1% of the UK total.

The table below gives max. winter counts, November – February, at the main sites, over 100 at least once. The National Level of Importance is 530.

		/1992	92/3	93/4	94/5	95/6	96/7	97/8	98/9	99/00	00/01
Glam.	Blackpill	119	45	46	20	89	89	45	10	3	
	Burry S.	1040	290	270	200	330	710	260	345	105	120
Carm.	Burry N.				146	262		96			
	TTG	123	125	59	103	100	350				
Angle.	Inland Sea	180						150	100		
Flint	Point of Air	46	400		100			70	45		60

Winter counts of over 100 also recorded: 169 at Burry Port Jan. 4th, 250 at Cefn Sidan Jan. 12th both Carmarthen and 1998.

Birds in Wales 1992-2000

Birds in Wales quote figures for Average Winter Peak Counts for the main sites for the period 1986-90. During the period under review there are no comparable figures available for the Dee. The Average Winter Peak Counts for 1996-2000 for the other sites are tabulated below, a decrease at both sites.

	1986-90	1996-2000
Burry Inlet	455	384
Swansea Bay (Blackpill)	111	47

Large numbers recorded annually on passage, counts of over 100 were: 100 at Point of Air and 110 at Burry S., both April 1992, 150 at Burry N. and 300 Dyfi, Ceredigion, both May 1993, 100 at Newton Beach, Glamorgan May 1st 1997, 260 Burry S. in September 1998, 200 at Inland Sea October 1998, 315 at Burry S. in October 1999 and in 2000, 121 GLWR Goldcliff, Gwent in May, 120 Burry N. in April, 207 at Ynyslas, Ceredigion May 7th, 300 Burry S. in September and 114 at Point of Air in September.

Inland records are unusual, during the period under review there were only 3 records all of singles in Brecon: at Brechfa Pool Apr. 13th 1993 and at Glasbury May 2nd 1993 & May 3rd 1997.

WHIMBREL *Numenius phaeopus* COEGYLFINIR

(C) Locally common in coastal counties on spring and autumn passage, scarce inland, small numbers occasionally over-winter. The spring passage predominates.

Birds in Wales states that there has been little change in the Whimbrel's distribution over the last century at coastal sites but an increase inland, where it is now a scarce but regular passage migrant. A larger passage is recorded in spring than in autumn, with the Gwent Levels being one of the most important sites at this time. E.g. 1,000 at Collister Pill on Apr. 30th 1976 and 739 on the Gwent Levels on May 5th 1991. During the period of this review the approximate passage periods were:

Year	Spring passage			Large flocks over 100:
1992	Mar. 27th	-	June 13th	Blackpill, 112 on Apr. 24th 1992, 107 at Kenfig Apr. 26th, 120 at Rhymney Est. Apr. 8th all Glamorgan
1993	Mar. 15th	-	June 22nd	105 at Goldcliff and 87 at St. Brides, Gwent Apr. 24th 1993, 132 at Berthlwyd May 8th, 110 at Kenfig on the 9th, 126 at Burry Inlet in May, 100 Gann, Pembroke May 2nd, 94 at Ynyslas, Ceredigion Apr. 30th and c100 at D/Glas May 10th
1994	Apr. 10th	-	June 11th	190 Peterstone Wentloog in late April & 180 there early May + 85 at Goldcliff, 177 Penclawdd Apr. 29th, 102 at Penclacwydd in April
1995	Apr. 9th	-	June 18th	235 Burry S. April, 114 Burry N. in May, 203 Penclawdd Apr. 29th, 104 in TTG in May, 133 Ynyslas Apr. 30th
1996	Mar. 23rd	-	June 28th	161 Burry N. in May
1997	Mar. 9th	-	June 6th	Counts from Burry S. 100 in April, 285 in May, 125 in June and 100 in July; 85 Burry N. in May, 135 at Brownslade, Pembroke June 6th
1998	Mar. 16th	-	June 18th	115 Castle Martin, Pembroke May 21st was the only significant count
1999	Mar. 17th	-	June 20th	140 at 3 sites along the Severn shore, Gwent, 175 at Burry S. in May & 100 in June and 160 at Burry N. in May
2000	Mar. 3rd	-	June 6th	125 at GLWR Goldcliff, Gwent in April and 102 in May, 330 at Burry S. in May

Year	Autumn passage			Large flocks:
1992	July 7th	-	Oct. 12th	96 at Berthlwyd, Glamorgan Aug. 15th
1993	July 7th	-	Oct. 22nd	220 passed Strumble Head, Pembroke Aug. 5th
1994	July 1st	-	Sept. 27th	300 Burry S. in July
1995	July 6th	-	Oct. 8th	73 Penclawdd Aug. 10th
1996	July 5th	-	Nov. 25th	105 Burry S. in July & 135 in August, 101 Blackpill Aug. 1st
1997	July 9th	-	Sept. 22nd	215 passed Strumble Head, July 12th – Sept. 12th, max. 26 on July 27th
1998	July 2nd	-	Nov. 19th	680 passed Strumble Head, July 16th – Sept. 13th, max. 370 on Aug. 5th
1999	July 3rd	-	Nov. 9th	190 at Burry S. in July. Only 180 logged passing Strumble Head, max. 65 on Aug. 30th
2000	July 2nd	-	Oct. 12th	230 at Burry S. in July and 175 in August

The total number of individuals recorded inland during the period were: Gwent (17), Glamorgan (7), Carmarthen (15), Brecon (52+), Radnor (10), Montgomery (21), Denbigh (1) and Flint (13-15).

Small numbers occasionally winter in Wales. During the period wintering records included: singles at Llansteffan, Carmarthen Jan. 16th and at Nevern Est. Pembroke Feb. 1st 1992. 2 wintered Red Wharf Bay, Anglesey 1994/5 and a late bird was at Bardsey, Caernarfon late Oct – mid. November 1994. Wintered Bardsey at both ends of 1996, one Skokholm, Pembroke Nov. 25th. Wintered Goldcliff / Redwick, Gwent in the early months of 1997, present at Skokholm, Bardsey and Bagillt Bank, Flint at both ends of 1997. 2 wintered Skokholm in the early months of 1998, singles at Dwyryd/Glaslyn in January, wintered at Bardsey & Bagillt Bank at both ends of the year, 2 at Blackpill Nov. 19th. Single at Skokholm Feb. 23rd 1999 with 1-2 there in November, 1 on Skomer until Nov. 20th, over-wintered at Bardsey, single Foryd Bay, Caernarfon Feb. 21st, 2 at Octel Amlwch, Anglesey Feb. 6th and present at Flint Castle Jan. 3rd & Feb. 12th.

In 2000 singles at Skokholm in both winter periods, at Skomer on Nov. 22nd, at Bardsey in the 1st winter period and 2 in Bull Bay, Anglesey Feb. 5th.

Individual of the American race, *hudsonicus* was at GLWR Goldcliff, Gwent May 6th – 7th 2000 (first Welsh record).

CURLEW *Numenius arquata*   GYLFINIR
(C) A fairly abundant and widespread but declining breeding species throughout most of Wales but local in Pembroke and the coastal fringe of south Wales. It is an abundant winter visitor to coastal areas, where small numbers are present all year round. Very small numbers winter inland.

The main wintering concentrations are on the Dee, Severn, Burry Inlet, Cleddau and Traeth Lafan. The total Welsh wintering population was estimated at 15-20,000, approximately 3% of the East Atlantic Flyway. The table below gives max. winter counts, November – March, at the main sites, over 1,000 at least once. The National Level of Importance is 1,200.

		/1992	92/3	93/4	94/5	95/6	96/7	97/8	98/9	99/00	00/01
Glam.	Burry S.	1098	1469	935	1175	940	1185	790	1235	1230	950
Carm.	Burry N.	1076	308	1550	939	694	1139	1218	1370	660	304
	TTG			995	4483*	478	694	820	609	666	304
Pemb.	Cleddau	1304	1541	1159	1732	1436	1283	1330	1169	1401	1167
Cere.	Dyfi	1033	909	914	984	768	892	681	623	873	541
Caern	Traeth Lafan	1220	1197	1276	1063	1412	1446	1044	522	1836	1610
Flint	Point of Air	1700	800	1429	1440	1110	1011	990	890	1145	789

* including 4,000 at Ginst Point.

The Average Winter Peak Counts for 1996-2000 are tabulated below.

	1996-2000
Burry Inlet	1,830
Carmarthen Bay (T/T/G)	653
Cleddau	1,324
Dyfi	767
Traeth Lafan	1,252

Large numbers were recorded in autumn, max. autumn counts, July – October, at the main sites, over 1,000 at least once:

	1992	1993	1994	1995	1996	1997	1998	1999	2000
Burry S.	1660	1555	1392	2050	1790	1765	1700	2185	1385
Burry N.	1500	1500	2442	2166	2056	2178	2064	2481	1576
Cleddau				1985	2734	1295	2393	2745	1152
Dyfi	1203	1100	902	932	1002	808	1016	958	1097
Traeth Lafan			2700		1500	1600	1200	4000	2205
Point of Air	1250	1600	1456	1460	1254	1470	1300	1778	1772

Inland reasonably large numbers were recorded regularly at Dryslwyn in Carmarthen max. 200 in January 1994, up to 85 at Glasbury in Brecon / Radnor and at Dolydd Hafren in Montgomery, usually about 100 but a max. of 300 on Apr. 13th 1993. Elsewhere large inland counts were: 300+ Lake Vyrnwy at the end of 1994, 71 Pont Glan y Wern, Clwyd Mar. 10th 1995.

Birds in Wales 1992-2000

Birds in Wales states that breeding numbers have declined over the past four decades. This trend probably continued into the 1990s. A summary of the max. number of breeding pairs reported during 1992-2000 for each county is shown on the map opposite.

In 1993 the RSPB conducted a breeding survey, using 1-km squares, producing a total of 10,700 pairs in Wales, Welsh Birds 2:35-42. These figures are thought to be far too high and are thus not a good indication of the Welsh breeding population. On the basis of the above counts for individual counties, the total Welsh population is unlikely to exceed 2,000 pairs (as published in *Birds in Wales*) and it is notable that all comparative counts note sharp recent declines.

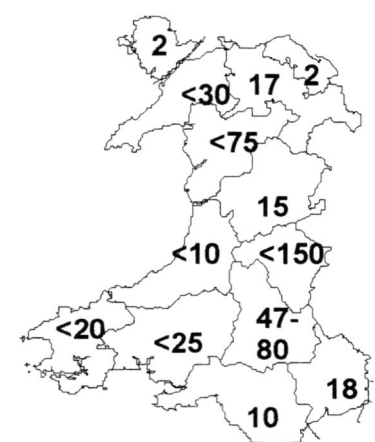

Results of the main surveys were:
Carmarthen: a survey of 15,000 ha of Mynydd Du in 1996 found none, compared to 3 pairs in 1978.
Pembroke: now breeds only on Skomer, with 8-18 pairs, probably declining.
Ceredigion: in 1994 it was considered that none were left breeding south of a line from Lampeter to Aberaeron, an area largely devoted to intensive dairying. In the north, in 1999, 12-13 pairs at Cors Caron, 13 at Cors Fochno and 10 other pairs noted (an incomplete survey). On the Dyfi SSSI, which lies mainly in Ceredigion but partly also in Meirionnydd and Montgomery, there were 34 pairs in 1996 (Cors Fochno lies within this site).
Brecon: in 1996 M.F. Peers surveyed 39 km² of upland commons and recorded 21 pairs, a density of 0.5 pairs per km² (*Welsh Birds* 1(5): 56-62).

Radnor: in 1997 A. Harris located breeding birds in 123 out of 268 1-km squares distributed throughout the county and holding 134-139 pairs *Welsh Birds* 2(2):7.
Meirionnydd / Denbigh: 38 pairs on Mynydd Hiraethog in 1994 and 3 there in 1996. This population has been declining steadily since 1977, when there were 64 pairs which had already declined to 42-45 in 1984-5.
Caernarfon: on the Llyn Peninsula in 1993, counts at 17 wader sites recorded 11-26 pairs, a decline of c76% compared to a similar census in 1986-7, when there were 69-82 pairs.
Anglesey: 2 pairs on Malltraeth RSPB. The whole area of this marsh (Cors Ddyga SSSI) held 16 pairs in 1985, arguing a marked decline.

SPOTTED REDSHANK *Tringa erythropus* *PIBYDD COESGOCH MANNOG*
(C) *An uncommon but regular passage migrant to coastal counties chiefly in autumn. Small numbers over-winter on the Dee, Burry and Cleddau. Rare inland.*
This species occurs regularly in small numbers usually less than 10, principally in the period August – October. The total number of individuals recorded in each month during 1992-2000 is shown in the following chart. Individuals were recorded wintering at several sites: Burry Inlet (Penclacwydd) in Carmarthen, Cleddau Estuary in Pembroke, Glaslyn in Meirionnydd, Conwy RSPB, Foryd Bay & Aber Ogwen in Caernarfon, Malltraeth & Red Wharf Bay on Anglesey, Point of Air, Oakenholt and Inner Marsh Farm in Flint. Approximate number of individuals wintering at each site is shown in the following table.

	/1992	92/3	93/4	94/5	95/6	96/7	97/8	98/9	99/00	00/01
Burry	5	3	11	10	1	1	3-4	5	3	4
Cleddau	2	5	6	1	1	6	5	4	5	
Glaslyn			1			2				
Conwy						1	4	2	1	
Foryd Bay	1	1		2	1				1	
Aber Ogwen	1	1	1	1	1				1	
Malltraeth		?					5			
Red Wharf Bay		?					1			
Point of Air								1		
Oakenholt			7	11	13	9	7	1		
Inner Marsh Farm	7	15	12	10	6	14	17	14	14	7

Large numbers were recorded on passage recorded at: 40 at Connah's Quay / Oakenholt Sept. 24th 1992, c75 there March – May & June – November 1993, 25 in October 1994, 25 on Oct. 13th 1995; 28 Penclacwydd September 1994, 26 in September 1996; a significant spring passage annually in April at Inner Marsh Farm, 18 in both 1997 & 1998, although the majority roost on the Cheshire side of the reserve.

Inland records of Spotted Redshank are very unusual and during the period under review there were singles at Dolydd Hafren, Montgomery Sept. 8th 1993, Sept. 5th 1999 and July 26th 2000, Hindwell Pool, Radnor Aug. 22nd 1995, Pwll Patti, Radnor Feb. 8th 1996 and at Glasbury Aug. 4th 1996.

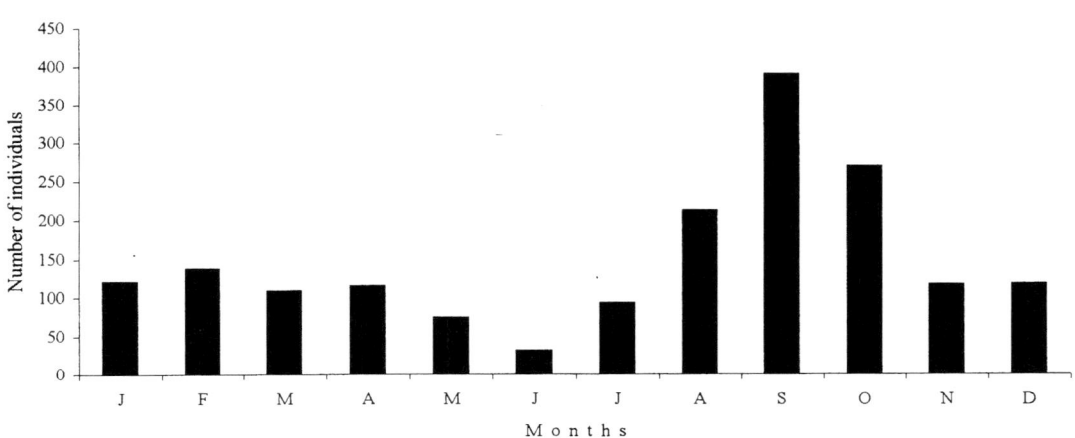

Monthly distribution of Spotted Redshanks 1992-2000

REDSHANK *Tringa totanus*   *PIBYDD COESGOCH*

(C) A scarce and declining breeding species, confined chiefly to salt-marshes but with a few pairs inland. Abundant wintering population.

Wintering birds return at the end of June, peaking in September / November and declining slightly thereafter. There is no evidence of spring passage. The table below gives max. winter counts, November – March, at the main sites, over 500 at least once. The National Level of Importance is 1,100.

		/1992	92/3	93/4	94/5	95/6	96/7	97/8	98/9	99/00	00/01
Glam.	Rhymney E.	950	770	442	800	900	400	520	850	800	460
	Taff / Ely	539	380	367	367	165	318	212	340		
	Burry S.	485	340	310	395	190	485	695	635	480	485
Carm.	Burry N.	340	308	460	833	397	855	733	613	489	186
	TTG	402	397	775	677	536	137	100	398	278	173
Pemb.	Cleddau	592	911	872	660	733	647	553	599	659	429
Caern	Conwy	500						245	300	300	250
	Traeth Lafan					334	622	336	195	1124	1270
Flint	Oakenholt	1300	450	1000	978		494	1100	450	350	395
	Point of Air	1000	530	780	830	570	585	465	240	569	450
	Bagillt Bank							1350	380		

Max. autumn counts, June – October, at the main sites, over 500 at least once:

	1992	1993	1994	1995	1996	1997	1998	1999	2000
Rhymney	650	200	435	600		800	276	550	1000
Ogwen								600	
Burry S.	405	430	490	570	450	350	160	510	220
Burry N.	700	831	1009	634	845	1051	1052	858	460
Cleddau								789	335
Oakenholt	2500	2000	2000	2000	3000	490	230	1550	2500
Point of Air	515	875	1450	850	350	490	230	180	1164
Bagillt Bank						1630	520		

Large counts elsewhere: 1,140 at Mostyn Dock, Flint August 1997.

Birds in Wales 1992-2000

Birds in Wales quote figures for Average Winter Peak Counts for the main sites for the period 1986-90. During the period under review there are no comparable figures available for the Dee or the Severn and the Taff / Ely estuary was lost to the Cardiff Bay Barrage. The Average Winter Peak Counts for 1996-2000 for the other sites are tabulated below. A very slight decline on the Burry and Traeth Lafan but a considerable drop in numbers on the Cleddau.

	1986-90	1996-2000
Burry Inlet	1,063	975
Carmarthen Bay (T/T/G)	-	290
Cleddau	1,442	638
Traeth Lafan	589	522

Birds in Wales reports a decrease in breeding Redshanks since the 1960s, in both range and number, with many suitable habitats being lost to drainage and land improvements. Coastal sites contain most pairs. A survey in 1991 found a total of 156-163 pairs on the coast at 13 sites, of which there were 20 prs. on the Dyfi. This compares to just 20 pairs inland (2 in Radnor, 4 in Montgomery, 2 in Meirionnydd, 6 in Ceredigion and 6 in Gwent).

During the 1990s the most important breeding site by far was the Dyfi SSSI, lying in Ceredigion / Meirionnydd / Montgomery. In 1996 a total of 87-92 pairs bred and 88 in 1999. In 1996 the RSPB also did a sample salt-marsh survey which found 48-53 pairs on 779 ha (6-7 pairs /km^2), Welsh Bird Report 1997 p38; 9-10 of these pairs also were in the Dyfi SSSI survey, so a total of 126-135 pairs.

Figures from the Dyfi suggest a dramatic increase (although this is probably due to a change in monitoring methods).

Elsewhere numbers decreased: up to 7 pairs breed Cors Caron, Ceredigion, c7 in total in Radnor and Montgomery, up to 15 pairs at Penclacwydd, Carmarthen, 10 in Gower, c10 in Gwent (these last 3 were all on salt-marsh but outside the RSPB survey). In Meirionnydd a further c12 pairs, (outside the Dyfi), Caernarfon and Anglesey. None seem to breed now on the Welsh side of the Dee.

In 2000, 11 prs. bred Gwent, 17 prs. Ynyshir, Ceredigion, 9 prs. Cors Caron, 3 prs. Penmaenpool, Meirionnydd, 2 prs. Morfa Madryn, Caernarfon and 2 prs. Inner Marsh Farm, Flint.

MARSH SANDPIPER *Tringa stagnatilis*   *PIBYDD Y GORS*
(B) A vagrant.

Three records were published in *Birds in Wales*, of individuals at Malltraeth, Anglesey June- July 1977, Oakenholt May 1990 and Penclacwydd, Carmarthen May 1990. Since then only two further records, the first on the River Clwyd at Rhuddlan, Flint Aug. 9th – 27th 1994. The second at RSPB Conwy, Caernarfon June 14th 1996.

GREENSHANK *Tringa nebularia*   *PIBYDD COESWERDD*
(C) Locally common in coastal areas on passage, chiefly in autumn. Small numbers over-winter on Anglesey, Caernarfon, Glamorgan and Pembroke. It is uncommon inland.

Monthly distribution of Greenshanks 1993-2000

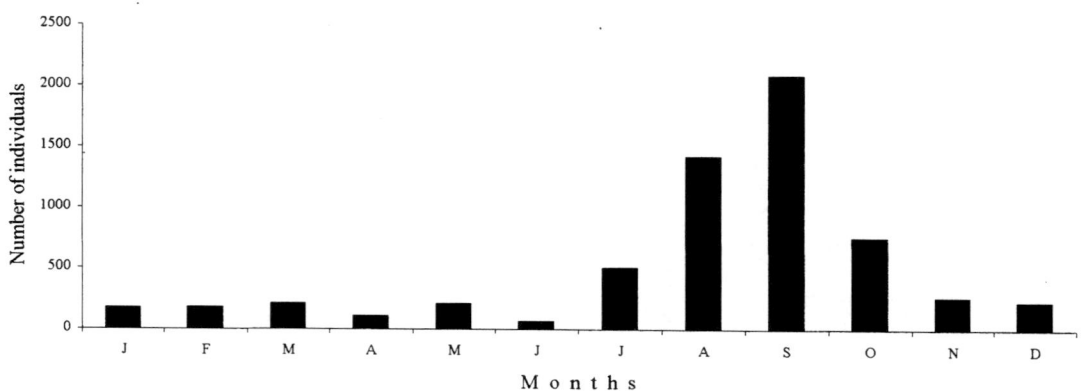

The total number of individuals recorded in each month is shown in the chart. This demonstrates a well marked autumn passage peak between July and October and far fewer spring migrants. The main sites for wintering Greenshanks were: the Burry Inlet & TTG Estuaries in Carmarthen, Cleddau in Pembroke, in Caernarfon Foryd Bay, Aber Ogwen and the Spinnies N. R., Menai Strait and Braint estuary on Anglesey and the Dee.

Large numbers occurred regularly at three sites, Penclacwydd, Cleddau and Inner Marsh Farm (occurred on both sides of the English / Welsh border but roosted on the English side). The following table summarises the peaks at these three sites. Elsewhere large counts included 35 at Oakenholt Sept. 6th 1992, 36 at Llandeilo, Carmarthen Sept. 25th 1998 and 32 at Foryd Bay, Caernarfon in September 1998. Greenshanks were recorded regularly inland in autumn, sometimes in small parties up to 7.

	Penclacwydd	Cleddau	Inner Marsh Farm
1992		11 in September	20 on Aug. 4th.
1993		28 at Sprinkle Pill Aug. 22nd	53 Aug. / Sept.
1994	54 in September	48 in Aug. & 47 in Sept.	51 Aug. / Sept.
1995		50 in Sept. & 37 in Oct.	45 in September
1996	56 in September	37 in July & 38 in Sept.	56 Aug. / Sept.
1997	59 in September	49 in September	39 in September
1998	58 in September	34 in September	85 in September
1999	54 in Aug. & 61 in Sept.	43 in Aug. & 41 in Sept.	56 Aug. / Sept.
2000	44 in September	37 in September	25 in September

LESSER YELLOWLEGS *Tringa flavipes*   *MELYNGOES BACH*
(B) A vagrant.
Nine records were listed in *Birds in Wales*, only two were recorded since. Single at Cemlyn, Anglesey May 7th – 13th 1999, was the first spring record in Wales and the first record for Anglesey. An individual wintered at Laugharne, Carmarthen from Nov. 12th 2000 – March 2001.

GREEN SANDPIPER *Tringa ochropus*   *PIBYDD GWYRDD*
(C) A regular passage migrant and winter visitor in small numbers to all counties.
Wintering numbers are relatively small and usually involve ones or twos in scattered inland and coastal sites. The table below gives approximate wintering numbers for each county. The first figure for the early months, the second for the end of the year.

	1992	1993	1994	1995	1996	1997	1998	1999	2000
Gwent		5 & 1	3 & 0			3 & 2		1 & 2	2 & 1
Glamorgan	5 & 2	12 & 6	10 & 2			8 & 4	9 & 8	7 & 14	16 & 8
Gower only	5 & 5	4 & 1	4 & 5	3 & 2	3 & 1	3 & 2	2 & 3	3 & 5	
Carmarthen	11 & 5	2 & 8	6 & 6	5 & 4	4 & 1	5 & 2	5 & 2	0 & 2	6 & 3
Pembroke	3 & 5	2 & 1	2 & 2	2 & 0	4 & 3	4-5 & 2	2 & 0	1 & 1	
Ceredigion	2 & 1	3 & 2	2 & 1	1 & 2	1 & 2	1 & 0	1 & 1	2 & 0	0 & 3
Brecon	1 & 2	1-2 & 2-3	2-3 & 1	6 & 1	0 & 1	2-3 & 5-7	2 & 3	2-3 & 2-3	2-3 & 1-6
Radnor			1 & 0	0 & 2	2* & 0	3 & 0	2 & 2	1-2 & 1-2*	1-2 & 1
Meiri.		1 & 0			1 & 0				
Caernarfon					1 & 1	1 & 1		1 & 1	
Anglesey		1 & 1		0 & 1					
Denbigh				1 & 2				0 & 1	
Flint							1 & 0		

* on the Brecon border. One at Llandewi Brefi wintered for 9 consecutive winters did not re-appear at the end of 1999 but did at the end of 2000.

As a passage migrant Green Sandpipers are mostly seen during the period July to September, peaking in August. The largest congregations in the period under review were: 9 at both Shotton June 29th and Inner Marsh Farm, Flint Aug. 19th 1992, 8 at Gobion, Gwent in August 1993, 7 at Hensol Lake, Glamorgan in autumn 1994, 10 at Burry Inlet in August & 7 on the Dyfi in September 1995. 10 at Gobion, Gwent & 9 at GLWR Goldcliff, both August, 7 at Cors Caron, Ceredigion in autumn and 7 at Screthrog, Brecon July 8th, all 2000. One summered at Teifi Marshes, Pembroke/Ceredigion in 1994.

WOOD SANDPIPER *Tringa glareola* PIBYDD Y GRAEAN
(C) A scarce passage migrant recorded in every county, principally in the autumn but with a scattering of spring records.

The bulk of records come from coastal sites. The numbers recorded annually in Wales continue to increase. The total number of individuals recorded in each month during the period 1992-2000 is shown in the chart, while the map shows the total number of records for each county in that period.

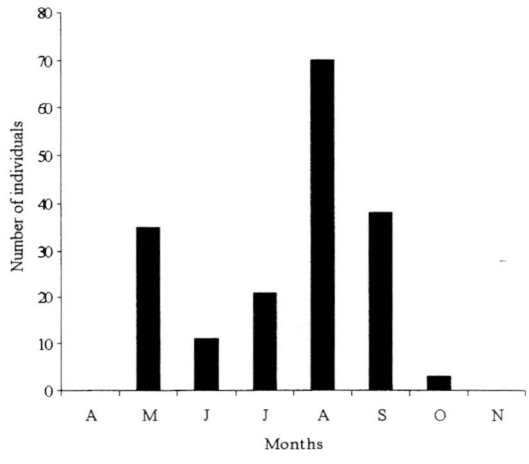
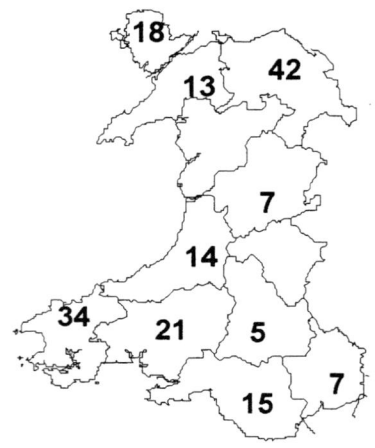

TEREK SANDPIPER *Xenus cinereus* PIBYDD TEREK
(A) A vagrant.
The first and only Welsh record was of an individual at Conwy RSPB, Caernarfon Apr. 29th – May 3rd 1999. The bird also visited Glan Conwy in Denbigh at times during its stay.

COMMON SANDPIPER *Actitis hypoleucos* PIBYDD Y DORLAN
(C) A summer visitor that breeds in all counties except Pembroke, albeit sporadically on Anglesey and Flint. Very small numbers over-winter.

The approximate wintering numbers in each county are tabulated below. The first figure is for the early months, the second for the end of the year.

	1992	1993	1994	1995	1996	1997	1998	1999	2000
Gwent		1 & 0	2 & 0			1 & 0			0 & 1
Glamorgan		1 & 1-2	2 & 3						2 & 3
Gower only				3 & 1	1 & 1	2 & 0	2 & 1	1 & 1	4 & 1
Carmarthen	0 & 3	0 & 1	2 & 0	0 & 1			1 & 0		1 & 0
Pembroke	5 & 4	4 & 0		1 & 3			1 & 3	3 & 2	3 & 0
Ceredigion		0 & 1	1 & 0			0 & 1-2	1 & 0		2 & 0
Brecon		1 & 0							1 & 1
Radnor					1 & 0			1 & 0	
Meirionnydd				1 & 1			0 & 1		
Caernarfon								2 & 0	0 & 1
Anglesey				1 & 0					0 & 1
Flint							1 & 0		

The largest concentration recorded during the period was of 45 on the Clwyd Estuary, Rhyl - Rhuddlan, Flint July 19th 1992.

Birds in Wales puts the total Welsh breeding population in the order of 1,000 pairs, of which fewer than 100 prs. in Gwent, 30 in Gower and 100-120 in Brecon. It also states that this species has been declining as a breeding species in Wales since the 1980s, possibly due to increased disturbance by tourists and anglers.

A survey of rivers in 1991 found 10-35 pairs/10km of the Tywi (total of 58 prs.), 107 prs. on the Wye, 16 prs on the Vyrnwy and 36 prs. on the Severn. Although the data for the last 9 years are incomplete, the estimate of 1,000 pairs breeding in Wales now appears to be on the high side. Data from the New National Atlas estimated the Welsh population as 664 pairs. BBS results 1994-1999 however suggest a population change of 71%. This would give a revised Welsh population estimate of 470 pairs.

A summary of other breeding records, during the period under review, arranged by county, are given below.

Carmarthen: 100 prs. estimated breeding in the county in 1992, 52 prs. on the Tywi, Carmarthen – Llandovery and 5 on the Cothi, 2 on the Cennen and one on the Sawdde in 1994.

Carmarthen / Brecon: 10 prs. found by RSPB survey of 15,000 ha of Mynydd Du in 1996.

Pembroke: a pair bred at Clarydale, the first confirmed breeding this century.

Ceredigion: 4 pairs on the Dyfi.

Brecon / Radnor / Ceredigion: 23 prs. on 150 km^2 of the Elenydd (70% of them in Ceredigion) in 1995.

Brecon: 9-12 prs. on the Usk, Pencelli – Talybont, 14 prs. on the Wye, Llyswen – Hay, 7 at Llyr-ffordd Fawr and 4 prs. on 4,000 ha of the Cnewr estate in 1996.

Radnor: 25 pairs in the Elan Valley but poor success rate.

Montgomery: 3-6 prs. Dolydd Hafren, 12 prs. Lake Vyrnwy, 2-3 prs. Llyn Coed y Dinas, 2 pairs Pwll Penarth, 2 prs. at Llanerfyl, 2 prs. at Cwm Llinau and 5 prs. on the Dyfi SSSI (also Meirionnydd & Ceredigion).

TURNSTONE *Arenaria interpres*   *CWTIAD Y TRAETH*
(C) An abundant wader on the coast, a passage migrant in spring and autumn but most numerous and widespread in winter. A few non-breeding birds are present throughout the summer. It is rare inland.

Birds in Wales 1992-2000

		/1992	92/3	93/4	94/5	95/6	96/7	97/8	98/9	99/00	00/01
Glam.	Cardiff	152									
	Blackpill	70	59	68	83	265	333	130	185	138	88
	Swansea Pier					125	129	85			
	Swansea Docks				141	122		450	145		
	Burry S.	145	137	210	265	220	95		159*	295	
Carm.	Burry N.	110	227	321	435	156	20		136	241	51
	Machynys			207	57	35					
	TTG		52	188	134						
Pemb.	Cleddau	97	107	127	85	61	60	178	107	45	77
Cere.	Aberaeron		80						61	105	107**

* Whiteford Point only.
** September.

Small parties forage along most rocky shores as well as on shingle beaches. The table above gives the max. winter counts, November – February, at the main sites, with over 100 recorded at least once. Other large counts included: in Glamorgan: 122 at Orchard Ledges in January 1994, 77 at Cardiff Heliport on Jan. 5th 1994, 79 at Porthcawl Jan. 6th; 82 at Foryd Bay, Caernarfon in January 2000 and 101 in December; 140 counted along the Clwyd coast in 1991/2 and 132 at Llandulas, Denbigh Feb. 2nd 1995.

Assuming that the Blackpill, Swansea Pier and Swansea Docks birds all belong to the same flock and by taking the largest figure produces an average winter peak of 247 individuals for the period 1996-2000. This is only slightly down on the 287 quoted in *Birds in Wales* for the period 1986-1990. The only other site which allows a comparison is the Cleddau, with an average winter peak of 90 during the period 1996-2000 compared to 133 in 1986-1990, quite a large reduction.

Birds depart their wintering grounds in early March. The table below gives max. spring passage counts, March - May, at the main sites, with over 50 recorded at least once.

		1992	1993	1994	1995	1996	1997	1998	1999	2000
Gwent	Goldcliff / Redwick						110	50		
Glam.	Cardiff	61								
	Rhymney	52								
	Kenfig	93								
	Blackpill	77	111	128	65	354	55	60	123	120
	Swansea Pier		111							
	Swansea Docks			152				45		
	Burry S.	2	95	15		125	15			140
Carm.	Burry N.	95	24	335	230	84	117	121		44
Clwyd	Clwyd Coast	160								

Large counts elsewhere: 50 at Sker, Glamorgan in March 1999, 54 at Port Talbot Harbour, Glamorgan in April 2000, 80 Gann, Pembroke Mar. 3rd 1999, 83 at Caernarfon Golf Course May 28th 1999, 100 Rhos-on-Sea, Denbigh Mar. 5th 1999.

The return passage starts at the end of July. The following table gives max. autumn passage counts, July – October, at the main sites, with over 50 recorded at least once.

		1992	1993	1994	1995	1996	1997	1998	1999	2000
Glam.	Cardiff	108								
	Rhymney	108								
	Kenfig	62								
	Swansea Pier		93					38		
	Swansea Docks			89						
	Burry S.	10	80	35	15	85		15	210	111
Carm.	Burry N.	366	392	391	352	149	158	244*		40

* at Burry Port Sept. 8th. Large counts elsewhere included 88 at Aberthaw Sept. 11th 1995 & 80 there in August 1999, 50 at Goldcliff, Gwent in August 1999. 147 at Blackpill in October 2000.

Inland there were 6 records: individuals at Llyn Tegid, Meirionnydd Jan. 29th 1992, at Llangorse, Brecon July 28th 1995 & Aug. 5th 1996 and at Llandegfedd Res. Gwent July 29th, Aug. 3rd and 23rd 1997.

WILSON'S PHALAROPE *Phalaropus tricolor*   LLYDANDROED WILSON
(B) A vagrant.
Additional record to the 8 published in *Birds in Wales*: an adult & 1st winter at Glan Conwy, Caernarfon Sept. 28th 1991. Since then there have been 2 records, at Point of Air, Flint Sept. 24th – Oct. 6th 1997 (this bird was also seen at Morfa Madryn, Caernarfon on Oct. 21st) and at the Broadwater, Tywyn, Meirionnydd Aug. 27th - 28th 1998.

RED-NECKED PHALAROPE *Phalaropus lobatus*   LLYDANDROED GYDDFGOCH
(B) A rare and sporadic visitor, March – December.
Birds in Wales quotes 37 individuals recorded in Wales, since then there has been 5 further individuals, 3 of which have been on Anglesey, the first at Llyn Traffwll May 27th – June 8th 1994, at Llyn Llywenan Sept. 8th 1995 and at Penmon Oct. 2nd – 6th 1998. In 2000 there was a female at Conwy RSPB, Caernarfon June 24th- 25th (the first for that county) and a juvenile at the Ogmore Estuary, Glamorgan Nov. 12th – 16th.

Breakdown of all Welsh individuals by month in the chart (2 undated records) and by county on the map:

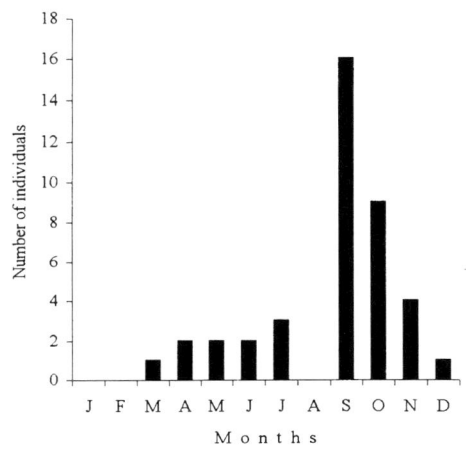
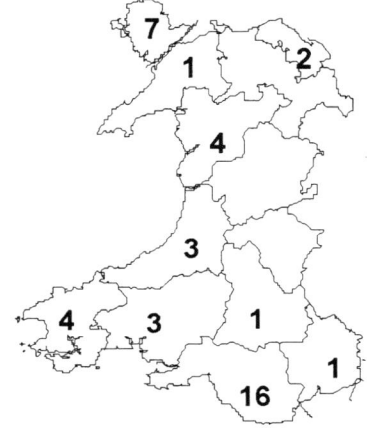

GREY PHALAROPE *Phalaropus fulicarius*   LLYDANDROED LLWYD
(B) An uncommon and irregular passage migrant, predominantly to coastal areas and in the period September – October.
The main site for the occurrence of this species was at Strumble Head, Pembroke, passage period late August to the end of November, annual totals of birds were:

1992	1993	1994	1995	1996	1997	1998	1999	2000
1	0	5	13	14	10	5	14	14

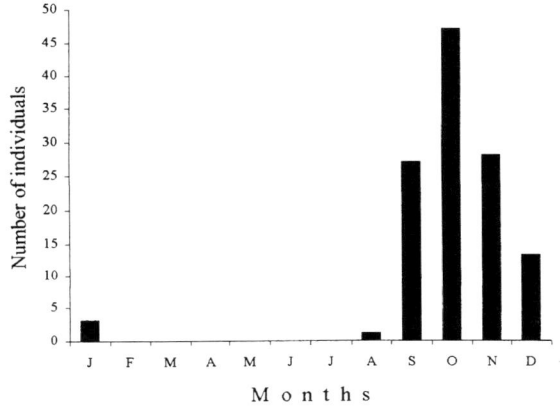

Autumn records elsewhere: 2 records from Gwent, 3 from Glamorgan, 2 from Carmarthen, 4 others from Pembroke, 15 from Caernarfon 10 of which from Bardsey, singles from Ceredigion, Anglesey and Flint.

The total number of individuals recorded in each month during the period 1992-2000 is shown in the chart.

There were only 15 winter records, all of singles at: Rhymney Est., Glamorgan Jan. 17th 1995, Burry Port, Carmarthen Jan. 7th – 10th 1996 and for Aberystwyth, Ceredigion Dec. 27th 1997 – Jan. 2nd 1998 & Dec. 18th 1998. In 2000 winter gales in December brought 11 individuals onto Welsh coasts, a single at Newgale, Pembroke on the 13th, 2 there 15th – 23rd with one remaining until the 25th, at Cei Bach, Ceredigion on the

6th, at least 4 in Aberystwyth 14th – 18th, 2 at the Broadwater, Towyn, Meirionnydd 23rd – 24th, a 1st winter at Porthmadog, Caernarfon on the 14th and an adult there 15th – 22nd.

There were two inland records, both of singles, the first at Llangorse, Brecon Sept. 13th – 16th 1992 – the first record for that county since 1966. The other record was at Llyn Coed y Dinas, Montgomery Dec. 19th 2000 – the first record for that county.

POMARINE SKUA *Stercorarius pomarinus* — *SGIWEN FRECH*

(C) A fairly common and increasing passage migrant to coastal areas, particularly in autumn.

	1992	1993	1994	1995	1996	1997	1998	1999	2000
Total	6	33	37	18	2	18	1	3	2
From	Apr. 30th	May 2nd	Apr. 23rd	Apr. 19th	June 9th	Apr. 24th	June 9th	Apr. 21st	May 7th
To	May 29th	June 15th	June 26th	Apr. 28th	June 30th	June 16th	-	June 3rd	-

Pomarine Skuas are recorded in Wales chiefly from headlands on Anglesey and Pembroke, mainly Strumble Head. Spring records are on the whole fortuitous, a summary of spring passage during the period under review is tabulated above.

One particularly early record, of an immature at Kenfig, Glamorgan Mar. 6th 1994. Large counts were: 17 off Mumbles, Glamorgan June 12th 1993, 10 at Pembrey, Carmarthen May 23rd 1994, 20 South Stack/Range, Anglesey in 30 minutes on June 3rd 1994, 14 there Apr. 19th 1995 and Peterstone Wentloog/Goldcliff, Gwent May 9th 1997.

Autumn passage noted at many coastal locations. Data from the main sites, Strumble Head, Pembroke, Bardsey, Caernarfon, Point Lynas, Anglesey and Point of Air, Flint are tabulated below:

Strumble Head

	1992	1993	1994	1995	1996	1997	1998	1999	2000
Total	65	16	55	41	50	14	19	149	51
From	Aug. 13th	Aug. 3rd	Aug. 23rd	Aug. 12th	Aug. 13th	Sept. 4th	July 15th	Sept. 11th	Aug. 3rd
To	Nov. 15th	Dec. 9th	Nov. 11th	Oct. 24th	Nov. 7th	Nov. 18th	Oct. 29th	Dec. 27th	Nov. 9th
Max.	20	3	6	4	11		3	27	6
date	Sept. 4th	Aug. 8th	Aug. 28th	Sept. 15th	Nov. 5th		Oct. 25th	Nov. 6th	Aug. 26th

Summary of the number of individuals recorded at Bardsey, Point Lynas and Point of Air:

	1992	1993	1994	1995	1996	1997	1998	1999	2000
Bardsey		2	2	3	2	2	5	11	6
Point Lynas	21	4	6*	4	2	1	7	8	7
Point of Air	10	4			4				

* includes a juvenile seen on Dec. 31st 1994.

Summary of autumn records elsewhere:

	1992	1993	1994	1995	1996	1997	1998	1999	2000
Glamorgan	2		4			2		3	1
Pembroke (other sites)		2		2	1	3		2	5
Ceredigion					1	1			
Meirionnydd	1								
Caernarfon (other sites)	5						1	1	1
Anglesey (other sites)	2		25	1	1		2		1
Flint (other sites)	1								

Most individuals have left Welsh waters by mid November but a few were seen in December, usually after gales. December records during 1992-2000 include: in Glamorgan 4 off Mumbles Dec. 31st 1994, singles off Port Eynon Dec. 4th & Kenfig Dec. 24th 1999; in Pembroke singles at Amroth Dec. 16th & Newgale Dec. 25th 1999, Strumble Head Jan. 3rd 2000; in Caernarfon singles off Porth Ysgaden Dec. 2nd, Porth Colmon Dec. 2nd with 4 on the 4th 1999 and Penmaenmawr Dec. 25th.

ARCTIC SKUA *Stercorarius parasiticus* SGIWEN Y GOGLEDD
(C) Passage bird off coastal counties; scarce in spring, more numerous in autumn, numbers recorded probably being closely correlated with the intensity of effort in sea-watching.

	1992	1993	1994	1995	1996	1997	1998	1999	2000
Total	15	54	47	16	16	16	24	11	13
From	Apr. 12th	Apr. 29th	Apr. 24th	Mar. 29th	Apr. 5th	Apr. 1st	Apr. 3rd	Apr. 10th	Apr. 17th
To	June 6th	June 12th	June 26th	June 11th	May 27th	May 29th	June 28th	June 28th	June 22nd

Max. 8 off Goldcliff, Gwent May 31st 1993, 8 off Cefn Sidan, Carmarthen Apr. 24th 1994, 16 from Glamorgan in 1998.

Winter records are very rare as virtually all individuals have left Welsh waters by the end of October. During the period under review there were five winter records, singles at Strumble Head, Pembroke Dec. 19th 1994, off Borth, Ceredigion Dec. 31st 1997 and presumably the same bird at Ynyshir on Jan. 4th 1998, Conwy, Caernarfon Jan. 7th 1998 and Penmon, Anglesey Mar. 7th 1999. The only multiple record was of 3 off Point Lynas, Anglesey Dec. 31st 1994. A few individuals are seen off the Welsh coast in spring, May – June. A summary is tabulated above.

Inland records few and far between: Gwent: 17 at Ynysyfro Res. Aug. 29th 1992 and singles at Llandegfedd Res. Sept. 13th 1993, Aug. 31st 1997 and Oct. 20th 1998.

Autumn passage was noted at many coastal locations, usually commencing in July with numbers building and reaching a peak in September before falling off during October. The table below summarises passage data from the main sites, Strumble Head, Pembroke and Point Lynas, Anglesey.

Strumble Head

	1992	1993	1994	1995	1996	1997	1998	1999	2000
Total	234	67	291	157	147	323	176	221	213
From	July 5th	July 10th	Aug. 12th	July 16th	July 6th	June 13th	July 23rd	July 5th	July 11th
To	Oct. 14th	Dec. 9th	Oct. 9th	Oct. 28th	Nov. 5th	Nov. 9th	Nov. 4th	Dec. 3rd	Nov. 3rd
Max.	45	21	38	15	26	52	18	22	17
Date	Sept. 3rd	Aug. 23rd	Sept. 12th	Aug. 25th	Aug. 25th	Sept. 6th	Sept. 10th	Sept. 30th	Sept. 6th

Elsewhere in Pembroke a large count of 10 from St. David's Head Oct. 3rd 1999.

Point Lynas

	1992	1993	1994	1995	1996	1997	1998	1999	2000
Total	72	21	41	40	24	140	30	22	20
From	July 1st	July 25th	Aug. 12th	Aug. 13th	Aug. 30th	July 15th	Aug. 16th	July 16th	Sept. 2nd
To	Oct. 25th	Dec. 4th	Oct. 3rd	Sept. 27th	Nov. 30th	Nov. 8th	Oct. 17th	Dec. 16th	Nov. 10th
Max.	10	8	15	10	9	19	8	5	
Date	Sept. 4th	Aug. 22nd	Sept. 14th	Aug. 28th	Oct. 5th	Sept. 7th	Sept. 13th	Oct. 4th	

At Point of Air, Flint figures are only available for 1992 and 1993. In 1992 a total of 128 individuals were logged passing between July 27th and Sept. 15th, max. 32 on Aug. 30th & Sept. 1st. In 1993 a total of 124 were logged between June 26th and Sept. 26th, max. 23 on July 31st. There were also 15 on Aug. 22nd 1998 and 7 on Dec. 7th 1999.

On Bardsey, Caernarfon 42 were logged between Aug. 24th – Oct. 26th 1999, max. 10 on Oct. 1st. In 2000, 88 were logged between Aug. 3rd and Oct. 31st, max. 17 on Aug. 22nd.

LONG-TAILED SKUA *Stercorarius longicaudus*   SGIWEN LOSTFAIN
(C) An uncommon visitor recorded chiefly along the western seaboard during westerly gales, spring records are very infrequent.

	1992	1993	1994	1995	1996	1997	1998	1999	2000
Total	7	1	39*	14**	11	19	25	52	11
From	Aug. 22nd	Oct. 2nd	Aug. 15th	Aug. 25th	Aug. 12th	Aug. 20th	Aug. 13th	Aug. 27th	Aug. 2nd
To	Sept. 11th	-	Oct. 23rd	Oct. 25th	Oct. 29th	Oct. 10th	Oct. 17th	Nov. 6th	Oct. 25th
Max.	3	-	8	3	3	7	9	13	-
date	Aug. 30th	-	Sept. 16th	Aug. 27th	Oct. 5th	Aug. 29th	Aug. 22nd	Oct. 3rd	-

*15 % adults, 79% of the records in September.
** 21% adults.

Birds in Wales states that up to 1991 there were only 91 accepted dated records, involving a total of 180 individuals in Wales. Since then a total of 179 were logged passing Strumble Head, the main Welsh sea-watching site. The number of individuals passing this site in each year, the passage period and the max. daily count are all summarised in the table above. The 1999 total of 52 individuals is the largest annual total to date but the maximum recorded in a day is still the 18 on Sept. 15th 1991. A single on Dec. 2nd 1996 and a juvenile on Dec. 3rd 1999 are the latest yet recorded in Pembroke.

Long-tailed Skuas were recorded elsewhere in 6 counties. All records were:
Glamorgan: Nash Point July 7th 1992, Porthcawl Sept. 10th 1994.
Pembroke: off Skokholm Oct. 4th 1995, a juv. from the Pembroke – Rosslare Ferry Aug. 24th 1996, St. David's Head Aug. 30th 1997, in 1999 2 on Oct. 3rd & an adult on the 30th, single Ramsey Oct. 25th 2000.
Ceredigion: Borth Oct. 3rd 1996.
Caernarfon: Bardsey immatures on Nov. 5th 1996 and Sept. 10th 1998, 3 on Oct. 10th 1997, in 1999 an immature Sept. 1st and one Oct. 2nd and in 2000 singles Sept. 12th & Oct. 30th and 3 on Sept. 6th & 7th.
Anglesey: at Point Lynas, 3 on Sept. 4th and a juv. Sept. 11th 1992, an adult Aug. 13th 1994, single Oct. 4th 1996, Aug. 22nd & 24th and an immature Sept. 13th 1998, in 1999 an immature Oct. 3rd, Oct. 26th & Dec. 15th (the latest yet recorded in Wales) and Oct. 2nd 2000.
Flint: 3 at Point of Air, Sept. 6th – Oct. 10th 1997.

Very few individuals are seen in spring and during the period there were only 3 records. In Pembroke at Skomer May 2nd 1995 and Jack's Sound Apr. 30th 1996 (the earliest Welsh spring record) and 2 at Llanddwyn, Anglesey on May 27th 1996.

GREAT SKUA *Stercorarius skua*   *SGIWEN FAWR*
(C) Regularly recorded on autumn passage along the west coast where small numbers are also seen in winter and spring. Numbers vary annually, the highest counts dependent on the strength and frequency of onshore gales.

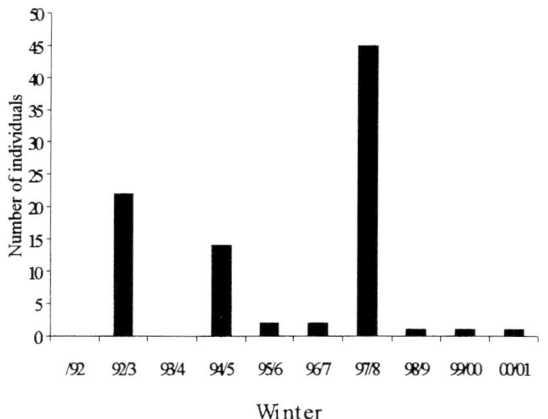

Winter

Although the vast majority of records come from the autumn months, in particular between August and October, smaller numbers are recorded in winter and spring. The chart opposite summarises the number of winter records, December – February, during the period 1992-2000.

Gales in December 1997 brought many individuals into Welsh waters. 26 were logged passing Strumble Head, Pembroke on the 26th, 3 the following day and singles there in early January 1998. A small party of at least 5 were seen along the Glamorgan & Gwent coast in early January 1998.

Summary of spring passage:

	1992	1993	1994	1995	1996	1997	1998	1999	2000
Total	3	19	8	3	3	17-22	6	4	3
From	Apr. 28th	Apr. 12th	Apr. 2nd	Apr. 27th	Mar. 2nd	Feb. 21st	Apr. 3rd	Mar. 26th	May 14th
To	June 6th	June 12th	June 26th	June 21st	May 31st	June 26th	June 11th	June 25th	May 30th

Autumn passage was recorded from many headlands but in particular off Strumble Head, Bardsey, Point Lynas and Point of Air. The following tables summarise the total number of birds logged, passage periods and the max. in a day for Strumble and the annuals totals from the other sites.

Strumble Head

	1992	1993	1994	1995	1996	1997	1998	1999	2000
Total	112	52	183	205	152	176	211	197	202
From	Aug. 5th	Aug. 3rd	Aug. 17th	July 21st	June 29th	July 23rd	July 11th	July 30th	July 11th
To	Nov. 11th	Dec. 19th	Nov. 15th	Oct. 28th	Nov. 7th	Nov. 19th	Nov. 12th	Nov. 29th	Dec. 14th
Max.	18	13	27	26	21	16	42	24	17
Date	Aug. 31st	Aug. 23rd	Sept. 11th	Sept. 24th	Sept. 30th	Aug. 29th & Sept. 13th	Oct. 17th	Sept. 20th	Sept. 2nd

Also from Pembroke, 10 off Skokholm in 1992, 8 off St. David's Head Oct. 30th and 10 in the Celtic Deep pelagic trips Aug. 15th & 28th 1999.

Totals from the other migration watch points:

	1992	1993	1994	1995	1996	1997	1998	1999	2000
Bardsey				12	22	25	14	27	24
Point Lynas	25	10	16 on Aug. 29th	28*	17	8 on Sept. 13th	26	19	26
Point of Air	6					10 - 12			

* a total of 53 off Anglesey in 1995 including these Point Lynas birds.

Large counts elsewhere included 17 off Cefn Sidan, Carmarthen during August 1993, 8 of which were on the 1st.

MEDITERRANEAN GULL *Larus melanocephalus* *GWYLAN MOR Y CANOLDIR*

(E) An increasingly common visitor, chiefly to coastal counties at all times of the year. Rare inland.

Since the first Welsh records in 1964 (in both Glamorgan and Caernarfon) this species has increased dramatically. *Birds in Wales* quote a minimum of 483 records in Wales during the period 1978 – 1987. Compare this to the total of 376 individuals recorded in Wales during 2000 alone. The true number of individuals occurring in each county is hard to assess due to individuals moving around from site to site and from county to county. The approximate annual totals, of individuals, for each county is shown in the following table.

Number in Pembroke nearly doubled in 2000. Some of this is probably due to a different way of calculating the total but some if it may be due to an increase in individuals just passing through the county. Evidence for this may be from Bardsey, where at least 30 individuals were logged passing the island in the autumn of that year.

	1992	1993	1994	1995	1996	1997	1998	1999	2000
Gwent		10	14	22-26	*70	18	26+	25+	24
Glamorgan	51	18	45	37+	40	76+	98+	100	101+
Carmarthen		6	7	3-5	9	9	10+	12	15
Pembroke		14	20	25-30	33	49	25-67	67	105
Ceredigion		12-16	12	8	10	9-13	20	20	57
Brecon			1						
Meirionnydd	4	1	3	3	1		13	9	20
Caernarfon	6+	1	1	14	2	8	27	29	48
Anglesey	5	4	6	17	5	2	5	5	6
Denbigh			6	24			4	5	
Flint		6			6	8-10	3	10	

*1996 figure of 70 for Gwent / E. Glamorgan.

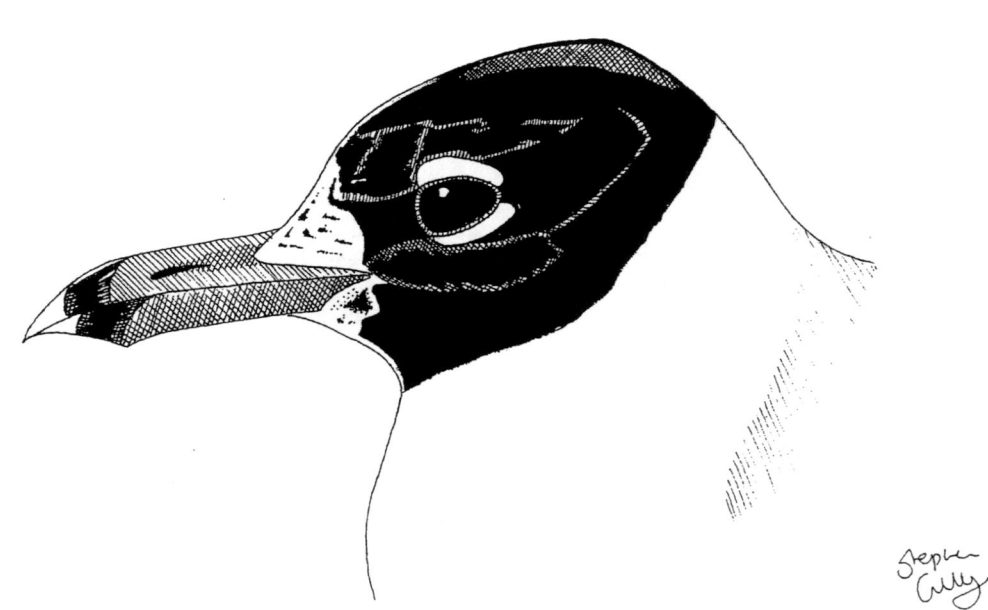

The site most favoured by this species in Wales is Blackpill, Glamorgan. Max. totals for each of the identifiable age-groups is tabulated below. The total number of individuals of each of the age in each month at Blackpill, during the period 1992-2000 is shown in the following chart.

	1992	1993	1994	1995	1996	1997	1998	1999	2000
Max. 1st year birds	-	-	-	13	3	10	5	6	11
Max. 2nd year birds	-	-	-	8	9	6	7	9	10
Max. adults	-	7	11	8	13	16	13	16	25
Total – at least	16	-	17	29+	25	32	45	43	36
Max. on one day	-	-	-	-	-	22 Sept. 19th	22 July 27th		27 Sept. 13th

One there on Aug. 12th 1999 had been ringed, pulli, at Newtown, Isle of Wight June 17th 1998.

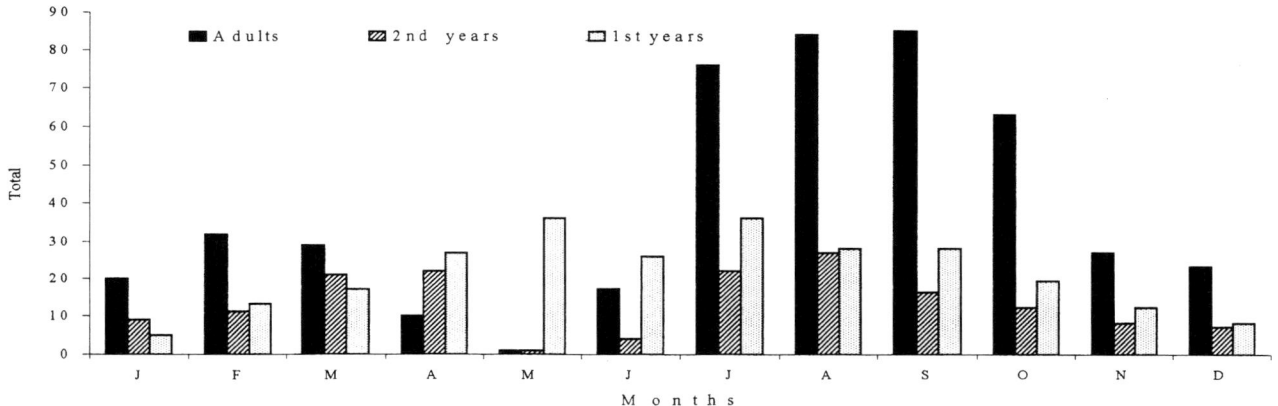

Analysis of age distribution at Blackpill clearly shows an influx of adults from June onwards (presumably this early influx is of non-breeding birds), most of which presumably pass through onto other sites. Numbers start dropping in March as adults move off towards their breeding grounds. Interestingly the number of 1st year individuals increases towards the end of the winter months, peaking in May & July. This may be due to the ease of identification amongst smaller numbers of Black-headed & Common Gulls in late winter / early spring. Numbers of 1st year individuals decrease in late summer as they moult into 2nd year plumage, causing a corresponding rise in that age group.

These results show some resemblance to those described by Hume (*British Birds* 69, 503-505) for data he collected during 1970-1975. He found a similar pattern for the early months but a higher proportion of adults were recorded. The data above are for a longer time span than Hume's and the numbers of birds involved is ten times that in the earlier study. There is also an influx of adults in late summer with individuals remaining throughout the autumn, whereas in Hume's, an influx August – October with birds departing thereafter.

Elsewhere, large congregations were: 5 summer plumaged adults at Llanrhidian, Glamorgan July 19th 1992, 6 wintered in Fishguard Harbour, Pembroke 1992/3 until April, 5 at Aberaeron, Ceredigion Sept. 29th 1993 (up to 12 at Inner Marsh Farm, Flint Aug. 29th 1998 were on the English side of the reserve). In 2000 4 adults were in Cardiff Bay, Glamorgan Feb. 19th, 4 adults Llanelli Dock, Carmarthen July 27th, 7 adults & 2 2nd summers at Penclacwydd, Carmarthen July 10th.

A total of at least 6 juveniles: at Goldcliff, Gwent July 25th 1993 in the company of an adult, at Blackpill a Hungarian ringed individual July 18th 1995 (which was subsequently present there, Mar. 19th – 20th & Apr. 2nd – 3rd 1996), July 18th 1996, on 9 dates in 1999, July 23rd – Sept. 27th, with 3 on Sept. 6th. One at the Gann, Pembroke, July 23rd 1994 had been colour-ringed 43 days earlier in Hungary (1743 km to the east).

Other colour-ringed individuals were at Aberystwyth, Ceredigion, a 2nd winter individual originally from Holland, in October 1993 and a 1st winter at Cemlyn, Anglesey in Jan. 1994 was originally ringed in Hungary.

In 2000 there were 7 records of colour-ringed individuals: a 1st summer at GLWR Goldcliff, Gwent May 17th, an adult at Peterstone Wentloog July 29th, adult marked as 1AM at Ogmore Est., Glamorgan Feb. 18th – 25th, an adult in Cardiff Bay, Glamorgan Jan. 23rd – Feb. 13th, another adult at Ogmore Est. Nov. 14th, an adult marked as 60X at Fishguard Harbour for its 4th winter in a row and another adult 67Z there & Nevern Est. All traceable individuals were from Holland / Belgium.

There were inland records: at Llandegfedd Res., Gwent 1998, 3 feeding inland in Pembroke 1998 as well as one at Llys y fran Res. A first summer at Llangorse Lake, Brecon Aug. 27th 1994 was the first record for the county and an adult was at Llyn Helyg, Flint Mar. 13th 1994.

Two adults were seen displaying at Inner Marsh Farm, Flint on the 5th & 8th June 1998 – unfortunately just over the border in Cheshire and a pair were also seen displaying at Cemlyn, Anglesey Apr. 8th & 12th 1998. An adult was in amongst the breeding Black-headed Gull colony at Llyn Helyg, Flint Mar. 19th 1999.

FRANKLIN'S GULL *Larus pipixcan*   GWYLAN FRANKLIN
(B) A vagrant.

A 2nd winter at Eglwys Nunydd Res. Glamorgan from Oct. 28th - Nov.1st 1998 and an adult at Blackpill July 2nd – 6th 1999 (quite possibly these 2 records relate to the same individual). These were the second & third records for Wales, the only previous record was of a second summer / adult at Aberdysynni, Meirionnydd Mar. 22nd 1986.

LITTLE GULL *Larus minutus*   GWYLAN FECHAN
(C) A fairly common, sometimes common passage migrant and winter visitor to the coast, scarce inland.

Birds in Wales states that this species was formerly rare but records have increased considerably in recent years. The table below summarises the number of individuals in each county during the period 1992-2000. The bar chart below shows the total number of individuals recorded in each month during the same period, excluding Strumble Head, Pembroke, where data is expressed for each passage period only (see the table below).

Large counts include 10 off Aberystwyth, Ceredigion on Jan. 16th 1993, 8 off Port Eynon, Glamorgan on Feb. 5th 1994, 51 off Ynyslas, Ceredigion on Jan. 21st 1997 and 65 immatures at Dwyryd / Glaslyn / Black Rock, Meirionnydd/ Caernarfon Mar. 7th – 8th 1999.

	1992	1993	1994	1995	1996	1997	1998	1999	2000
Gwent		5				3-4		8	7
Glamorgan		25	19-21	21	6	13	19	7	5
Carmarthen		2	1	4	1				4
Pembroke		25	28	29	53	42	17	38	24
Ceredigion		18	22	12	6	73	10	15	17
Brecon		6	1	6	6		12	2	6
Radnor								2*	
Meirionnydd	1		1	3			2	6	2
Caernarfon	14	2	3	5	6	9	5	65	18
Anglesey	7	6	1	7	11	3	2	2	5
Denbigh				5					
Flint						233		9	2

* Radnor/Brecon.

Data of autumn passage from Strumble Head, Pembroke not included in the above table but summarised below:

	1992	1993	1994	1995	1996	1997	1998	1999	2000
Total	6	20	29	24	35	33	8	36	21
From	Sept. 3rd	Nov. 11th	Oct. 15th	Aug. 18th	Oct. 13th	Aug. 27th	Aug. 24th	Aug. 24th	Sept. 24th
To	Dec. 5th	Dec. 9th	Dec. 31st	Nov. 6th	Nov. 16th	Dec. 27th	Oct. 25th	Dec. 12th	Dec. 14th

Monthly distribution for the month of February is exaggerated due to the large passage of 230 past Point of Air, Flint Feb. 13th – Apr. 3rd 1997, max. 62 west on Feb. 18th (large numbers were also recorded in NW England).

Birds in Wales quotes a marked spring passage, mainly in May, during the period 1973-1975. Data from the period under review suggests that this passage has ceased. This data also shows fewer numbers in November and December than mentioned in *Birds in Wales*. It has been proposed that this is due to the Irish Sea wintering population having shifted its main wintering area to the north and are therefore less likely to be storm driven in to Welsh waters during these months.

Inland Little Gulls are scarce but almost annual at Llangorse Lake, Brecon (mainly in April / May and September) with single records coming from Talybont, Brecon, Glasbury, Brecon/Radnor, Llandewi Brefi, Ceredigion / Carmarthen and at Glaslyn, Montgomery. The total number of individuals recorded at Llangorse Lake in each month during 1992-2000 is shown in the following chart.

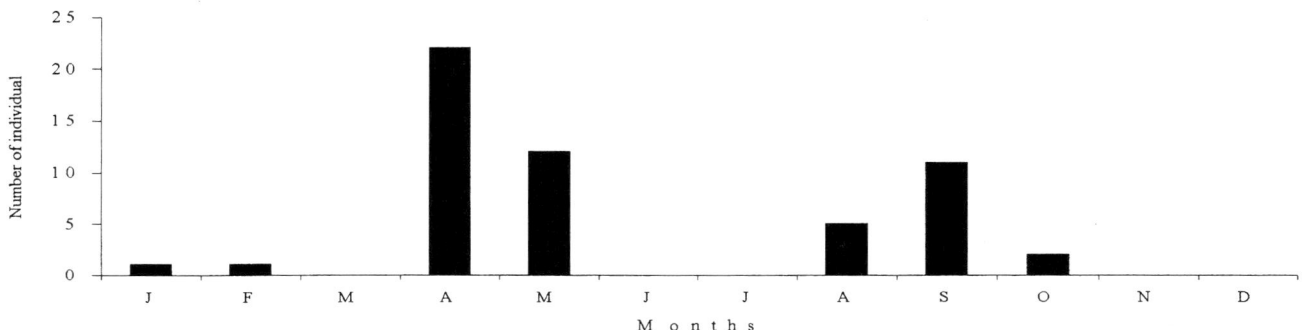

Monthly distribution of Little Gulls at Llangorse 1992-2000

SABINE'S GULL *Larus sabini*   GWYLAN SABINE
(C) An uncommon visitor to the coast in autumn, numbers vary according to the frequency of westerly gales.

In Wales this species occurs close inshore as a result of Atlantic storms as it moves south in winter. Peak numbers are in September and the first half of October. *Birds in Wales* quotes a total of 295 individuals recorded in Wales up to the end of 1991. Compare this to a total of 147 during the period of this review. This increase in records probably only reflects the increase in sea-watching and improved identification and optical aids rather than any change in occurrence. The table below summarises the total number of individuals recorded in each county during the 1992-2000. The data from Strumble Head clearly shows the importance and dedication of sea-watchers from that site.

A 1st winter was seen off Strumble Head, Pembroke Jan. 3rd 1999 – this was the 1st January record for Wales.

The chart above clearly shows the autumn passage and a significantly larger number of juveniles than adults. 1997 was the most productive year to date with a total of 56 individuals recorded in Wales. The largest total previously was 37 during 1983.

	1992	1993	1994	1995	1996	1997	1998	1999	2000
Gwent			1*						
Glamorgan		1	3*		2	1	1		
Carmarthen							1		
Pembroke - Strumble Head	5		6	8	5	42	4	7	17
Pembroke - other sites					3				
Ceredigion						2	3		
Meirionnydd						2			
Caernarfon	1				2	8	2		1
Anglesey	3		6	1	1	1	3		3

* a 1st summer was seen at Rhymney Great Wharf, Glamorgan July 14th and subsequently in Gwent, another passed the Range, Anglesey May 26th 1995.

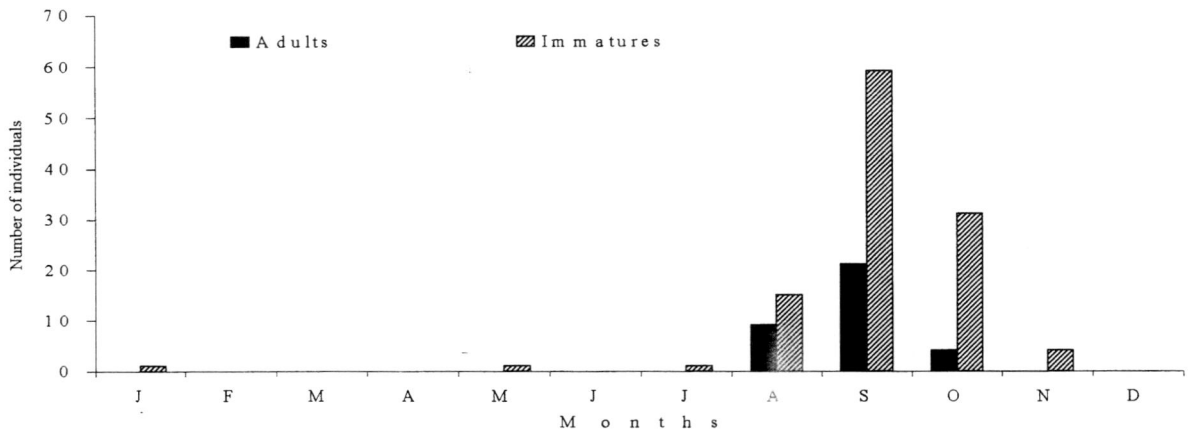

Monthly distribution of Sabine's Gulls 1992-2000

BONAPARTE'S GULL *Larus philadelphia*

GWYLAN BONAPARTE

(B) A vagrant.

Four records in this period compared to just 3 previously. The Welsh total is now 7 records of 8 birds.

A first summer at Inner Marsh Farm, Clwyd from June 1st-14th 1994, a first year at the Taff Est., Glamorgan from Apr. 15th-21st 1995, a juvenile off Strumble Head, Pembroke on Sept. 28th 1997 and finally 2 first year birds, the first multiple record in Wales, at Beddmanarch Bay, Anglesey on May 16th 1998, with one remaining on the 17th.

The 1990 record at Inner Marsh Farm, Flint although originally accepted by BBRC and published in *Birds in Wales* was later rejected on review by BBRC.

BLACK-HEADED GULL *Larus ridibundus*

GWYLAN PENDDU

(C) A locally common breeding resident, scarce only in south and south-east Wales during the summer months. A common and widespread passage migrant and winter visitor.

		/1992	92/3	93/4	94/5	95/6	96/7	97/8	98/9	99/00	00/01
Gwent	Llandegfedd Res		10000	15000							
Glam.	Blackpill	6320	5200	6460	4578	5590	5450	4500	3900	3934	5040
	Ogmore Est.						3,000	3,500	2,523	2450	500
	Burry S.	4360	1500	1838	2535	1215	1840	3930	2395	1935	370
Carm.	Burry N.		893	675	1764	661	799	270	450	175	275
Pemb.	Cleddau								4100	1524	1821
Brecon	Llangorse Lake	5-10,000	1200	3000	2000	3000	2000	2000	800	1700	

In winter the Black-headed Gull frequents a wide range of habitats, both coastal and inland. The 1973 BTO Winter Gull Roost counted a total of 10,635 for the whole of Wales. A comparable survey in January 1993 produced a total of 85,737. An 8 fold increase. The table above gives max. winter counts, November - March at the main sites, where over 1,500 were counted once.

Large roosts elsewhere: in Glamorgan: 10,000 Rhymney estuary winter 1994/5, 1,000 at Ponticill Res. in December 1996, 1,320 at Jersey Marine in December 1996; 1,000 at Llys y fran, Pembroke February 1998 with 500 in November. In 2000 there were counts of 2,000-3,000 at GLWR Golcliff, Gwent Mar. 9th, 2,220 Cardiff Bay, Glamorgan in January, 1,000 on the Teifi Est., Cere./ Pemb. in January & February and 2,045 at Alaw Est. Nov. 17th.

Birds in Wales states that passage birds are encountered from August onwards. Max. summer counts, July – October from the main sites are shown in the following table.

		1992	1993	1994	1995	1996	1997	1998	1999	2000
Glamorgan	Ogmore estuary						3000	2600		726
	Blackpill	3550	2710	3060	3370	4460	2660	3920	3047	4560
	Burry S.	6890		6744	5320	6700	6980	6040		5560
Carmarthen	Burry N.		2773	1136	1794	1785	1747	1680		2111
	T/T/G		3200	3018	1555	1823	1608	1705		853
	Upper Loughor	1152	1022	1474	987	1829	2148	1696	2138	1460

Large counts elsewhere included: 1,000 Eglwys Nunnydd Res., September 1993, 2,000 Rhymney Estuary July 1994 both Glamorgan, 2,347 on the Cleddau Est., Pembroke in September and 4,800 Point of Air, Flint August 1997.

Birds in Wales states that between 1919 and 1973, the number and size of Welsh colonies had gradually increased to a total of 64 colonies in 1973, numbering 7956-9172 pairs. Since then the Welsh population has been in decline and by 1991 the total Welsh breeding population was fewer than 4,000 pairs.

During the period of this review this decline appears to have continued. The largest figures for each county as quoted in Welsh Bird Reports are tabulated below. Although not from complete surveys and data lacking from 2 counties, what data there is suggests by the end of the century there were fewer than 3,000 pairs breeding in Wales. There are no firm reasons for this decline but drainage, agricultural improvement, acidification and possibly mink may all have had an effect.

	1973 survey		Present population estimate	
	Colonies	Pairs	Colonies	Pairs
Gwent	12	1710 - 1883	8	200
Glamorgan	None	60 prs. in 1991	2	30-80
Carmarthen	None		1	0-15
Pembroke	None		none	-
Ceredigion	13	1739-2054	8	200
Brecon	3	152-162	4-5	80-240
Radnor	9	251-265	3	300
Montgomery	12	1710-1883	5	260
Meirionnydd	6	1434-1514	?	?
Caernarfon	3	481-698	1	100
Anglesey	5	849	5	800
Denbigh	8	1181-1442	?	?
Flint	3	305	1-2	?

RING-BILLED GULL *Larus delawarensis*   *GWYLAN FODRWYBIG*

(C) *An uncommon but regular visitor to coastal areas, chiefly in the period from mid-February to the end of May.*

	1992	1993	1994	1995	1996	1997	1998	1999	2000
Gwent					2				
Glamorgan	25	8	10	12	12	10	11	12	7
Carmarthen	3		1	3-4	1				
Pembroke	3	3	3	10	3	7	6	3	1
Ceredigion				1		1			
Meirionnydd				1					
Caernarfon	3	1			2	1		2	2
Anglesey		1		1	5				1
Denbigh	1	1	1	1					
Flint						1			

Since the first British record, at Blackpill on Mar. 14th 1973, the number of records of this species has increased dramatically. With so many individuals in south Wales it is impossible to produce exact totals for each county. The table gives the total number of individuals recorded in each county during the period 1992-2000.

Birds in Wales 1992-2000

The monthly distribution of individuals in Wales, clearly shows that December – April are the best months to see this species, with a defined peak in February. The site most favoured by this species continues to be Blackpill in Glamorgan. The total number of individuals in each age group in each month during the period is plotted below. The main adult arrival is clearly in December, peaking in February, with numbers dropping through to April and few adults present during the summer months. Second years follow a similar monthly distribution to adults but are less pronounced. The same can be said for first years except that they remain longer into the summer months.

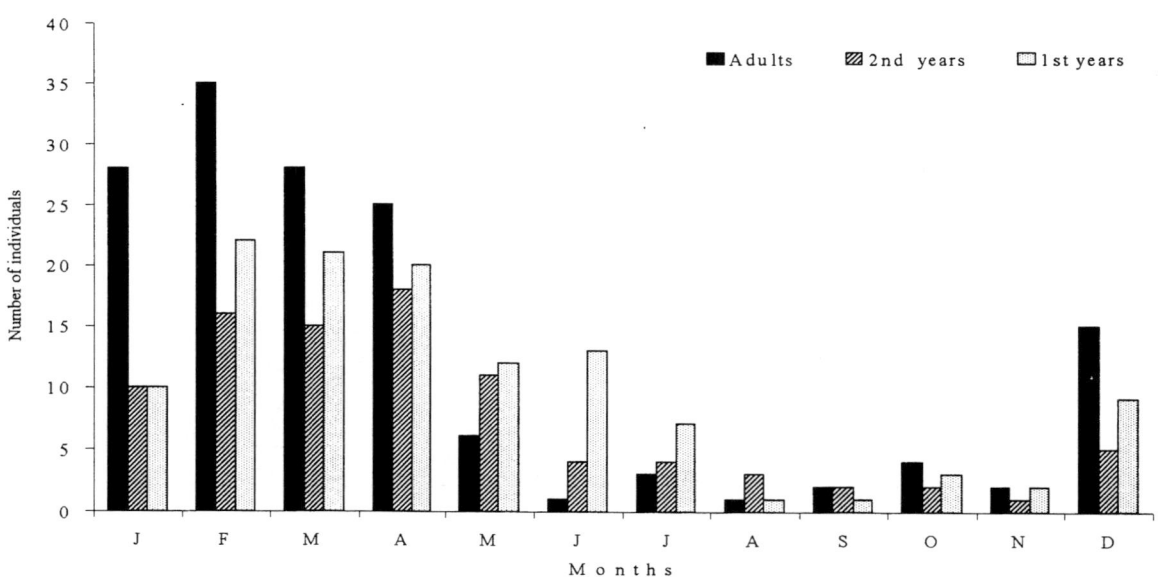

Age Distribution of Welsh Ring-billed Gulls 1992-1999

COMMON GULL *Larus canus*   *GWYLAN Y GWEUNYDD*
(C) A numerous and widespread winter visitor with small numbers recorded on passage.

	91/2	92/3	93/4	94/5	95/6	96/7	97/8	98/9	99/00	00/01
Llandegfedd Res.		2000	1500					9000		
Blackpill	2050	1800	2800	2340	3150	2290	2400	2220	1480	1022

In Wales the Common Gull's winter distribution is mainly coastal although sizeable flocks roost on inland reservoirs, particularly Llandegfedd. The table above gives max. winter counts, November - March at the main sites, Llandegfedd Res. in Gwent and Blackpill in Glamorgan, where over 1,000 counted regularly.

Large counts elsewhere included: in Gwent 2,500 at Dingestow in January 1999, 800 at Monmouth in February 1999 and 1,000 there Dec. 12th 2000; in Carmarthen 975 at T/T/G estuary in February 1994 and 3,000 on Nant y Caws land fill site, Carmarthen Jan. 23rd 1998; 1,000 at Amroth, Pembroke feeding on large shoals of small fish on Dec. 9th 1999; 2,000 Teifi Est., Cere./Pemb. in January & February 2000; 925 at Morfa Aber, Caernarfon in July 2000; 1,000 at Llandulas, Denbigh Feb. 11th and Mar. 19th 1992; 900 at Plas Uchaf Res. Denbigh Mar. 14th 1997; and in Flint 1,500 at Gronant Jan. 4th 1992 with 800 there Jan. 28th 1998, 3,500 at Bagillt Bank Feb. 6th 1995 and 1,200 at Dyserth on Jan. 16th 1999. The only site with consistently large summer / autumn counts, over 1,000, was Burry S., Glamorgan:

1992	1993	1994	1995	1996	1997	1998	1999	2000
620	1034	1201	1585	3210	1220	970	2700	5060

LESSER BLACK-BACKED GULL *Larus fuscus*   *GWYLAN GEFNDDU LEIAF*

(C) Breeding summer visitor, passage migrant and winter visitor.
The number of breeding pairs in each county during the period of this review is summarised below.

Gwent: Roof-top nesting was reported at 4 sites (3 of which in Newport) in 1993 but many more thought to breed at Ebbw Vale Steel works and Bryn Mawr. In 1997 bred at Uskmouth Power Station and 30 pairs bred at Dunlop Semtex Pool. In 1999 16 pairs bred at Dunlop Semtex and 8 at Whitehead Steelworks, Newport and in 2000 a total of 25-30 prs. bred at the two sites.

Glamorgan: 2,000 pairs bred on Flat Holm in 1992 with 2,500 pairs there in 1994. *Birds in Wales* quote 1,100 prs. in 1969 increasing to 4,055 in 1975 but decreasing since then to a low of 1,400 prs. in 1988. Figures suggest the population is increasing again. Summary of pairs at urban / industrial sites (incomplete figures): 108 in 1994, 36 in 1996 and 53 in 1998.

Pembroke:

	1992	1993	1994	1995	1996	1997	1998	1999	2000
Caldey	198	224	265	377	452	502	483	587	
St. Margarets	8	18			27		82	c50	
Skokholm	3017, 305 y	4652, 700 y	4017, 1000 y	3000, 300 y	3137, 750 y	2937, 600 y	3269, 800 y	2894, 0.14 y/pr.	2419
Skomer	16-18000, <1000 y	20200	19090,	15500, <1000 y	14400	14343	12065, 0.16 y/pr.	12028, 0.42 y/pr.	10007
Ramsey		203	294	273	413	344	262	315	
Bishop's & Clerks		52	77	65	113	92	72	72	

Single counts were: 16 prs. at Flimstone & 11 prs. Broad Haven to New Quay in 1992, with 22 prs. and 15 prs. respectively in 1993, 46 prs. Stackpole to Linney Head in 1995 (an area which includes Flimstone and Broad Haven to New Quay), 166 prs. at Middleholm and 157 on the mid & south coasts in 1998, 22 prs. at Herbrandston and 20 prs. at Grassholm in 1999. The total population for the county in 1996 was 18,882 prs, 18,632 in 1997 and 16,600 in 1988.

Birds in Wales states Skomer's population as 5,559 in 1969 increasing to 17,500 in 1988. The population peaked in 1993 at c20,000 prs. after which it has dropped substantially to 12,000 prs. The Skokholm population was put at c2,000 prs. in 1969, rising to 4,186 by 1988. As with the Skomer population, numbers increased until 1993 but have dropped since to around 3,000 mark, with minor fluctuations.

Ceredigion:

	1992	1993	1994	1995	1996	1997	1998	1999	2000
Cardigan Island	2400	3517	4700	3180	3933			2763	1654
Penrhip/Llangranog			4+	3	11				
Penmoelcilian				6	13				

One pair nested on a roof top in Aberystwyth in 1999.
The Cardigan island colony has been steadily increasing according to *Birds in Wales*, from 1,500 prs. in 1990 – although some of this increase may be due to a change in survey methods.

Brecon: a juvenile still being fed by adults at Llangorse Lake early August 1994.

Meirionnydd: 80 pairs at Llyn Trawsfynydd in 1994, 230+ adults there in 1997. Population increased from only 20 prs. in the early 1990's as stated in *Birds in Wales*. A single nest in the Herring Gull colony at Trefeddian quarry, Aberdyfi in 1996 and breeding on roof-tops in Tywyn was reported in 2000.

Caernarfon:

	1993	1994	1995	1996	1997	1998	1999	2000
Bardsey	220	402	396	420 0.56 y/pr.	375 0.49 y/pr.	233	478 123y	
Ynysoedd Gwylan		19	15		5			

Anglesey: 386 at Bodorgan Head in 1996, 173 at South Stack in 1998, 132 in 1999 – increasing since 1991 (fewer than 20 before this) and 6 at Range in 1999. Thought to be a major predator at the Cemlyn Tern colony. No data from the main colonies of Puffin Island and Skerries.
Flint: One pair bred Oakenholt 1998.

Large counts, over 1,000 away from the breeding grounds in the summer include: 1,418 at Rhaslas Pond, Gwent Aug. 25th 1997 with 1,489 there in August 1999, up to 2,000 at Rhymney Great Wharf, Glamorgan July 1994 & 2,800 there in August 1999 and 1,000 Frainslake, Pembroke Sept. 14th 1996.

The number of individuals wintering in Wales has been increasing since the 1960s. The BTO Winter Roost Count in January 1993 produced a total of 11,605 for the whole of Wales. The table below gives max. winter counts, October – March, at the main sites, those where over 700 counted once. The principle site in Wales is Llys y fran Res., in Pembroke (not too far from the main breeding islands). Numbers there topped 10,000 in 1988.

		91/2	92/3	93/4	94/5	95/6	96/7	97/8	98/9	99/00	00/01
Gwent	Llandegfedd Res		5000		1000						
	Silent Valley							1800			
	Dowlais						12000				
Glam.	Taff / Ely								1,000		900
Pembroke	Llys y fran Res.	2300	8000	7000	7500	5000		12000	3800	12000	
Brecon	Llangorse	100	1250 Sept.	660	830	860	820 Sept.	860	1280 Sept.	1100 Sept.	1050 Oct.

HERRING GULL *Larus argentatus* — *GWYLAN Y PENWAIG*

(E) *Formerly an abundant resident in coastal areas, now much reduced on the islands and a scarce breeding bird on the mainland coast. Frequent in many inland areas, breeding on several freshwater sites.*
Breeding data incomplete but breakdown of the number of pairs at some of the main sites, by county is summarised below.

Gwent: Bred on rooftops in Newport and at Ebbw Vale Steel works in 1993, single pairs in 1997 at Tintern Abbey, Dunlop Semtex and Usk. In 1999 pairs bred at Newport Steelworks, Monmouth and Llanwern and in 2000 40 prs. bred at Pirelli Newport, 10y counted at Denny Island and birds probably bred at Llanwern and Dunlop Semtex.

Glamorgan:

	1992	1993	1994	1995	1996	1997	1998	1999	2000
Flat Holm	200		300						
South coast sites		9	2-3	6-7 at 3 sites	12 prs at 5 sites	5 prs. at 3 sites		10 prs. at 2 sites	
Urban rooftops			38	12+	25	69	13+	4	6+
Port Talbot				12		64	113	74	94

Pembroke:

	1992	1993	1994	1995	1996	1997	1998	1999	2000
Caldey	876	1045	1297	1377	1457	1334	1471	1637	
Stackpole - Giltar				48*	135	131	57*	170	227
Castle Martin coast				192*	260	267	316	276	302
St. Margaret's	138	242			197	224	315	353	
Skokholm	363	450	382 (sample of 105 nests 138 y)	382 (103n – 70y)	317 1.35y/pr	336	387 0.78y/pr	330 0.87y/pr	309
Skomer	525	530	568	448	401 1.21y/pr	361 1.34y/pr	299 0.75y/pr	375	367
Ramsey	c400	148	184	167	174	195	220	250	
Bishop's & Clerks		48	92	77	203	128	134	133	

* incomplete counts.

Single counts: 75 prs. Thorn Island in 1994, 937 prs. Cemaes – Port Stinian, north coast in 1997, 218 prs. on the Angle peninsula, 170 Barafundle to Gilton and 33 at Grassholm in 1999. County wide survey in 1997 found 4425 prs. at 11 sites. A small number breed on roof tops in Tenby.

Birds in Wales stated the Skomer population to be 660 prs. in 1945, increasing to 2,350 prs. in 1979 followed by a decrease to 1,409 prs. in 1981 and an increase to 8,000 prs. The population decreased from then on mainly due to a large botulism outbreak. On Skokholm a similar story, 270 prs. in 1934, increasing to 1,400 prs. by 1973. Since then numbers have gone done to around the 300-400 mark, with the exception of 1993. On Skomer the population continued to decrease, possibly due to food competition with the Lesser-black Backs.

On Caldey the population was put at 3,857 prs. in 1975 which crashed to a mere 802 prs. by 1991. Since then its population has almost doubled to the 1999 figure of 1,637 prs. On Ramsey there has been a steady decrease since 1,003 prs. in 1973 although numbers increased towards the end of the 1990s.

Ceredigion:

	1992	1993	1994	1995	1996	2000
Cardigan Island	800-900	400			720	60
Penmoelcilian					515	
Graig Du Aberaeron				12-15		
Complete survey					1,876	1,303

Roof top nesting recorded in Aberystwyth in 1995 and 1998. In 1999, 50 pairs nested on Aberystwyth rooftops (41 prs. in 2000), 4 at Borth and 85 prs. on the coast between Borth and Llanrhystud.

Meirionnydd: 30 pairs at Trefeddian quarry, Aberdyfi and 152 at Tywyn in 1996. In 2000, 31 prs. bred at Aberdyfi, 249 at Tywyn and 52 at Barmouth Quarry.

Caernarfon:

	1993	1994	1995	1996	1997	1998	1999	2000
Bardsey	300	371	426	494	502	522	508, 485y	534, 597y
Ynysoedd Gwylan	270	271	231		295			
Conwy RSPB					3 prs., 6y		2	1

300 individuals noted Great Orme in July 1998. 2 pairs bred at Manweb Caernarfon in 1999. 2-3 prs. on Caernarfon roof-tops in 2000.

Anglesey: 485 on South Stack and 58 on the Range in 1999. Bred at Cemlyn, Llyn Eli in 2000.

Flint: One pair nested BHP's Point of Air terminal in 1997.

In 1993 the BTO Winter Roost Survey counted 29,065 for the whole of Wales. The table below gives max. winter counts, October - March at the main sites, where over 2000 counted once. Other large counts were: 2,000 at Bryn Pica Landfill sites, Glamorgan Dec. 17th 2000, 7,500 off Amroth, Pembroke feeding on shoals of small fish in December 1999 and 2,800 at Bangor Hills, Caernarfon Feb. 2nd 2000.

		91/2	92/3	93/4	94/5	95/6	96/7	97/8	98/9	99/00	00/01
Glamorgan	Burry S.	2520	1190	1860	470	975	1085	995	6385	1210	
	Rhymney				2500					2000	
	Ogmore								650	3252	
	Blackpill	580	520	346	606	640	569	1246	1316	2801	2798
Carmarthen	Pendine				5000						
	Burry N.								6700		
Denbigh	Kinmel Bay								5000	1800	
	Pensarn								2200	1550	
	Gronant							2500			

Large counts, of over 1,000 during the summer months, away from the breeding colonies included: in Glamorgan: 1,000 Whiteford Point in June 1993, 2,000 on the Rhymney Estuary in July 1994, 1,640 Burry S. June 1995 with 1,275 there in August 1996, 1,330 in June 1997, 1,045 in May 1998, 2,390 in September 1999 and 4790 in July 2000, 4,000 at Ogmore est. in September 1999 and 1774 at Blackpill in August 1999 & 1,636 in July 2000; in Carmarthen: 15,000 Cefn Sidan July 20th 1994, at T/T/G 1,880 in August 1996 and 1,259 in August 1998; 1,800 at Dysynni, Meirionnydd in September 1997; 1,700 at Briant Est. Anglesey June 3rd (all 1st & 2nd summer individuals).

YELLOW-LEGGED GULL *Larus cachinnans* — *GWYLAN GOESMELYN*

(C) Unusual visitor, chiefly to the coast.

Although BOU have not separated this sub-species from Herring Gull, there is the possibility that it will be given full species status in the future. *Birds in Wales* quote no records until 1983, since then numbers of records continue to increase in south Wales, with a distinct peak in late summer to winter. The number of records are probably correlated to the amount of time spent examining gull flocks. The main site for this gull is Blackpill in Glamorgan.

Pair RSPB Conwy may have attempted to nest, Feb 27th – Aug. 30th 1998. A breakdown of individuals by county:

	1992	1993	1994	1995	1996	1997	1998	1999	2000
Gwent	3	3	3	7	5	7	9	12	
Glamorgan	7	1	12	11	12-14	20	14	16	22
Carmarthen	3-4	4	5	5			1		2
Pembroke		2	5	6		3	1	3	2
Ceredigion	1								
Brecon					2	1	2	3	1
Caernarfon	1	1		1	1	4	6	7	1
Anglesey	1		1	1					
Denbigh			1						
Flint							3	4-5	

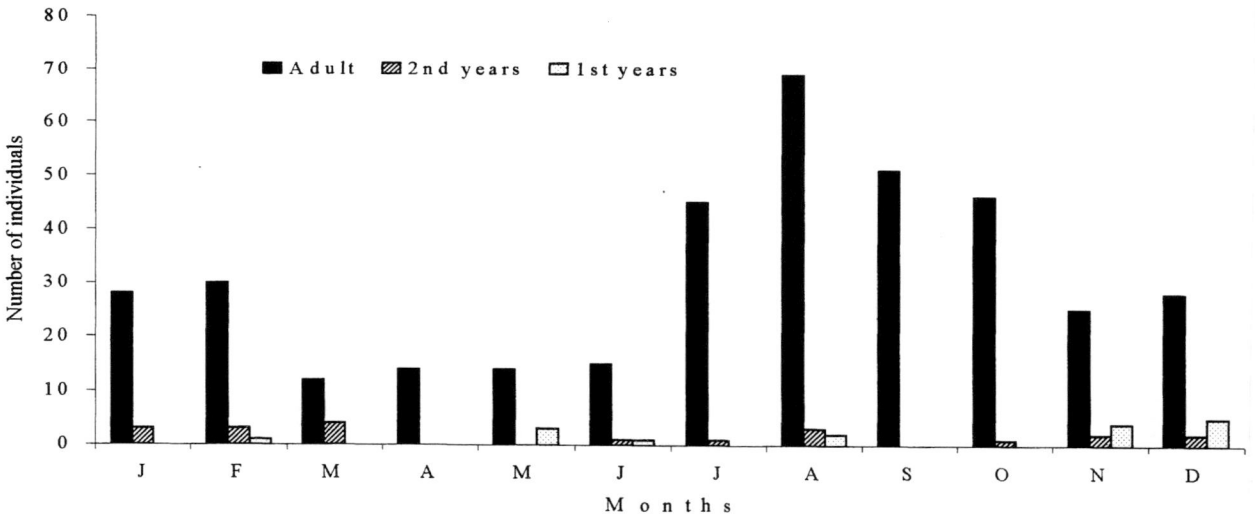

Age Distribution of Yellow-legged Gulls 1992 - 2000

Birds in Wales 1992-2000

ICELAND GULL *Larus glaucoides*   *GWYLAN YR ARCTIG*

(C) An uncommon and irregular winter visitor to coastal areas, formerly rare.

Up to 1983 this white-winged gull was rare in Wales. There were significant influxes during 1983 & 1984 (when 40+ individuals were recorded) and since then this species has been recorded annually, principally in the southern counties. As yet it is still to be recorded from the inland counties of Brecon, Radnor and Montgomery. The table below shows the number of individuals recorded in each county during the period 1992-2000, while the total number of individuals in each age group is shown in the chart.

There were two accepted individuals of the race *kumleini* recorded, the first & second records for Wales. The first a first winter at Llys y fran Res., Pembroke Feb. 1st – 5th 1998, the other an adult at Blackpill Dec. 15th 1998 until Mar. 2nd 1999, then assumed the same returning individual there Sept. 9th until Feb. 20th 2000, returning again Nov. 24th – Dec. 11th.

	1992	1993	1994	1995	1996	1997	1998	1999	2000
Gwent		1	2				3		
Glamorgan	1	1	7	3+	4	10	9	11	10
Carmarthen			1	1		1-2	5		
Pembroke	1	5		4	1	1	2	4	6
Ceredigion		2		1			3	3	4
Caernarfon			1			2	5	3	2
Anglesey		2	1			1		1	1
Denbigh	1		2	2					
Flint	1			1	1	1			

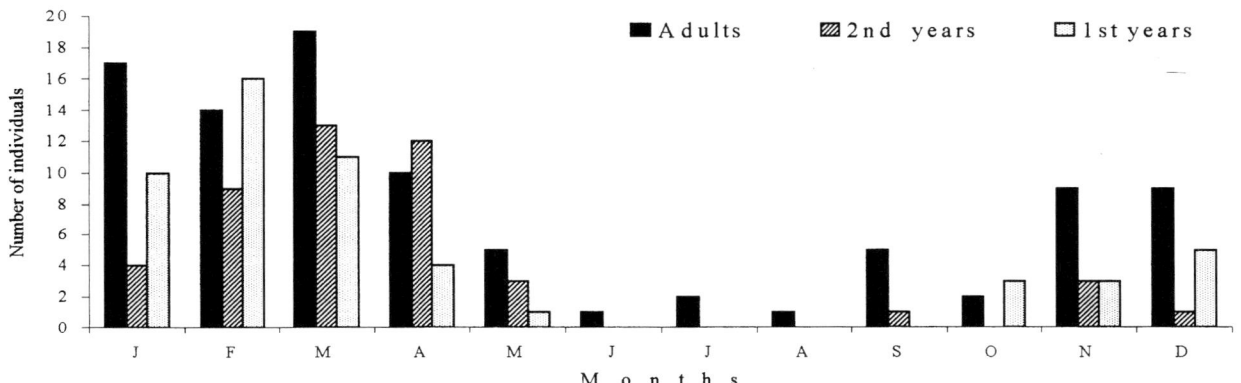

Age Distribution of Iceland Gulls 1992-2000

GLAUCOUS GULL *Larus hyperboreus*   *GWYLAN Y GOGLEDD*

(C) An uncommon but regular visitor to coastal areas of Wales, chiefly in winter but occasionally in summer.

In general this species is more likely to be encountered in Wales than Iceland Gull. Most records are from the coast or estuaries and as with the latter, it is still to be recorded in Brecon and Radnor. The only record so far from Montgomery was of a 2nd winter at the top of the Dyfi estuary in October 2000. The table below summarises the number of individuals recorded in each county during 1992-2000.

The monthly distribution of different aged individuals plotted in the following chart, clearly shows that the best month to see this species in January – March. Several individuals have summered.

Birds in Wales 1992-2000

	1992	1993	1994	1995	1996	1997	1998	1999	2000
Gwent	1	1				1		1	
Glamorgan	4	3	10	5	2-3	5	4	6	3
Carmarthen			1		1		1		
Pembroke	2	1	4	3	2	3	6	4	2
Ceredigion	5-6	4		1	1	1-2	2	4	1
Montgomery									1
Meirionnydd	1						1		
Caernarfon	1-3				3	1	2	1	1
Anglesey		2	2	1				1	1
Denbigh	5	3	3	9					
Flint						1	3	1	1

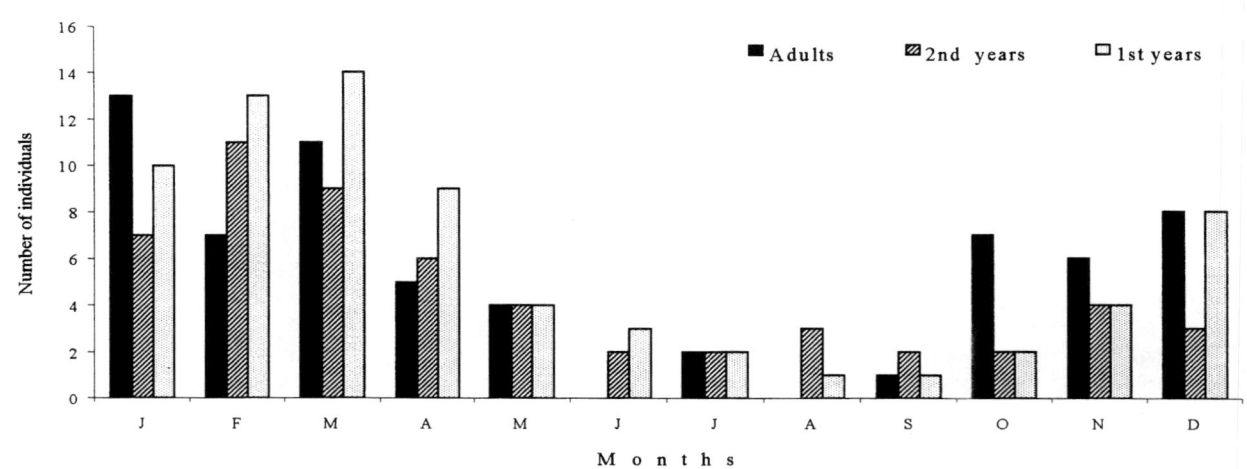

Age Distribution of Glaucous Gulls 1992-2000

GREAT BLACK-BACKED GULL *Larus marinus* GWYLAN GEFNDDU FWYAF
(C) Common and widespread in coastal areas in winter; locally an uncommon breeding bird.
Breeding data incomplete but breakdown of the number of pairs at some of the main sites, by county:

Gwent: 23 nests Denny island in 1993.

Pembroke, main sites (over 20 pairs):

	1992	1993	1994	1995	1996	1997	1998	1999	2000
St. Margaret's	46	31			48	69	80	65	
Skokholm	25	23	32	37 33y	37 52y	39 59y	46 1.1y/pr.	49	53 1.13y/pr.
Skomer	40	53	68	64	58 1.14y/pr.	54 1.14y/pr.	53 1.1y/pr.	65 0.96y/pr	61 1.52 y/pr.
Middleholm		31			34				
Bishop's & Clerks		12		31	32	34	42	34	

County total put at 321 prs. in 1997, with only 9 prs. on the south coast in 1998.

Birds in Wales states that the Skomer population has been decreasing since 235 prs. in 1962 (intensive control measures killed over 800 adults in 1960's & 1970's). The population now appears fairly stable in the region of 50-70 prs. On Skokholm the population was at its highest in 1949 when there were 72 prs. Numbers have decreased since then due to control measures but over the last 8 years the population is increasing and has almost doubled. The St. Margaret's population was put at 169 in 1969, it decreased dramatically to 58 prs. (Seabird register of 1985-87) and has been fluctuating since around the 65 prs. mark.

Ceredigion: 15 nests along the coast in 1996. A pair bred on the chimney stack of the Old College, Aberystwyth in 1998 – the first record of roof-top nesting so far in Wales. 12 prs. Cardigan Island and 4 prs. on the mainland in 2000.

Caernarfon:

	1994	1995	1996	1997	1998	1999	2000
Bardsey	2	1 pr., 1y	2 pairs, 3y	3 prs., 6y	2	3 prs., 5y	3 prs., 7y
Ynysoedd Gwylan	51	50		51			

Anglesey: Single pairs nested at Llyn Alaw and Valley lakes in 1997, total of 4 pairs in 1998. In 1999 1 pair at Ynys Llanddwyn, 2 prs. at South Stack and one at the Range.

Peak counts during summer / autumn, June – November, at Blackpill & Burry S. in Glamorgan and T/T/G in Carmarthen:

	1993	1994	1995	1996	1997	1998	1999	2000
Blackpill	112	116	106	117	379	81	106	54
Burry S.	295	177	185	675	295	180	160	115
T/T/G	142	134	212	110	109	109	78	33

Other large counts were of 130 at West Helwick, Glamorgan Oct. 24th 2000, 200 on Skomer, Sept. 11th 2000, 155 on the Dyfi, Ceredigion/Meirionnydd in July 1996, 360 in Bangor Harbour, Caernarfon Feb. 2nd 2000, 120 at Pensarn/Abergele/Towyn, Denbigh on Oct. 6th 1999, 200 at Rhyl/Kinmel, Denbigh on Oct. 11th 1999 and 132 at Point of Air, Flint in October 1997.

ROSS'S GULL *Rhodostethia rosea*   GWYLAN ROSS
(B) A vagrant.
An adult at Aberystwyth, Ceredigion on Dec.30th 1994, followed by presumably the same bird at Porthmadog, Caernarfon from Jan. 14th-16th and 22nd 1995. The second record for Wales, the other was an adult in Fishguard Harbour, Pembroke in February 1981.

KITTIWAKE *Rissa tridactyla*   GWYLAN GOESDDU
(C) A breeding species in coastal areas from Glamorgan to Caernarfon, except Meirionnydd. Large numbers occur offshore on passage.
Birds in Wales quotes 13 colonies in 1959, with an additional 5 by 1991. Since then only one new colony at Mumbles Pier, Glamorgan. Breeding data incomplete but breakdown of the number of pairs at some of the main sites, by county:

Glamorgan:

	1992	1993	1994	1995	1996	1997	1998	1999	2000
Devils Truck	100*	40	4+	28	29	38	10	27	26
Thurba Head			60, 26y						
Worms Head				60 adults	128			94	
Mumbles Pier			54	70	85	92	88	81	169, 59y

*Devil's Truck colony of 100 pairs in May 1992 was deserted by July – observation proved that it was not due to the presence of Peregrines.
Birds in Wales quotes the Worm's Head colony as 114 prs. in 1991, a slight decrease since then, presumably some of these birds have moved to Mumbles Pier.

Pembroke:

	1992	1993	1994	1995	1996	1997	1998	1999	2000
Castle Martin coast	427 0.19y/pr	394 0.44y/pr	403 0.56y/pr	310 0.36y/pr	278 0.23y/pr	204 0.41y/pr	153 0.31y/pr	147 0.85y/pr	148 0.9y/pr
St. Margaret's		170		162	123*	77	38	39	
Skomer	2172 0.47y/pr	2478 0.65y/pr	2512 0.9y/pr.	2295 0.94y/pr.	2262 0.45y/pr.	1959 0.68y/pr	2092 0.8y/pr.	2156 0.95y/pr.	2257 0.78y/pr
Grassholm	50	50			39	25	22	22	
Ramsey	361	416 1.16 y/pr.	468 70% hatched 1.56 y/pr.	489 77% hatched 1.28 y/pr.	341 40% had young	474 0.85y/pr	450 0.9 y/pr.	471	459

Of the 123 pairs on St. Margaret's island in 1996 only 2 eggs were known to hatch and no young were fledged. Similarly of the 39 pairs on Grassholm in 1996 only one young was seen. All the 25 pairs on Grassholm in 1997 failed.

A complete census of the county produced figures of 3,043 prs. in 1996 and 2,739 prs. in the following year.

In 1999, 147 counted at Elegug stacks and the largest breeding numbers on Skomer for a decade but the lowest numbers at St. Margaret's in 30 years. The breeding successes of all the colonies along the south coast has been poor for many years. It is thought that the Skomer population could be a "sink" for the population as a whole, as this species is the most mobile of the gulls.

The Seabird Register of 1989 put the county total at 9,120 prs in 20 colonies (an increase of 29% since 1979). On Skomer the population had built up from 1,500 prs. in 1945 to 2,466 prs. by 1991. A slight decrease since then, with around 2150 prs. annually since 1995. The colony at Stack Rocks (Elegug stacks, part of the Castle Martin peninsula between Stackpole and Linney Head) had increased from 87 prs. in 1959 to 486 prs. by 1990, which then decreased by 25% by 1993. Since then the population effectively crashed.

Ceredigion:

	1992	1993	1994	1995	1996	1997	1998	2000
Cardigan Island	3*	1	0	0				
Ynys Lochtyn	struggling	declining		12	13	1	12	
New Quay Head		250+		467	477	306	395	375

Caernarfon:

	1993	1994	1995	1996	1997	1998	1999	2000
Bardsey	200	199	220	198 / 1.2 y/pr.	205 / 0.78 y/pr.	149 / 111y.	243 / 304y	278 / 312y
Great Orme	1030	1015	1210	1448	1258	No count	652	1147
Little Orme	687	856	734	401	711	299	661	582

The 1999 figure on Bardsey was the best number in 20 years. Historical figures for the Great Orme show fluctuations between a low of 70 and a high of 1400 prs. with a complete cycle of increase followed by decrease of 20 years [1399 prs. in 1989 followed by the usual decline to the 1999 figure].

The major colony at Carreg y Llam has not been counted recently.

Anglesey: Only 8 nests on South Stack in 1998, only one in 1999 - a dramatic decline from the 246 nests in 1992. 28 adults seen on ledges at Ynys Moelfre in 1999.

Large passage noted in 1999, 6,000 off Bardsey on Nov. 2nd, 3,000 off Point Lynas, Anglesey on the 6th, 10,000 in 4 hours off Strumble Head, Pembroke on Dec. 3rd and 12,000 passed there on Oct. 31st 2000.

Inland there were numerous records: a large scale "wreck" in the beginning of 1993 brought several birds to Glamorgan, Ceredigion, Meirionnydd, others at Llandegfedd Res., Gwent in January & February and to Talybont Res., Brecon in January.

Otherwise singles at: Llandegfedd Res. July 14th 1996, Mar. 18th and Aug. 4th – 12th 1997; in Brecon one picked up Oct. 27th 1992 and released at Llangorse Lake, another there Apr. 5th 1995; Llanddewi Brefi, Ceredigion Mar. 28th 1995.

IVORY GULL *Pagophila eburnea*

(B) A vagrant.

A first winter at Aberthaw, Glamorgan on Jan. 2nd 1998. *Birds in Wales* quote 6 possible records, all bar the most recent one, a 1st winter at Burry Port, Carmarthen Oct. 10th – 12th 1988, are insufficiently documented as to establish their true identity and may in fact refer to other white-winged gull species. The above Aberthaw record is therefore only the 2nd for Wales.

GWYLAN IFORI

GULL-BILLED TERN *Gelochelidon nilotica*

(B) A vagrant.

One at Penrice, Hangman's Corner and Knelston, Glamorgan from July 4th-5th 1993, one at Penclacwydd WWT and Machynys Pond, Carmarthen from July 4th-Aug. 17th 1996 which was also seen on the Glamorgan side of the estuary on July 6th and at Whiteford Sands on Sept. 14th. This bird was later seen at Tacumshin, Wexford, Ireland. These constitute the 9th and 10th records for Wales.

MORWENNOL YLFINBRAFF

CASPIAN TERN *Sterna caspia*   *MORWENNOL FWYAF*
(B) A vagrant.
One at Kenfig Pool, Glamorgan on Apr. 11th 1989, one on Skomer, Pembroke on May 28th 1994, another at Kenfig, Glamorgan on Aug. 6th 1997 (also seen at Eglwys Nunnydd Res. on the 8th), and one on Bardsey, Caernarfon on May 28th and June 1st 1998. The 6th to 9th records for Wales.

SANDWICH TERN *Sterna sandvicensis*   *MORWENNOL BIGDDU*
(C) A summer visitor and passage migrant to coastal areas in spring and autumn, scarce inland; it is a regular breeder on Anglesey.
There is only one breeding colony in Wales, on Anglesey. Numbers there have been consistently within the range of 450-1000 pairs since 1985. The table below summarises the number of pairs and the number of young fledged at this site, during 1992-2000.

	1992	1993	1994	1995	1996	1997	1998	1999	2000
Pairs	500	564	400+	650	650-700	450	460	604	450
Young fledged		168	350-400	506	770	600+	450	550	283

Inland records are unusual and during the period under review individuals were seen at: Llandegfedd Res., Gwent Aug. 7th and 27th with 2 on the 10th 1997; Llanishen / Lisvane Res., Glamorgan Aug. 10th 1997; at Llangorse, Brecon: single Apr. 29th 1995, 2 on June 1st 1996, 2 on May 27th 1997, 2 on May 11th 1998 and in 2000, 2 on Apr. 17th – 18th and a single Sept. 17th; 2 Elan Valley, Radnor on Apr. 10th 1999; at Dolydd Hafren, Montgomery 2 on Apr. 24th 1995. Those in Radnor and Montgomery were the first county records.

The only large record of spring passage was of 500 at Dwyryd / Glaslyn estuary, Meirionnydd Apr. 25th – 26th 1995.

Birds in Wales states that the summer influx takes place from mid July onwards, peaking in late July to early September. The main localities are the various estuaries of north Cardigan Bay, the Dee and Burry Inlet. These are of considerable importance as nursery areas. The table below summarises the max. autumn passage counts at these sites.

Other large counts of passage birds included: a mixed flock numbering 2,500 Sandwich/Common/Little were off Point of Air Aug. 27th 1997. 480 were counted at Morfa Madryn & Morfa Aber on Aug. 26th 2000. 248 past Bardsey, Caernarfon on Aug. 24th 1998 and 160 on Sept. 15th 2000. A total of 673 were logged passing Strumble Head, Pembroke autumn 1998, max. 79 on Sept. 11th, 1183 in autumn 1999, max. 131 on Sept. 30th and 1,059 in autumn 2000, max. 106 on Aug. 28th.

Winter records are very rare. During the period under review there were singles at: Port Eynon, Glamorgan Nov. 22nd 1997, Pwlhelli, Caernarfon Jan. 26th 1998 and from Anglesey at Beddmanarch Bay Feb. 9th – 10th 1996, at Beaumaris Jan. 18th 1997, at Penmon, Anglesey Nov. 2nd 1999, with 3 at Llanfairfechan, Caernarfon on Nov. 3rd 1999.

	1992	1993	1994	1995	1996	1997	1998	1999	2000
Burry Port, Carmarthen				52 Sept. 28th		158 Aug. 29th	120 August		88 Aug. 22nd
Ynyslas, Ceredigion	<50	415 Sept. 2nd	585 Aug. 6th	694 Aug. 31st	362 Sept. 3rd	190 July 19th	200 Aug. 30th	550 Aug. 31st	140 Sept. 3rd
Dysynni, Meirionnydd		200 Sept.	150 August	60 September	100 Sept.	150 September	200 July	200 Sept. 11th	
Mawddach, Meirionnydd			70 Sept.	90 August	150 August	110 August			
Point of Air, Flint	1,000 July 20th	940 August		605 July 31st	900 Aug. 14th	920 July 27th	1,050 July 21st	300 July 31st	300 Sept. 2nd

Birds in Wales 1992-2000

ROSEATE TERN *Sterna dougallii*   *MORWENNOL WRIDOG*

(E) Confined as a breeding species to Anglesey, where there is less than 10 pairs; elsewhere it is occasionally recorded on spring and autumn passage along the coasts.

	1992	1993	1994	1995	1996	1997	1998	1999	2000
Colonies	1	2	2	1	1	2	2	1	1
Pairs	7	21	20	10	1	3-4	6	3	2
Young fledged	10	36	21 at one site	8	1	2**	1	2	0

This species is only just maintaining its foothold in Wales. It is sad to think that only 30 years ago, in 1969, there were 202 prs. breeding in Wales and that this level of population (130-251 prs.) was maintained until 1986. The table above summarises the number of colonies and productivity in Wales, during the period. Birds present at another site in 1995 but did not breed, a single individual was also at Cemlyn, Anglesey in May 2000. In 1997 one bred with a Common Tern, producing one hybrid offspring.

Roseate Terns are recorded on passage in very small numbers, mainly from coastal sites although during the period under review there were 3 inland records, all from Llangorse Lake, Brecon, of 3 on May 16th 1993 and singles May 7th & 10th 1999 (only 3 previous records). The total number of individuals recorded in each month during the period is shown in the following chart.

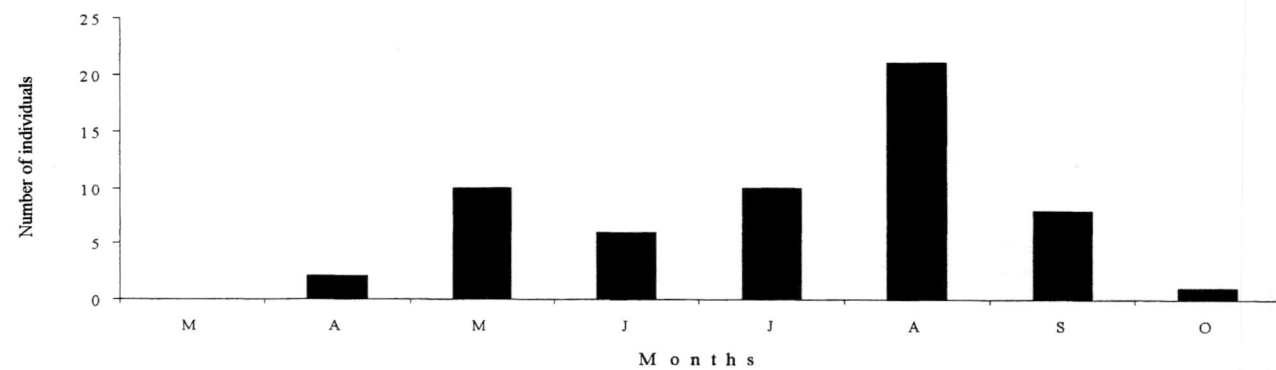

Monthly distribution of passage Roseate Terns 1992-2000

COMMON TERN *Sterna hirundo*   *MORWENNOL GYFFREDIN*

(C) A summer visitor which breeds on Anglesey and Flint. It is a spring and autumn passage migrant to all coastal areas and is the commonest of the Sterna species inland.

Birds in Wales states that this species has increased from 302 prs. in 1969 to 470's in 1978-1991. This increase has continued although the actual number of colonies has decreased. The table summarises the number of colonies and productivity in Wales, during the period:

	1992	1993	1994	1995	1996	1997	1998	1999	2000
Colonies	7	6	5	6	7	6	5	4	4
Pairs	485	403-427	473-478	593	454-474	553	553	638-757	638
Young fledged	572 at 5 sites of 365 prs.	475-500	516-536			555	828	291-321 at 3 sites of 205 prs.	1.3y/pr.

It must be noted that the above data for 1994 & 1999 are incomplete since the 2 species, this and Arctic, were not differentiated at some sites.
In 1995, 421 pairs at Shotton reared 400+ young but of the 124 prs. in 3 colonies on Anglesey only 63 young were reared, with one colony deserted by mid-May.
In 1996 240-260 pairs at Shotton reared 567 young, the best productivity since 1992 but the 6 colonies on Anglesey only reared 154 young (predation, poor weather and egg collectors took their toll).
330 pairs at Shotton in 1997 reared 403 young, while the 5 colonies totaling 223 prs. on Anglesey reared 152. In 1998 there were 380 prs. at Shotton – 585 young, a 6th colony completely failed before census.

In 1999 there were 433 – 552 prs. at Shotton, where 1045 eggs were laid but many failed at the early young stage / late incubation due to the poor June weather. Many re-laid but the offspring were very thin and most did not survive.
Interesting records include 3 at Kilpaison, Pembroke Mar. 30th 1996, with Arctics off Aberystwyth Oct. 29th & 30th 1996, a late bird at Machroes, Caernarfon Nov. 8th 1998.
Large counts include 305 Point of Air on July 22nd and 530 there on Aug. 8th 1998 and a count of 672 on the Dee Webs count in August 2000.
The monthly totals of Common and Arctic Terns logged passing through Llangorse Lake, Brecon are plotted in the following chart, clearly showing the larger totals of the former species.

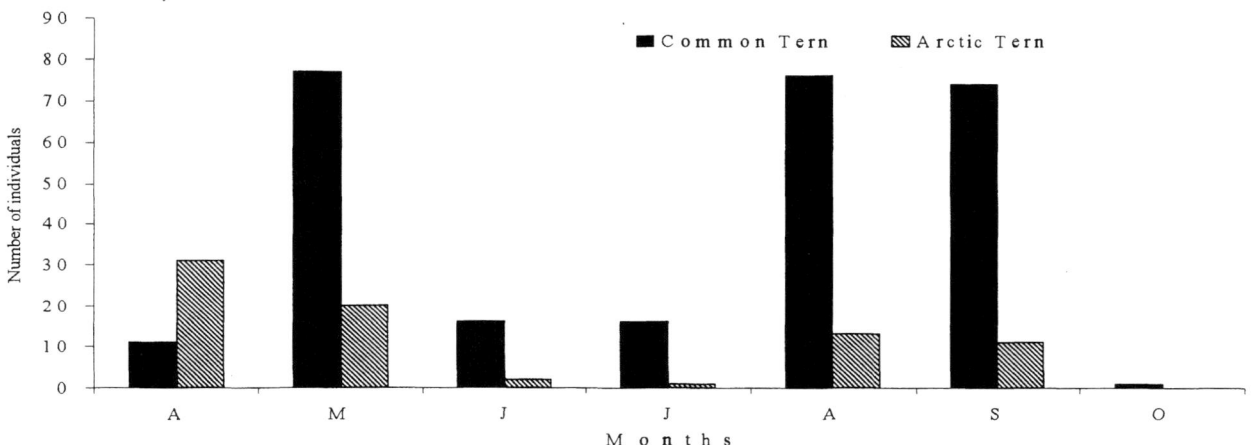

Monthly distribution of Common and Arctic Terns at Llangorse Lake 1992-2000

ARCTIC TERN *Sterna paradisaea* MORWENNOL Y GOGLEDD

(C) The commonest of the tern species breeding in Wales. It is confined to the Anglesey coast for breeding. Away from there, it is a regular passage migrant to coastal areas, generally scarcer than Common. Scarce inland.

	1992	1993	1994	1995	1996	1997	1998	1999	2000
Colonies	4	4	3	3	3	4	4	3	3
Pairs	1,113	934-957	1,073-1078	1,277	1,202	1,257	1,305	1430	1705
Young fledged	335 from 3, holding 813 prs.	765-815	1,030-1,080	1,050	1,200	1,966	925-1,800	1828-1958	1.3-1.5y/pr.

According to *Birds in Wales* the Welsh breeding population had increased from 436 pairs in 5 colonies in 1969 to an average of 889 pairs in 3 colonies during 1988-1992. Figures from the period under review clearly show that this increase has continued. The table above summarises the number of colonies and productivity in Wales, during the period 1992-2000.

Large numbers were recorded during passage: 1,100 past Sker Point, Glamorgan in 30 minutes. Apr. 26th 1995 and 150 at Cefn Sidan, Carmarthen May 12th 1998.
Late records were from Aberystwyth, Ceredigion Oct. 27th and Bardsey, Caernarfon Oct. 31st and Nov. 3rd, all 1996. A winter record of a single at Sandy Water Park, Carmarthen Jan. 5th 1992 is very unusual.

FORSTER'S TERN *Sterna forsteri*   *MORWENNOL FORSTER*
(B) A vagrant.
During the period under review there were 3 accepted records. A 1st winter at Fishguard, Pembroke on Jan 10th - 11th 1994, a single in Bangor Harbour, Caernarfon from Jan 20th-24th 1995 then moved to Caernarfon and Foryd Bay from Feb. 3rd - 9th and a 2nd winter also in Bangor Harbour Dec. 2nd – 18th 2000. Records difficult to assess due to returning wintering individuals, but likely that as few as five birds involved in the records since the first in Wales in 1984.

LITTLE TERN *Sterna albifrons*   *MORWENNOL FECHAN*
(C) A summer resident which breeds only at one colony in Flint, otherwise a spring and autumn passage migrant to other coastal areas, rare inland.

	1992	1993	1994	1995	1996	1997	1998	1999	2000
Pairs	52	45	77	65	78	80+	35-40	86	75
Young fledged	100+	45	120-140	35-40	120-140	9	20	111	57

The Gronant colony has been wardened by the RSPB since 1975, when there were only 15 prs. The colony continues to rise although there was a large decrease in 1998. In 1993 foxes predated 25 nests despite electric fences. 1996 was the most successful yet but heavy rain, high tides and foxes had drastic effects in 1997 & 1998. The table above summarises the number of breeding pairs and productivity in Wales, during the period 1992-2000.

This species is sometimes recorded on passage: total of 83 in Pembroke in 1992 of which 15 past Skomer on Sept. 3rd, 19 on the 4th and 23 on the 7th, a single there Oct. 26th 1996. Large counts included 350+ at Gronant July 15th 1996, 175 at Point of Air Aug. 8th 1998 and 111 on the Dee Webs count in June 2000, all Flint.

There have been very few inland records. During the period under review there were singles at Llangorse Lake, Brecon Apr. 22nd & 28th 1994, Apr. 17th & May 19th 1996, Aug. 30th 1997, 4 there on May 16th 1998 and singles May 11th 1999 and Sept. 15th 2000. Singles at Llyn Coed y Dinas, Montgomery Apr. 14th 1996 and Llyn Helyg, Flint May 24th 1998.

WHISKERED TERN *Chlidonias hybridus*   *CORSWENNOL FARFOG*
(B) A vagrant.
One at Llyn Traffwll, Anglesey on June 18th 1993 moved to Cemlyn Bay from 19th-23rd, and one at Llandegfedd Res., Gwent from July 15th-28th 1994 were the only records during the period under review. These constitute the 6th and 7th records in Wales, ending a run of 16 blank years.

The previous 5 records were from Llan Bwlch-Llyn, Radnor in April 1956, Glandyfi, Ceredigion in July 1965, 2 at Blackpill, Glamorgan in May 1974, at Eglwys Nunnydd Res., Glamorgan in September 1974 and at Shotton, Flint in July 1976.

BLACK TERN *Chlidonias niger*   *CORSWENNOL DDU*
(C) A fairly common passage migrant, chiefly to coastal counties in autumn but also recorded on inland waters; spring records are usually fewer in number but there are substantial influxes in some years.

Black Terns are now a regular occurrence in singles and small groups, especially in the southern counties, with widespread movements in some years. The map opposite summarises the total number of individuals recorded in each county during 1992-2000. The sea-watching headlands of Pembroke and Anglesey account for the majority of records. The high total in the former is exaggerated by the total of 335 individuals recorded in 1997.

There are two distinct passage periods, spring passage (April – June) and more pronounced autumn passage (July – November). The following tables for spring and summer summarise passage periods and the number of individuals involved in each year during the period. The total number of individuals recorded in each month during 1992-2000 is shown in the chart below.

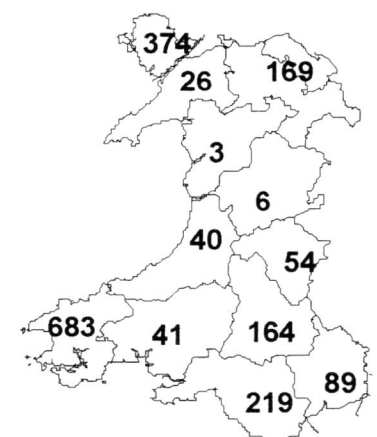

Spring passage:

	1992	1993	1994	1995	1996	1997	1998	1999	2000
From	Apr. 20th	Apr. 29th	Apr. 24th	Apr. 11th	Apr. 23rd	May 3rd	Apr. 5th	Apr. 21st	Apr. 27th
To	May 18th	June 10th	May 31st	May 23rd	May 19th	June 6th	May 14th	Apr. 29th	May 7th
Number	26	64	24	8	16	17	12	3	71

Max. counts: 31 at Kenfig, Glamorgan at dusk May 12th 1993 and 26 at Sker May 7th 2000.

Autumn passage:

	1992	1993	1994	1995	1996	1997	1998	1999	2000
From	Aug. 7th	Aug. 18th	July 26th	June 26th	Aug. 1st	Aug. 5th	July 30th	July 21st	Aug. 27th
To	Oct. 2nd	Sept. 18th	Oct	Oct. 13th	Nov. 11th	Sept. 20th	Sept. 20th	Oct. 11th	Oct. 23rd
Number	232	34	95	88	59	477	106	194	306

Birds in Wales 1992-2000

The majority of autumn passage birds were recorded passing of the watch points of Strumble Head, Point Lynas and Point of Air. Maximum counts from these sites include: 58 Strumble Head, Pembroke Sept. 11th 1992, total of 327 logged on 15 dates Aug. 17th – Sept. 20th 1997 of which 238 were Aug. 27th – 29th, a total of 147 Aug. 15th – Oct. 11th, max. 51 on Aug. 25th 1999 and a total of 87 Aug. 27th – Oct. 23rd 2000, max. 45 on Sept. 24th; a total of 143 logged passing Point Lynas, Anglesey Sept. 2nd – Oct. 12th 2000, max. 70 on Sept. 20th and 69 at Point of Air, Flint Aug. 27th, with 22 on the 29th 1997.

Late bird was seen at Ynyshir, Ceredigion Nov. 24th 1994. Black Terns are recorded almost annually at Llangorse, Brecon, monthly totals of 5 in April, 22 in May, 2 in July, 67 in August, 74 in September and 3 in October. These mirror the totals for the whole of Wales, as shown in the following chart.

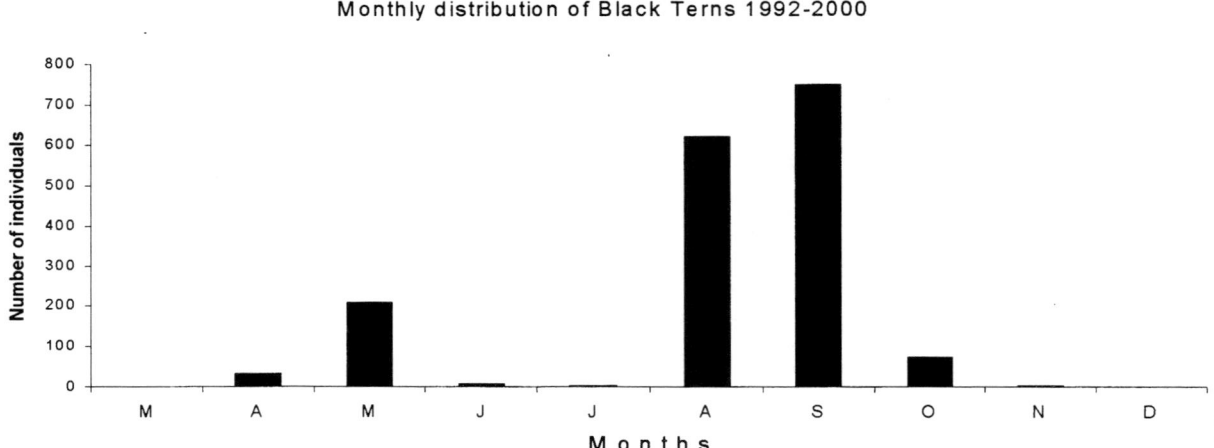

Monthly distribution of Black Terns 1992-2000

WHITE-WINGED BLACK TERN *Chlidonias leucopterus* CORSWENNOL ADEINWEN
(B) A rare passage migrant.
Distribution of all Welsh records by age (2 unaged records: Shotton Lake Oct. 4th – 17th 1970 and Llangorse Aug. 23rd 1910) and by county:

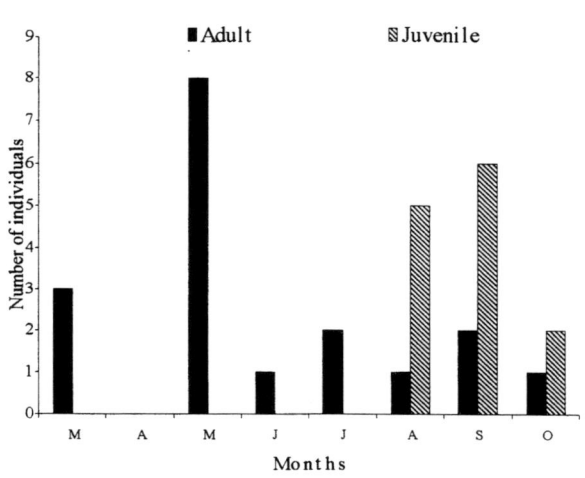
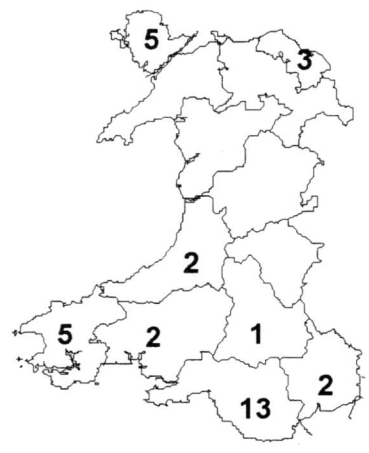

24 individuals quoted in *Birds in Wales*, with an additional record of an adult at Llandegfedd Res., Gwent on May 29th – 31st 1991. Since then there have been 8 more individuals recorded in Wales. Three adults were at Kenfig Pool, Glamorgan on May 18th 1992, an adult past Strumble Head, Pembroke on Oct. 15th 1997 and an immature at Llyn Traffwll, Anglesey from Sept 8th-14th 1998. In 1999 juveniles at Aberaeron, Ceredigion Sept. 13st and at Strumble Head Aug. 1st and a juvenile at Llandegfedd Res., Gwent Sept. 22nd 2000.

GUILLEMOT *Uria aalge*   *GWYLOG*

(C) *Resident breeding species in south-west and north-west counties, moving out to sea to moult after breeding and returning to the vicinity of colonies as early as December.*
Breeding counts of birds on ledges (not pairs) from various sites are organised below by county.

Gower: Worms Head: 100 in 1995, 177 in 1996, 50 in 1997, 47 in 1999 and 240 in 2000.

Pembroke (productivity expressed as young per site):

	1992	1993	1994	1995	1996	1997	1998	1999	2000
St. Margaret's		842		791	334	379	375	534	
Castle Martin coast	7,178	7,163	7,271	8,077	6,902	8,375	9,452	10,658	9250
Skokholm	478	684	643	684	509	613 0.95y/s	774 0.94 y/s		996 0.87 y/s
Middleholm				183	212	214	253		
Skomer	8,032 0.72y/s	8,792 0.72y/s	8,427 0.72y/s	9,995 0.79y/s	9,174 0.77y/s	9,721 0.77y/s	10,899 0.77y/s	12,135 0.61y/s	13,852 0.65 y/s
Ramsey	2,169	2,091	2,397	2,487	2,640	2,904	3,235		
Grassholm	293	310		670	685	712	876		

Counts at smaller colonies include 15 at Caldey in 1993, 9 there in both 1997 & 1998, 7 at Bishop's & Clerks in 1997 and 8 the following year. The Castle Martin peninsula is also known as Stackpole to Linney Head, includes Castle, Flimston and Crickmail & Mewsford Points. Census of the whole coast found 20,570 in 1996 (896 on the north coast), 23,785 in 1997 (8,379 of which were along the south coast) and 26,954 in 1998 (9,452 along the south coast).
During the Sea Empress disaster of 1996 1,591 oiled victims were picked up, of which 1,104 were dead. Breeding data for the following summer showed a massive decrease in colonies in the areas affected by the oil (south Pembroke).
Birds in Wales quote that the figures for Skomer have been steadily increasing since 3,920 counted by Operation Seafarer, with the last estimate of 5,000 pairs in the 1980s. Skokholm's population has nearly doubled in the last 10 years from the 300 birds counted in the 1980s. On Ramsey the population has risen from the 1500-1700 to today's figures, while St. Margaret's population rose from 118 in 1969 to 842 in 1993 but has declined by nearly half in the late 1990s.

Ceredigion:

	1992	1993	1994	1995	1996	1997	1998	2000
Cardigan Island	17	5	9	9	2	3	1	16
Penmoelcilia				132	216	215	247	1017
Ynys Lochtyn				310	378	325	467	
Dolyfrau				210	252	217	311	
New Quay Head				2062	2900	2744	2949	4235

Caernarfon:

	1993	1994	1995	1996	1997	1998	1999	2000
Bardsey	280	282	269	280	270	237	399	629
Ynysoedd Gwylan	150	171	142		172			152y
Great Orme	1520	1935	1740	684	1875	-	622	1512
Little Orme	665	980	865	726	927	732	444	603

Birds in Wales states that the Bardsey population has been steadily increasing from the 20 prs. in 1901 to 50-60 in 1922 and to 400 birds in 1989. Numbers have taken a dip since then but appear to be increasing again. The Great Orme population has shown the same sort of fluctuations as for Kittiwake, a similar story at the Little Orme where there was only 251 in 1969, increasing to 840 in 1986.

No data available from 2 of the biggest colonies at Carreg y Llam (the largest in Wales) and Cilan Head.

Anglesey: 3,071 counted at South Stack in 1992, 3,086 there in 1998 (little change from 1996 & 1997 figures), 3,414 in 1999 and 3,266 in 2000. Numbers have increased here, the population was put at 1363 in 1976.

Large movement recorded off Strumble Head, Pembroke in 1999, total of 34,000 on 5 dates Nov. 5th – Dec. 3rd.

RAZORBILL *Alca torda* LLURS

(C) *A breeding resident on the cliffs and islands of north-west and south-west Wales. Some disperse south to the Portuguese coasts in winter with a proportion entering the Mediterranean; part of the population remains in Welsh waters.*

Breeding counts of birds on ledges (not pairs) from various sites are organised below by county.

Worms Head, Gower: 12 nests in 1996, 25 in 1997, 83 in 1999 and 80 in 2000.

Pembroke:

	1992	1993	1994	1995	1996	1997	1998	1999	2000
St. Margaret's		156			100	91	71	115	
Castle Martin coast	673	638	641	621	823	971	1083	1013	997
Caldey	23	31	41	105*	16	13	12	10	
Skokholm	731	804	818	891	941 75% fledging rate	1,073 0.84 y/pr.	1,011 0.87 y/pr.	1180	1246 0.81y/pr
Middleholm				243	253	256	242		
Skomer	3,135	3,676 0.56y/pr	3,085 0.54y/pr	3,393 0.72y/pr	2,934 0.64 y/pr.	2,931 0.67 y/pr.	3,337 0.66 y/pr.	2938 0.56y/pr	3896**
Ramsey	979	936	1088	1267	1317	1469	1472	1731	1490
Bishop's & Clerks				87	116	138	226		
Grassholm		25		0	5	17	11	21	

* the high count of 105 was uncharacteristic – thought to be a fortuitous day.
** the highest ever count for the island.

Castle Martin peninsula is also known as Stackpole to Linney Head, includes Flimston, Broad Haven to New Quay, Mewslade & Crickmail Points, Castle and St. Govan's. Total county population censused post-Empress, at 7,198 in 1996, 7,759 in 1997 and 8,327 in 1998. It must be noted that counts are now based on the number of individuals, not apparently occupied sites. *Birds in Wales* states the Skomer population to be 2,100 prs. in 1963 and as 3,248 individuals in 1983. Ramsey as 600-1,000, Skokholm 700, Middleholm at 170, St. Margaret's at 200, Stackpole / Flimston (Elegug stacks) at 400.

Ceredigion:

	1992	1993	1994	1995	1996	1997	1998	2000
Cardigan Island		12	20	25	40			425
Penmoelcilia				233	225	160	74	
Ynys Lochtyn				52	48	33	18	
Dolyfrau				65	82	57	34	
Carreg Newydd				24	40	48		
Carreg Wynt					17			
New Quay Head				261	314	286	168	321

Caernarfon:

	1993	1994	1995	1996	1997	1998	1999	2000
Bardsey	250	108 ad 142n	89 194n	105 198n	96 271n	233n	800 ad	535 prs.
Ynysoedd Gwylan	70	61			69			82 prs.
Great Orme	189	212	180	257	187		196	
Little Orme	86	88	39	46n	34	36	47	

Bardsey population is expanding dramatically. 1300 in the Seabird Register of 1985-87, including birds at Carreg y Llam and Cilian Head from which there have been no counts in the period.

Anglesey: at South Stack, count of 626 adults in 1998 and 527 in 1999, similar to 1996 & 1997 figures. 1-2 nested at Porth Wen in 1999.

The Sea Empress disaster, in Pembroke/Carmarthen in February 1996 produced 341 casualties, 92 % of which were dead.

Large movements included: past Strumble Head, Pembroke, 20,000 in 11 hours, Oct. 25th 1992, 8,000 in 7 hours on Oct. 10th, with 11,000 on the 12th 1997, 4,000 in 4 hrs. off Strumble Head on Dec. 3rd 1999 and 12,000 in 7 hours on Oct. 2nd 2000; 2,000 in Cardigan Bay off Ynyslas, Ceredigion on Oct. 12th, 1,400 on the 19th 1992 and 2,500 past Bardsey Nov. 2nd 1999.

BLACK GUILLEMOT *Cepphus grylle* *GWYLOG DDU*
(C) A very scarce breeding resident in Anglesey and rarely elsewhere in north Wales. Occurs sparingly off the coasts of Wales outside the breeding season.
The only breeding sites are on Anglesey, at Fedw Fawr, Porth Eilian & Puffin Island and Ynys Gywlan Fawr, Caernarfon. A summary of available data:

	1992	1993	1994	1995	1996	1997	1998	1999	2000
Fedw Fawr (birds)	11	6	6	11	12	11	8-9	8	13
Porth Eilian						1 pr.	1 pr.	3prs., 2ad + 4 juv. on July 20th	
Puffin Island									One on eggs in June
Ynys Gwylan Fawr									1 pr.

Individuals from the two main sites could be seen off Traeth Llugwy, Penmon and Point Lynas, all Anglesey and Penmaenmawr, Caernarfon all year.

Other records from this county include singles at Ynys Llanddwyn Apr. 10th & Sept. 16th 1994. Records from Caernarfon in 1999 include 1-2 Bangor, singles at, Llanfairfechan, Great & Little Ormes and from Porth Iago July 14th 2000.

Other records outside the breeding season were of singles in Caernarfon at Bangor Harbour Jan. 24th 1995, Porth Dinllaen Jan. 2nd 1999, Llandudno / Penrhyn Sept. 28th 2000 and 3 Pwllheli Harbour Feb. 24th – Mar. 3rd 1997. At Bardsey individuals were seen on Oct. 24th 1993, June 13th 1994 and June 15th 1997.

The only other records from the breeding season were from the Castle Martin peninsula, Pembroke where individuals were present Jan. 1st – Apr. 14th 1993 and June 17th – 19th 1994.

Passage: singles in Glamorgan: at Port Eynon & Mumbles June 20th 1993 and Mumbles June 5th 1995; in Pembroke: at Strumble Head: Apr. 25th, Aug. 30th Sept. 3rd 1992, Aug. 28th & Sept. 20th 1999, July 11th and Nov. 3rd 2000, at Fishguard Harbour Apr. 7th 1993, May 20th 1994, Jan. 26th 1997 and Dec. 30th 1998 and at Skomer June 8th 1993; at Morfa Harlech, Meirionnydd Oct. 26th 1992, at Gronant, Flint Mar. 22nd 1992.

LITTLE AUK *Alle alle*   CARFIL BACH

(C) Scarce winter visitor in irregular numbers offshore on the western sea-board. Most sightings in late autumn and early winter; often associated with westerly gales.

	1992	1993	1994	1995	1996	1997	1998	1999	2000
Total	10	8	7	36	16	8	2	4	2
Daily Max.	5	3		13		3			
Date	Feb. 16th	Oct. 25th		Sept. 24th		Nov. 9th			

The majority of records were from Strumble Head, where annual totals for each year during the period 1992-2000 are shown in the table above.

The following chart shows the total number of living individuals recorded in each month during the period, while the map shows the total number recorded in each county.

Dead individuals found: Rhosilli, Glamorgan Dec. 13th 1997, in Pembroke at Broad Haven N. Jan. 16th 1994, Skokholm over the winter of 1993/4, Freshwater West Dec. 11th 1997; Ceredigion at Borth Nov. 26th 1992, Aberystwyth Jan. 28th 1993 and Ynyslas Jan. 17th 1994; Trearddur Bay, Anglesey Jan. 13th 1994.

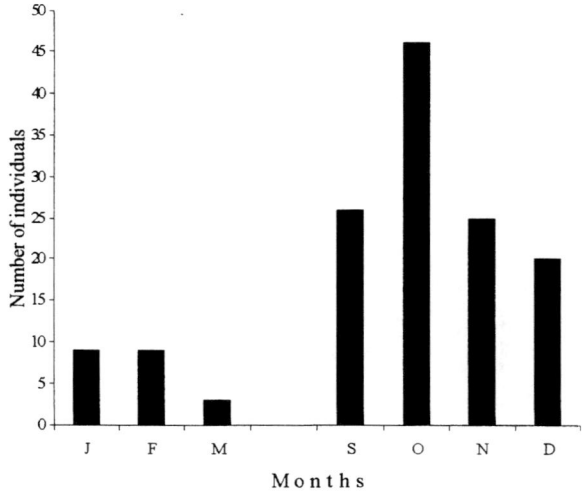
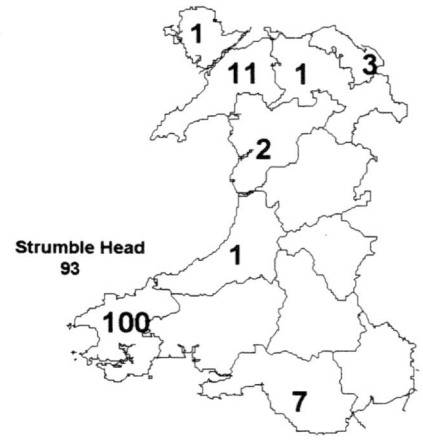

PUFFIN *Fratercula arctica*   PAL
(C) *A breeding bird of offshore islands and a few western headlands. Formerly bred in large numbers.*
Breakdown of breeding data for some key sites, a combination of numbers of breeding pairs and numbers of individuals.

Pembroke	1992	1993	1994	1995	1996	1997	1998	1999	2000
Skokholm	2505	2145	3320	2700	3332	3250 0.63 y/b	2774 0.66 y/b	3083	3092
Skomer	5-6000 prs.	11000 0.69y/b	6-6500 prs. 0.8y/b	10473 0.78y/b	9141 0.83 y/b	9049 0.64 y/b	9235 0.79 y/b	9213 0.7y/b	10614 0.75y/b
North Bishop	30b	28b	8b	12b	22b	31b	24b	27b	

Main sites only: figures for Skomer and Skokholm censused in April and June/July – the table below only shows the April count. b represents burrows.
County population censused post-Empress, at 12,342 individuals in 1997 and 12,043 in 1998. The Seabird Register of 1985-87 put the county population as 14,000 prs., of which 7,400 prs. were on Skomer and 5,000 individuals on Skokholm. Data from Skomer suggest a stable population with perhaps a hint of an increase. Note that counts are now of adult birds present in April, a pre-breeding peak attendance count. This should not be compared to the estimates of 6000 – 7000 pairs which have been postulated several times in the past.

Caernarfon: at Ynysoedd Gwylan, 390+ in 1993, 524 burrows in 1999 and 1113 burrows in 2000, a two fold increase. Individuals were seen carrying fish at Bardsey in 2000, the first time since 1953.

Anglesey: South Stack – 25 in 1995, 56 in 1997, 17 birds in June 1999 and 20 birds in July 2000. 4 birds at Puffin Island / Penmon Point Apr. 9th 1999 and 40+ individuals were seen on Puffin Island in June 2000 (post rat eradication).

STOCK DOVE *Columba oenas*   *COLOMEN WYLLT*
(E) *Breeding resident occurring widely but not numerously up to approximately 305 m; most common in lowland areas. Appears to be getting scarcer across most of Wales, probably as a result of a change in farming practices and a decline in arable land.*

Using data from the National Atlas produces a Welsh population estimate of 15,300 pairs. Although there is little concrete data, this species appears to be getting scarcer throughout Wales and a population estimate of around 10,000 pairs is more realistic.
By 2000 this species was described as fairly common in Gower, locally common in Glamorgan but scarce & local in Ceredigion and becoming scarcer in Montgomery. Appears to be decreasing Radnor and in Pembroke where has largely disappeared from the coast. In Meirionnydd it is a scarce breeder, confined to the east of the county, particularly the Dee valley.
Systematic breeding information is available only from Brecon, where at least 40 pairs were in the Usk & Wye valleys in 1995, in 1998 51 pairs were found in mixed farming areas in the SE & 6-7 pairs north of Mynydd Epynt, while in 1999, 47 pairs located in the SE (30 prs. in 2000) with only 5 pairs to the north of Mynydd Epynt (16 prs. in 2000). The population was estimated at 200-300 pairs in this county in 1990 but has decreased since to probably no more than 100 pairs.
In 1999, few large flocks recorded, with most counties recording low numbers. Distribution of the significant flocks suggests a strong link with the remaining areas of arable or mixed farming. At Inner Marsh Farm, Flint wintering numbers have fallen sharply probably linked to the conversion of arable land into pasture. Only one record from Skokholm, Pembroke where 60 pairs bred in the 1970's (N.B. breeding only recorded there 1967-1984). *Birds in Wales* quote a record of 400 at Pontiets, Carmarthen in 1980 – this is now thought to be unlikely. Winter flocks of 60 or more were:

1992	two flocks of 60+ in Gwent at the year's end; 80 at Llanhamlach, Brecon on December 31st
1993	67 at Bonvilston, Glamorgan on Jan. 3rd and 150+ in the Hendre Lake area, Cardiff on Oct. 16th; 160+ at Talgarth, Brecon in Jan. and Feb. and c.200 here in Nov., with 62 at Llanhamlach on Feb. 7th
1994	300+ at Talgarth, in Jan. and Feb., 150 at Garthbrengy on Dec. 24th, both Brecon
1995	100 at Aberthin, Glamorgan on Nov. 18th; up to 60 at Oak Hill, Pembroke in Jan. and Feb.; 300 at Garthbrengy, Brecon on Mar. 12th; c. 200 at Dolydd Hafren NR, Montgomery in December
1996	68 at Martletwy, Pembroke on Jan. 21st
1997	70 at Llanfilo, Brecon on Feb. 2nd and 80 at Trefeitha on Oct. 20th
1998	60 at St. Nicholas, Pembroke on Dec. 14th
1999	80-100 at Dingestow, Gwent Oct. 10th, 96 at Leckwith Mar. 7th & 107 at Great Porthamel, Dec. 5th, both Brecon.
2000	92 at St. Arvans, Gwent Jan. 22nd, 100 at Leckwith, Glamorgan Jan. 2nd, 70 at Great Porthamel, Jan. 2nd and 100 at Dolydd Hafren, Montgomery during January – March.

Birds in Wales 1992-2000

WOOD PIGEON *Columba palumbus* — YSGUTHAN
(D) Abundant breeding species: passage migrant and winter visitor.
An abundant species occurring numerously throughout all of Wales. Using data from the National Atlas produces a Welsh population estimate of 225,600 pairs.

COLLARED DOVE *Streptopelia decaocto* — TURTUR DORCHOG
(D) Now a widespread and common resident, having first appeared in Wales in 1959.
Since the 1st record in Wales in 1959, this species has shown an explosive expansion of range and is now distributed all over Wales. Using data from the National Atlas produces a Welsh population estimate of 13,000 pairs. Results from BBS, 1994-1999 suggests a population change of 125%, producing a revised population estimate of 16,250 pairs.

TURTLE DOVE *Streptopelia turtur* — TURTUR
(B) Very rare summer visitor in eastern Wales with numbers declining further; passage migrant, mainly spring, in small numbers.

As a breeding bird, this species has been declining since the 1970's and is now restricted to Gwent, where a few pairs were recorded most years, with breeding confirmed in some years although there were none in 1998, 1999 and 2000.

Turtle Doves are recorded annually on passage or in some cased summering. The total number of non-breeding individuals recorded in each county during the period 1992-2000 is shown on the map opposite.

Pembroke accounted for the highest number of records, of which 124 came from its islands. Likewise Bardsey in Caernarfon.

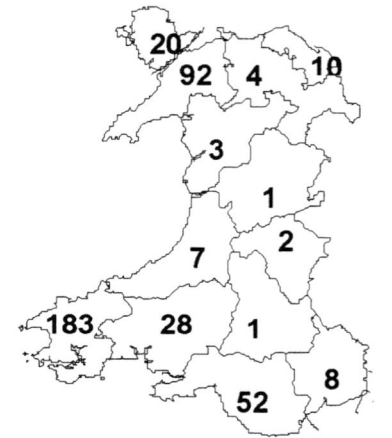

CUCKOO *Cuculus canorus* — COG
(E) Formerly a locally common breeding summer visitor to all counties but appears to be declining and far from common; also recorded on passage.

This species has been declining in Wales since the 1940's, particularly in lowland and valley bottoms. Very little comparative data from the period under review. In Brecon, it has declined sharply and is now confined to hill grazings and "rhos" land, where Meadow Pipits remain numerous. Few recent figures but up to 18 males recorded recently on Mynydd Epynt MoD, 16 males on 820 ha of rhos and common in the Llanwrtyd area in 1998 (2 males/km^2) and 12 males on 39km^2 of common in central Brecon in 1995 (0.3 males/km^2). In 2000 there were 19 males counted on the Epynt and 10 on 513 ha of bogs and commons in the north of the county – a 37% decline since 1998.

There were 14 males in 7 1-km squares distributed in 3 10-Km squares in Radnor in 1997.
By 2000, described as scarce and local in Ceredigion and Montgomery, decreasing in Glamorgan, Radnor, Anglesey and Clwyd. Still fairly common in Gower and abundant in Meironnydd.

Using data from the National Atlas produces a Welsh population estimate of 1,800 pairs. Results from BBS, 1994-1999 suggests a population change of 69%, producing a revised population estimate of 1,200 pairs.

YELLOW-BILLED CUCKOO *Coccyzus americanus* — COG BIGFELEN
(B) The Yellow-billed Cuckoo is a vagrant, which has been recorded three times in Wales, all in the 19th century. It breeds in North, Central and South America.
One near Porthclais, Pembroke on Oct. 30th was the fourth record for Wales and the only 20th century record. The other three had all been found dead, at Stackpole, Pembroke 1832, Aberystwyth, Ceredigion in October 1870 and at Craig y Don, Anglesey in November 1899.

BARN OWL *Tyto alba* TYLLUAN WEN
(E) A breeding resident in all counties, population appears to be stable in many counties after a prolonged period of decline.

Barn Owls were formerly a familiar sight throughout lowland Wales but has declined over the whole of its range during this century. The last comprehensive survey, 1982-85 reported 25 sites in Gwent, 36 in Glamorgan, 34 in Carmarthen, 31 in Pembroke, 78 in Ceredigion, 17 in Brecon, 12 in Radnor, 20 in Montgomery, 30 in Meirionnydd, 47 in Caernarfon, 68 on Anglesey, 24 in Denbigh and 16 in Flint.

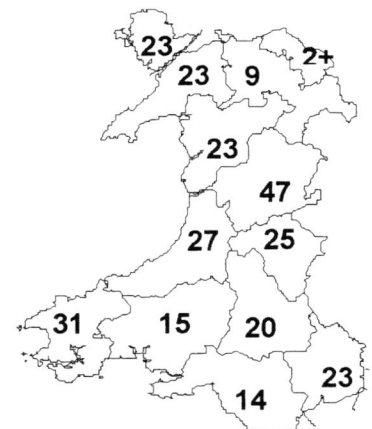

The map opposite shows the maximum number of pairs / occupied sites in each county, during the period 1992-2000. The data are incomplete, do not represent detailed surveys and for some counties probably reflect observer coverage. Even so the populations in Gwent and Brecon appear to be at the same levels as in the 1982-85 surveys, those for Montgomery and Radnor have increased (although this is probably due to better coverage), while those in Glamorgan, Carmarthen, Pembroke, Ceredigion, Meirionnydd (may be due to poorer coverage), Caernarfon, Anglesey, Denbigh and Flint have decreased significantly.

The survey in 1982-85 put the total Welsh breeding population at 462 prs., a decline from the estimate of 1416 prs. in 1932. The figures suggest that this decrease has continued.

LITTLE OWL *Athene noctua* TYLLUAN FACH
(E) Locally common breeding resident in most eastern counties, scarcer and probably decreasing in the west. Greatly under-recorded.

Breeding records are very incomplete, and give little real indication of population trends. The following gives the number of pairs confirmed or probably breeding, with the total number of sites from which birds were recorded during the year given in brackets:

	1992	1993	1994	1995	1996	1997	1998	1999	2000
Gwent	(6)			7 (5)		7 (22)		6	11 (20)
Glamorgan	(2)			(3)	2 (7)	4 (9)	3 (22)	4 (7)	4 (14)
Carmarthen	0 (7)	1 (4)	3 (3)	0 (5)	3 (2)	3 (5)	1 (3)	(3)	
Pembroke				5	4	5 (7)	3-5 (6)	4 (5)	3 (2)
Ceredigion					1				
Montgomery							(3)	(4)	1	(5)
Brecon	2 (8)				(8)	3 (19)	(12)	(5)		(4)
Radnor							(3)	(1)	(10)	
Meirionnydd					(2)		(3)	(1)	(1)
Caernarfon			5	6	7	4 (5)	3 (6)	3 (5)	4-5 (7)	4 (2)
Anglesey	(3)			3 (7)	(5)		(4)	Several (11)	2 (6)
Denbigh							(4)	(8)	(12)	
Flint	(1)								

Recorded at a total of 35 sites in Brecon during 1992-99, the bulk in the Wye and Usk valleys and nearby. Declining, it has now disappeared from the area north of Epynt and contracting in its heartland. Thus in 20 sites recorded in 1996 in the SE of the county no more than 15 were occupied in 1999. This may be as a result of wet summers. Described as scarce and only in coastal areas of Ceredigion, thinly spread in Montgomery and fairly common on the main river valleys of Gower but scarce elsewhere in that county. This species is probably declining in Radnor and in Meirionnydd it is mainly a coastal species, with very few individuals inland and appears to be declining, with several unoccupied territories in the SW.

In *Birds in Wales* it was suggested that the total Welsh breeding population at the time was likely to be in the region of 2,000 prs. on the basis of surveys in the 1980's and estimates used in atlas surveys of individual counties & individual studies (500-1,500 prs. in Gwent, 100 prs. on the Gower, 50 prs. in Pembroke, 25 prs. in Brecon and 54 on Anglesey). The tabulated figures suggest that there has been a substantial decline. By totalling the highest figure of sites for each county during the last 8 years, this species is present at less than 150 sites. Even allowing for poor observer coverage in some counties, a total Welsh breeding population of twice this amount, 300 prs. is probably an over estimate.

TAWNY OWL *Strix aluco* *TYLLUAN FRECH*
(E) A common and widespread breeding resident in areas of woodland in all counties but possibly declining.

Using data from the National Atlas produces a Welsh population estimate of 2,160 pairs compared to the 5-6,000 quoted in *Birds in Wales*. The difference may be real and due to significant declines or may just reflect the method of extrapolation.

The only systematic counts were of 15 prs. Dinas / Gwenffrwd, Carmarthen in 1994 and 12-13 territorial males along the R. Irfon from Llanwrtyd to Abergwesyn, Brecon in 1995. Paper published in *Welsh Birds* (1(3): 52-58).

LONG-EARED OWL *Asio otus* *TYLLUAN GORNIOG*
(C) A scarce resident in some counties. Also a scarce winter visitor and passage migrant.
This is an elusive species but the true number of breeding pairs is probably less than the 30 prs. as suggested by *Birds in Wales*. Records of breeding or probable breeding were:

1992	a pair at Mynydd Rhiw Gregan, Glamorgan and one in Dyfi Forest, Meirionnydd
1993	4 breeding pairs at three sites in Gwent and a pair at Mynydd Rhiw Gregan, Glamorgan; a pair possibly bred in Montgomery.
1994	one pair nested in Gwent, but the young were predated.
1995	two pairs nested in Gwent; probable breeding in Ceredigion; one pair bred in Brecon.
1996	one pair bred in Radnor; possible breeding in Clwyd. A check of 50 1-km squares in Radnor during mid-March produced responses from 6.
1997	one pair bred in Ceredigion, one pair in Montgomery and one in Denbigh
1998	one pair reared three young in Gwent and two pairs raised 4 young in Denbigh.
1999	three pairs in Gwent, at least 4 young fledged, one pair bred in Ceredigion, 5 pairs in north Radnor, one pair fledged 2 broods in Montgomery, one pair bred on Bardsey, Caernarfon – the first breeding record for the island, 1-2 pairs in Denbigh, one of which fledged 2 young.
2000	two pairs in north Ceredigion, pairs bred at 4 sites in Radnor, 1 pair on Bardsey (2y found dead later) and 1 pair Denbigh.

Winter roosts of five or more were up to 12 birds at Clytha, Gwent during Jan. – March 1998; up to 9 at Wentloog Levels, Glamorgan in Jan. – March 1992, with 5 here in December 1993; 5 at Euroclad, Glamorgan in Jan. 1993; 5 near St David's, Pembroke in Feb. 1993 and 7 at Rhodiad y Brenin, Pembroke in Jan. 1997, with 5, presumed the same, at Hendre Farm.

SHORT-EARED OWL *Asio flammeus*   *TYLLUAN GLUSTIOG*
(C) Scarce breeding resident. Winter visitor and passage migrant in small numbers to all counties.

	1992	1993	1994	1995	1996	1997	1998	1999	2000
Skomer	5, 6y	13, 32y	8-9, 25-30y	2, 1y	6, 13-14y	6, 13y	5, 5y	4	1-2, 3y
Abergwesyn							1	1	
Elenydd						2		6	1
Llanbrynmair	3-5	3		1 probable		2		2	
Lake Vyrnwy			1			1	1		4
Berwyns		2 possible			3		1 possible	1 pr., 3y	1

A RSPB survey of 15,000 ha of Mynydd Du, Brecon/Carmarthen found one summering individual in 1996.

Birds in Wales 1992-2000

Birds in Wales quotes the Welsh breeding population to be 20 prs. in 1988/9 and in the range 15-40 prs. in 1992 (at least 23 prs.). During the 1990's breeding took place at Skomer in Pembroke, Abergwesyn in Brecon, Elenydd in Radnor, Llanbrynmair Moor & Lake Vyrnwy in Montgomery, on the Berwyns (Meirionnydd / Denbigh).

The table above summarises the number of pairs at each site. Overall there has not been any change in the total Welsh breeding population, with great fluctuations in some years. Of interest is the increase in mid-Wales, although this is perhaps just pairs moving around.

Away from the breeding sites there were records in all months of the year, the highest numbers from October to February. The map opposite shows the total number of individuals recorded in each county during the period 1994-2000. The largest numbers came from Skomer Island, Pembroke where there were 10 in November 1993, 12 in December 1996, 20 in December 1997 and 8 in October 1999.

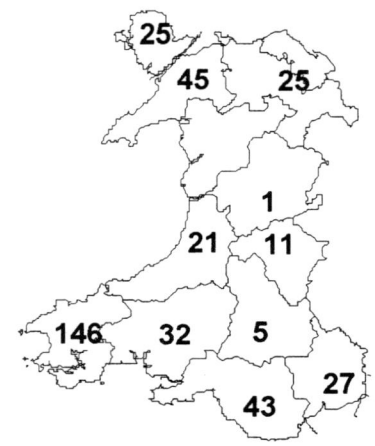

NIGHTJAR *Caprimulgus europaeus*   *TROELLWR*
(E) Widespread but scarce summer visitor. The breeding population has increased over the past ten years and is now mainly dependent on clearfell areas in forestry plantations.

The table below summarises the number of territorial males recorded in each county during the period 1992-2000:

	1992	1993	1994	1995	1996	1997	1998	1999	2000
Gwent	28	22	19	19		5+		8	5
Glamorgan	33-42	14	26	13	12+	18	13	18	18
Carmarthen	5	3	1	6	4	2-3	3	9-10	
Pembroke	1		1			1	1		
Ceredigion	5				2-3	2	1-2	2	2-4
Montgomery	7	3	bred	5-7	14-16	11-13	7-9	13	9
Brecon	14	8	4	10-13	12	1	7	2	3
Radnor	11		2					2	
Meirionnydd	41	bred			3				2-4
Caernarfon	20	1	5	2		1	2	2	5
Anglesey									
Denbigh	26	7	8	4	3 (*)	4+	11+	8	7-8
Flint	2								

(*) One male here was bigamously paired, thus 4 females.

Birds in Wales quotes a total Welsh population of 193 prs. in 1992 (as shown on the map) compared to 57 in 1982 (both results of National Surveys). Records for years other than 1992 are incomplete, but there are indications of a decline in some counties. Whether these declines are real or are due to clear-fell areas being planted causing birds to move to new, yet undiscovered sites, is not known.

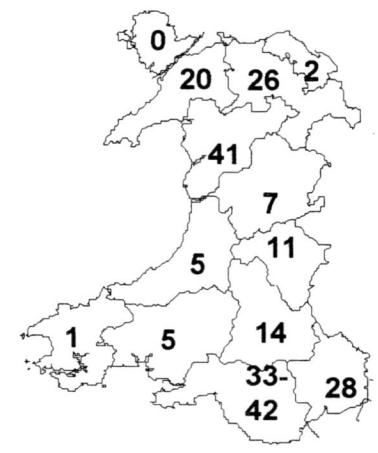

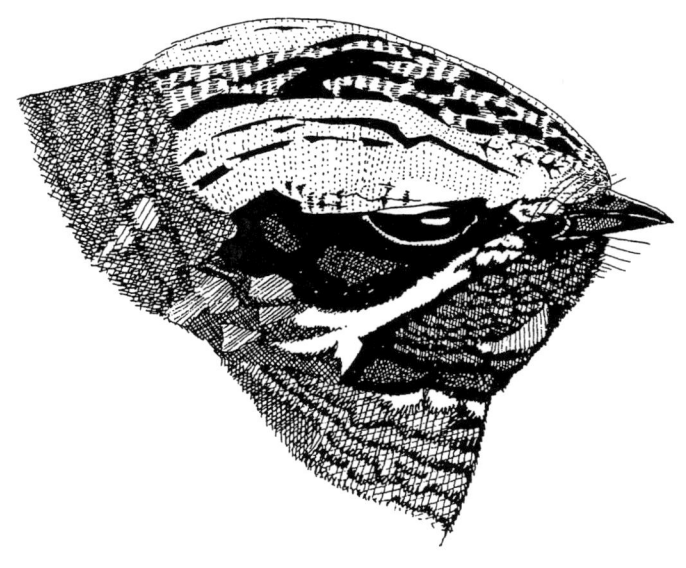

SWIFT *Apus apus*
(C) A widespread and plentiful summer visitor.

GWENNOL DDU

Swifts are familiar and conspicuous summer visitors to towns and villages throughout Wales. Spring gatherings of 400 or more were: in Glamorgan: 2000+ at Eglwys Nunydd Reservoir on May 12th 1994, 1050 on May 22nd 1997, in 1999, 1600 on May 13th with 500 on the 26th and in 2000, 3000 on May 18th, 1000 on the 23rd and 800 on June 1st & 9th, 1,000 at Kenfig on May 18th, 400 at Llanishen/Lisvane Res on May 20th and 750+ at Clyne Common on May 29th, all 1995; 750 at Pembroke on May 1st, 1996; 400 over Llangorse, Brecon on May 23rd.

In autumn there were large numbers recorded: 500 at Skirrid Fach, Gwent during June and July; in Glamorgan: 400 at Eglwys Nunydd Reservoir on Aug 3rd 1994, 400 on July 11th 1998 and 400 on Aug. 17th 1999, 2000 at Kenfig Pool on July 20th 1995, 600 at Port Talbot Docks July 4th 1999 and 1,000 at the Ogmore estuary on July 25th 1999.

Using data from the National Atlas produces a Welsh population estimate of 8,500 pairs. Results from BBS, 1994-1999 suggests a population change of 153%, producing a revised population estimate of 13,000 pairs.

ALPINE SWIFT *Apus melba*
(B) A rare visitor to Wales.

GWENNOL DDU'R ALPAU

Additional record to the 27 individuals published in *Birds in Wales*, of an individual at Rhuddlan, Flint on Apr. 19th 1988. Since then there were records of singles: in Glamorgan at Aberthaw on May 19th 1996 and Eglwys Nunnydd Res. on May 13th 1999 and May 4th 2000; in Carmarthen at Llanelli on Apr. 26th 1998; in Pembroke at St. David's May 30th and Dinas Cross Aug. 25th both 1999; in Ceredigion at Borth on Apr. 9th 1993; in Caernarfon over the Conwy river on Apr. 21st 1996 and at Rhuddlan, Flint on July 20th 1999.

Breakdown of Welsh records of Alpine Swift, by month is shown on the chart and by county on the following map.

Birds in Wales 1992-2000

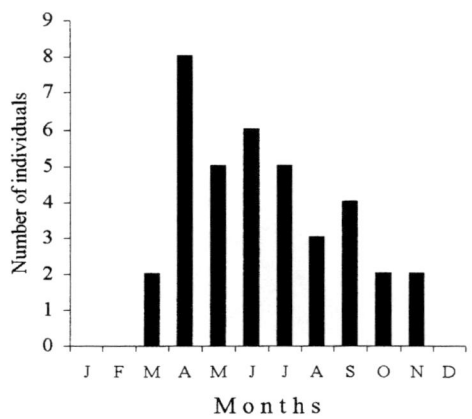
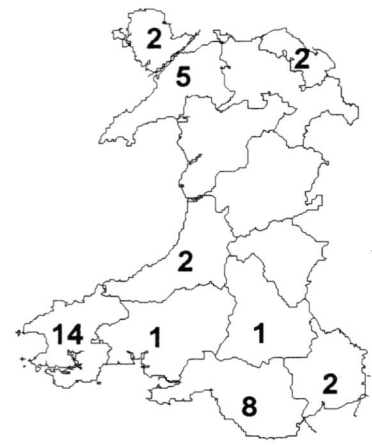

KINGFISHER *Alcedo atthis*
(C) Resident in lowland areas.

GLAS Y DORLAN

Birds in Wales stated the Welsh breeding population to be 400 prs. in a good year. The map opposite shows population estimates for each county (data from British Birds' Rare Breeding Bird Panel and county recorders).

Unfortunately there are virtually no contemporary data for comparison. The only systematic counts were from Brecon, where there were 4 pairs on the River Usk between Pencelli and Talybont in 1994, with 8-10 pairs with young between Fennifach and Talybont in the late summer of 1994 and 7 pairs between Brecon and Fennifach in 1996. Overall county population estimated at 40-50 pairs in 1990 but is thought to be no more than 40 now. In Radnor, a survey of the River Wye between Newbridge and Hay on Feb. 4th 1996 found 15 birds. 4 pairs on a 7 km stretch of the River Severn in 1995 & 1996 around Newtown, Montgomery. The population of Carmarthen is estimated at 50 – 60 pairs.

Described as scarce in Gower, scarce in Ceredigion but widespread in Montgomery along the Vyrnwy, Camlad and Banwy rivers. Distribution is local but probably increasing in Glamorgan. A sporadic breeder on Anglesey.

An all Wales survey is long overdue.

BEE-EATER *Merops apiaster*
(B) A rare visitor to Wales.

GWYBEDOG Y GWENYN

Birds in Wales records 29 individuals in Wales. Since then there have been a further 5. Singles were recorded in Pembroke at Skomer on May 23rd 1993 and at Skokholm on May 13th 1997, in Ceredigion at Llanfihangel-y-Creuddyn on June 26th 1993, in Caernarfon at the Great Orme on May 16th 1997 and at Pen-y-Sarn, Anglesey May 25th – 28th 1993.

Breakdown of all Welsh individuals by arrival date (2 undated records from Pembroke) is shown by the following chart and by county in the map.

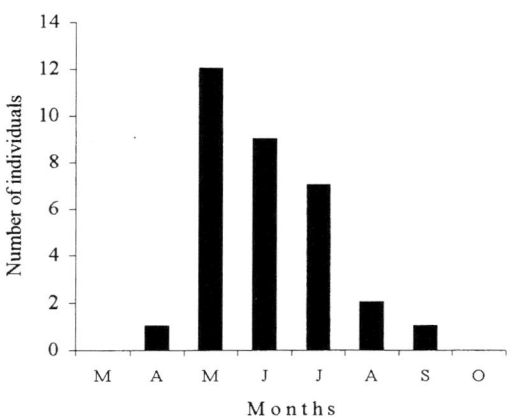
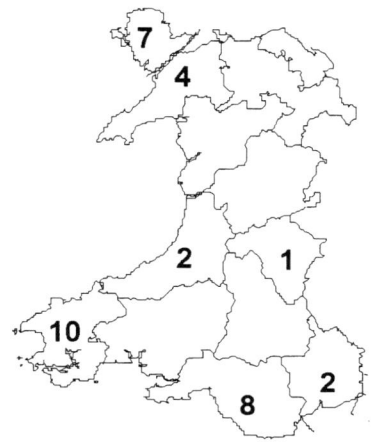

HOOPOE *Upupa epops*   *COPOG*

(B) An uncommon but regular passage migrant to Wales.

A pair reared three young in Montgomery in 1996, the first proven breeding in Wales. One bird returned to this site in 1997 but did not stay. *Birds in Wales* summarised all the Welsh occurrences of this species up to and including 1986. Since then there has been a further 155 individuals recorded. A breakdown of records in the period 1987-2000, by month (not including the 1996 breeding record or the subsequent returning adult in the following year) is shown in the following chart and the total number of individuals recorded in each county (including all records) is shown on the map. Pembroke accounted for the majority of records, 60, of which 6 were on Skomer, 4 on Skokholm and 4 on other islands but the majority were scattered all along the coast.

The pattern of occurrence is similar to that described in *Birds in Wales*, where 162 individuals were recorded in the period 1967-86, compared to the 155 in the last 14 years, 1987-2000.

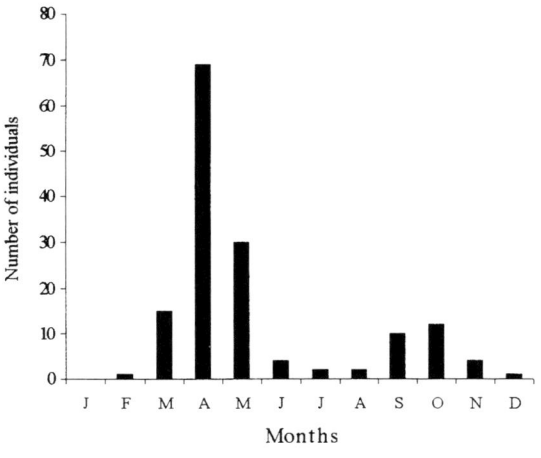
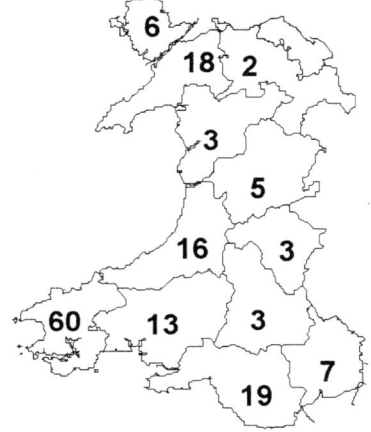

WRYNECK *Jynx torquilla*   PENGAM

(B) An uncommon but regular passage migrant in Wales, chiefly to coastal counties in autumn.

Birds in Wales quote a total of 141 individuals recorded in Wales during the period 1969 – 1991, mainly in autumn, particularly September and a less obvious spring passage. In the period under review, 1992-2000, there were 93-96 individuals recorded, including 2 records from Brecon, at Carngafallt May 31st & June 1st 1997 and at Halfway Forest Sept. 11th 2000. The majority of records came from Pembroke, 34 of which came from the islands.

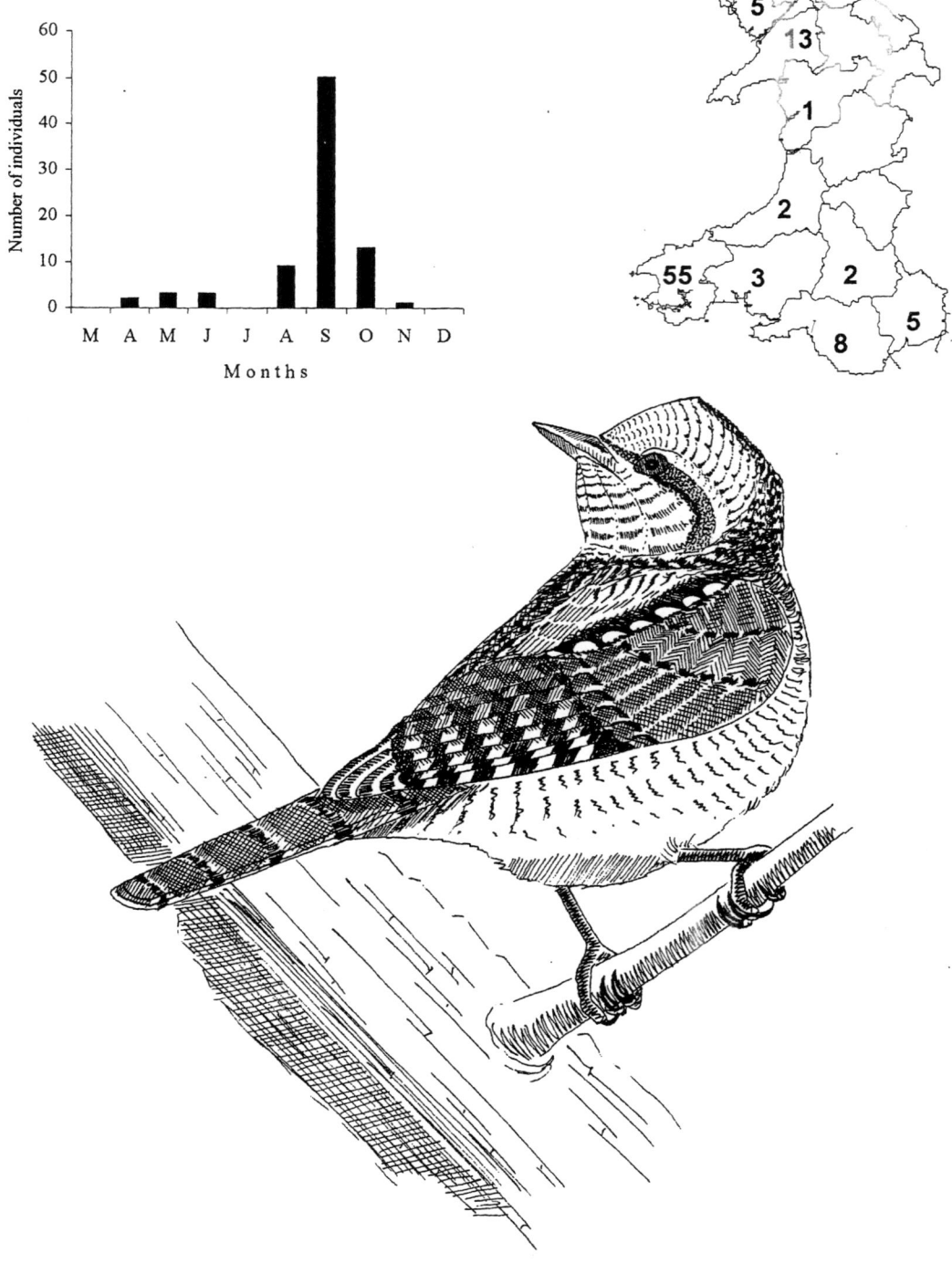

GREEN WOODPECKER *Picus viridis*   *CNOCELL WERDD*

(E) *Once a common breeding resident throughout lowland Wales except on Anglesey. It now appears to be doing well in eastern counties but declining in most western counties, including Montgomery and Radnor.*

Birds in Wales compares the distribution of breeding Green Woodpeckers using the original Breeding Atlas and the New Breeding Atlas. Stating that there appears to be little change in the overall distribution of the species but a reduction in confirmed breeding records from Pembroke, Caernarfon and Anglesey, suggesting a slight retraction in range and a total Welsh breeding population in the range 2,500 – 5,000 pairs.

During the period of this review, 1992-2000 there has been a further reduction in the number of breeding pairs in Wales. The map opposite shows population estimates for the number of breeding pairs in each county. The range figures were provided by County Recorders, the other figures are from the number of pairs reported in Welsh Bird Reports 1999 – 2000. Admittedly these are not all full surveys. By totalling these up an estimate of the Welsh population is 1,000 pairs.

The map clearly shows that the stronghold for this species in Wales is in the east, particularly Gwent and to a lesser degree Glamorgan and Radnor.

In Gwent a breeding survey in 1995 recorded birds in 300 1-km², in 170 tetrads (the 1985 Gwent Atlas estimated a county population of 350+ pairs).
Comments by County recorders on the distribution of this species in Welsh counties, for the end of 2000 were:
Glamorgan: described as widespread and common in Gower.
Carmarthen: mostly in old parkland sites near wooded upland valleys and to the north east of the coalfield.
Pembroke: 50 tetrads where this species was found during the Pembroke Breeding Atlas were sampled in 1998 and individuals were found in 34, mainly in the south of the county. Declining.

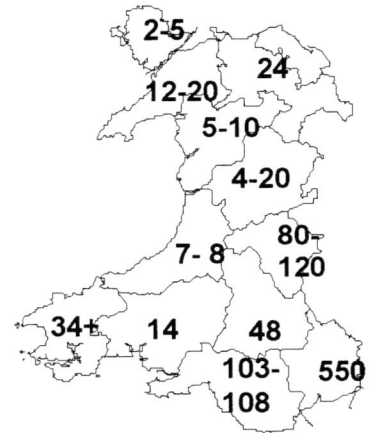

Ceredigion: very scarce and declining.
Radnor: disappearing, especially in areas recently colonised by Goshawk.
Montgomery: recorded in small numbers in the Severn and Vyrnwy valleys in the east of the county. Appears to be declining. Meirionnydd: declining, possibly due to a reduction in grazing within woodlands and the agricultural improvements of older pastures.
Caernarfon: Widely distributed but probably not declining.
Anglesey: absent from much of the island.

GREAT SPOTTED WOODPECKER *Dendrocopos major*   *CNOCELL FRAITH FWYAF*

(D) *A common breeding resident in wooded parts of Wales but scarce on Anglesey and the Llyn peninsula.*

Using data from the National Atlas produces a Welsh population estimate of 2,700 pairs. Results from BBS, 1994-1999 suggests a population change of 130%, producing a revised population estimate of 3,500 pairs.

Birds in Wales 1992-2000

LESSER SPOTTED WOODPECKER *Dendrocopos minor* — *CNOCELL FRAITH LEIAF*

(B) A scarce and local breeding resident in all counties, principally distributed in the lowlands and along the main river valleys.

There are few records of confirmed breeding for this species, which is almost certainly under-recorded.

Max. number of sites from which birds were recorded is shown on the map opposite.

No data is available for Flint. The data for Radnor is a population estimate although the species is thought to be declining rapidly.

This species as described was very scarce in Gower, similarly in Ceredigion and Glamorgan in 2000. In Carmarthen this species was found at a total of 39 sites in the period and it is estimated that the county population may be around 100 pairs.

In Brecon the population was estimated at 30-50 pairs in 1990 but has declined since. As riverine alders appear to be important, the decline of this tree due to disease may be bad news in the long term for this woodpecker.

Using data from the National Atlas produces a Welsh population estimate of 400 pairs. Although the data are incomplete a figure of 300 pairs seems more realistic.

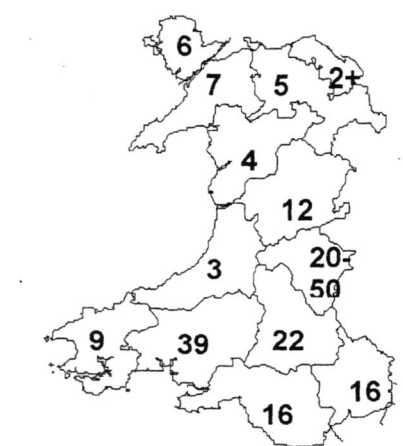

SHORT-TOED LARK *Calandrella brachydactyla*   EHEDYDD LLWYD
(B) A rare visitor to Wales. It breeds from Iberia, southern France and north-west Africa east to Manchuria.

Birds in Wales quote 12 records, 7 of which were from Skokholm, Pembroke, 4 from Bardsey, Caernarfon and a single mainland record at Llanfairfechan, Caernarfon. Since then singles were recorded in Pembroke at Dale airfield on Aug. 19th-25th 1994, 2 on Skokholm Oct. 10th – 16th and a single at Strumble Head on Oct. 14th-27th 1995; at RSPB Conwy, Caernarfon on May 12th – 13th 2000; in Anglesey at South Stack on May 25th 1993.

A breakdown of all Welsh records by arrival date is shown in the graph.

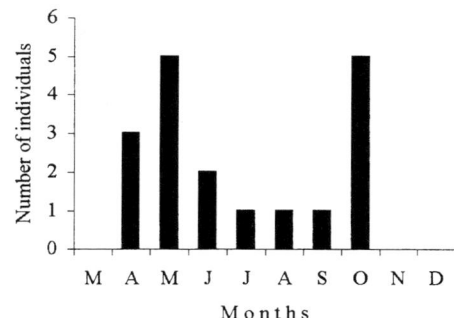

WOODLARK *Lullula arborea*   EHEDYDD Y COED
(B) Formerly resident but now extinct as a breeding species.
The annual total of individuals recorded in Wales is shown in the graph.
There have not been any confirmed breeding records in Wales since a pair bred in Radnor in 1981. All records were:

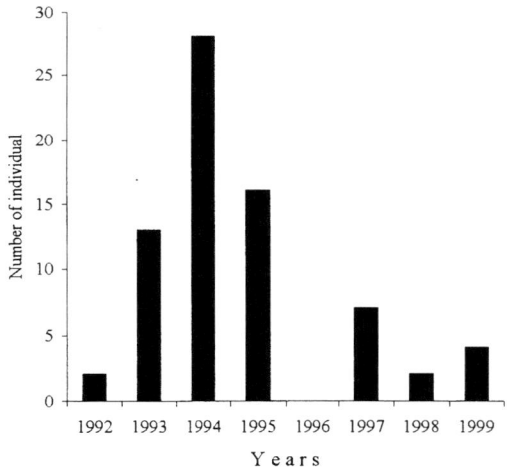

Gwent: all records came from one site, Dingestow, where there were on 6 Oct. 31st and Nov. 1st 1993, 10 on Oct. 25th-31st 1994, 7 on Nov. 25th 1995 and a single Oct. 11th & 15th 1999.

Glamorgan: 4 at Whiteford Point Oct. 17th-29th 1993, 10 at East Moor Farm, Scurlage on Nov. 3rd, 8+ at Clyne Golf Course on Dec. 21st 1994 with 10 here on Feb. 8th 1995 (presumably the same birds), 7 at Mewslade Farm on Jan. 2nd 1995, singles at Cardiff Heliport on Feb. 8th and Rhoose Quarry May 15th 1997, at Rhiw Tor Cymru Garn on Oct. 21st and at Middleton Dec. 28th 1999.

Carmarthen: 5 at Broad Oak on Oct. 21st 1997, with one still there on the 23rd.

Pembroke: singles at: Skomer on Oct. 18th 1992 & Oct. 11th 1993, at Strumble Head on Oct. 31st 1992, Nov. 18th 1995 & Nov. 11th 1998, 2 at Skokholm on Oct. 23rd 1993 and at least one at Porthclais Nov. 18th 1998 & Jan. 29th – 20th 1999 (presumed to be the same bird).

Caernarfon: a single at Bardsey on Oct. 26th 1995.

SKYLARK *Alauda arvensis*   EHEDYDD
(C) A widespread but declining breeding species; abundant passage migrant in autumn and winter.

A widespread breeding bird. Using data from the National Atlas produces a Welsh population estimate of 142,000 pairs. Estimates from CBC data put the Welsh population at 120,000 – 131,000.

Breeding information: 334 found on Mynydd Epynt, Brecon in 1999, a increase of 43% since 1996 and in only 27% of sites monitored at Llanwrtyd Wells. 445 males on 39km^2 of common in central Brecon in 1995, 11.4 prs/km^2. 7.9 prs/km^2 found on mixed farmland in SE Brecon. In 2000 300+ were counted on the Epynt at densities up to 15 prs/km^2, 40 prs. On 703 ha of bog in the Llanwrtyd Wells area – a decline of 29% since 1998, 15 prs/km^2 on Pantylln Moors, a 40% decline since 1993. 620 – 900 pairs on the Elan Valley estate, Radnor in 1999, in some places up to 30/km^2. 80 pairs at Lake Vyrnwy, Montgomery in 1999 – a low density but typical of heather moorland.

Birds in Wales 1992-2000

Winter flocks of 200 or more were:
- 1992   c.1000 at Croesgoch Pembroke on Jan. 11th
- 1993   480 at Wernffrwd Glamorgan on Jan. 13th and 200 in stubble at Llandeilo-Abercywyn Carmarthen on December 26th
- 1994   350 at Llangwm Pembroke on December 21st
- 1995   1504 at Sker Point Glamorgan on Nov. 3rd; 200 at Porthmadog Caernarfon on December 24th
- 1996   430 at Llanrhidian marsh Glamorgan on Jan. 22nd; 276 at Wernffrwd Glamorgan on the 24th; 200 at Talbenny Pembroke on Nov. 15th
- 1997   200 at Wernffrwd Glamorgan on Feb. 13th; 421 at Wernffrwd on Nov. 16th; 560 at Whiteford Point Glamorgan on December 15th; 200 between Gwbert and Mwnt Ceredigion on December 1st; 300 on arable near Inner Marsh Farm Flint on Oct. 13th
- 1998   200 at Wernffrwd Glamorgan on Jan. 30th; 300 at Sarnau Ceredigion on Jan. 7th-9th
- 1999   400 at Talbenny, Pembroke Jan. 17th, 500 at Freshwater West, Pembroke in maize stubble November & December, 200 Abermenai Point, Anglesey Dec. 28th
- 2000   2000 on weedy maize stubble at Gupton/Freshwater West, Pembroke in January and 500 there in December, 200 at Crug Glas, Pembroke Jan. 1st and 200 in the salt-marsh at the Braint Estuary, Anglesey on Jan. 22nd

Passage flocks included 200 at Skomer, Pembroke on Oct. 10th and 200 at Castle Head, Pembroke on Oct. 19th 1995.

SHORELARK *Eremophila alpestris*   *EHEDYDD Y TRAETH*
(B) *Regular annually at one North Wales site in very small numbers, otherwise a scarce and irregular winter visitor and scarce passage migrant.*

The number of individuals seen in each winter is shown in the following graph. A breakdown of individuals by county in each winter (none in 1992/3 or 2000/01) is tabulated below:

	1993/4	1994/5	1995/6	1996/7	1997/8	1998/9	1999/2000
Pembroke					1	3	
Ceredigion			1				2
Meirionnydd				1			
Caernarfon		3		1	4	1	
Anglesey	1		1		1		1
Flint			4		4	35	
Total	1	3	5	3	10	39	3

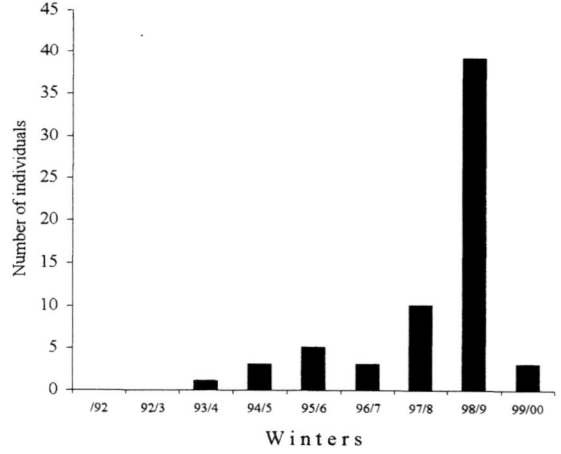

The main wintering area was at Gronant / Point of Air, Flint, where there were 4 during January – Apr. 17th 1995, 4 Nov. 14th – December 1997, before moving to Mostyn Dock in February 1998, 5 increasing to 24 by December 1998 and to 35 by January 1999 with 3 remaining until Apr. 16th.

Two other sites had multiple records: Castle Martin, Pembroke: single Feb. 7th – Mar. 21st 1998, 3 Nov. 21st 1998 – Apr. 17th 1999; and Cemlyn, Anglesey: singles Nov. 28th 1993 – Mar. 27th 1994, Mar. 16th – Apr. 22nd 1997, with 2 there Nov. 1st 1998.

SAND MARTIN *Riparia riparia*   *GWENNOL Y GLENNYDD*
(C) Locally common breeding visitor, most numerous in eastern counties, increasing. Passage migrant in spring and autumn.
Although Sand Martins breed in all Welsh counties, their distribution is uneven, absent from the uplands and with only small, scattered colonies in the west. The table below summarises the number of breeding pairs in each county during the period.

	1992	1993	1994	1995	1996	1997	1998	1999	2000
Gwent		292	400-410			160+		255	
Glamorgan		370	375-380	569	413-418	263-268	386-391	481*	785-790**
Carmarthen	1441	714	2089	3000		245	101		
Montgomery	92	125		180	410	160	170+		
Brecon		187	150	510	270	381	264	330	435***
Meirionnydd								20+	
Caernarfon								63	182
Anglesey							10+	19-20	30
Denbigh		50							
Flint					269	51			100

* figures for Gower only, where a 43% increase compared to 1998.
** figures for Gower only, a 62% increase on 1999 (637 along the Loughor, 76 Tawe, 22 Clydach, 21-26 Neath, 24 Afon and 5 Llan).
*** a 32% increase.

2915 holes were counted along the Tywi, Cothi & tributaries, Carmarthen in 1995, a 40% increase on 1994 figures. 190 nest holes were counted at Abergavenny Castle Meadows, Gwent in 2000.
The Brecon population is estimated at around 400 pairs, of which 350 were on the Usk (Aberbran – Talybont) in 1995 before declining to 127 in 1999, a further 100 pairs on the Usk down to Crickhowell, 100-120 pairs on the Wye, 20-40 pairs on the Irfon and 20-40 pairs Builth – Llanwrtyd. In 2000 there were 75 prs. on the Wye at Glasbury, 40 on the Irfon, 40 at Talybont Res. and 280 along the Usk between Aberbran and Ashford.
Using data from the National Atlas produces a Welsh population estimate of 12,600 pairs. Results from BBS, 1994-1999 suggests a population change of 115%, producing a revised population estimate of 14,500 pairs.

Passage numbers of 500 or more were:
1994   600 at Llandegfedd Res., Gwent on April 1st and 1000 there on 2nd-5th and 16th-17th; 500+ at Eglwys Nunydd Res. Glamorgan on April 7th and 1200 there on the 25th. In the autumn there were 2000 at Kenfig Pool, Glamorgan on Sept. 1st; 1000 at the Teifi Marshes Ceredigion/Pembroke in July and 1000 at Llangorse Lake, Brecon on July 11th.
1995   500 at Skomer, Pembroke on April 30th; 800 at Llangorse Lake on May 16th. In autumn there were 500 at the Teifi Marshes on August 7th; 1000+ at Cors Fochno/Ynyslas Ceredigion on July 31st and 500 at Llangorse Lake on July 28th
1996   800 at Eglwys Nunydd Res. on April 17th; 800 at Llangorse Lake on April 16th and 29th. In autumn 1060+ at Upper Loughor, Glamorgan on July 17th and 1000-1500 at Llangorse Lake on August 14th
1997   1000 at Llys y Fran Res., Pembroke on April 12th; 800 at Llangorse Lake on April 17th and 1000 there May 4th-6th. In autumn 500+ at Peterstone/Wentloog, Gwent on August 1st and 500 at Llangorse Lake in mid August
1998   600 at Llangorse Lake on April 23rd and in autumn 600-700 on July 20th
1999   2000 at Llangorse June 25th and Sept. 7th, 800 there on Aug. 12th. 700 at Eglwys Nunydd Res. Aug. 14th with 600 there on the 26th.
2000   500 at Uskmouth, Gwent July 23rd, 700 at Eglyws Nunydd Res. July 8th & 600 on Aug. 31st, 500 roosting at Kenfig on July 23rd, 2000 at Llangorse May 3rd, 500 at Glasbury, Brecon on July 6th and 800 at Ynyshir, Ceredigion Apr. 22nd.

SWALLOW *Hirundo rustica*   *GWENNOL*
(C) A widely distributed and common summer visitor and passage migrant.
On of the few species to breed in each 10-km square in Wales. Using data from the National Atlas produces a Welsh population estimate of 58,000 pairs. Results from BBS, 1994-1999 suggests a population change of 137%, producing a revised population estimate of 80,000 pairs.

Winter records are rare. Records during December – February included: one at Point of Air, Flint on Feb. 5th 1992; one picked up freshly dead at Abergavenny, Gwent on Jan. 20th, 1993; one (thought to be the same bird) on Cardigan, Ceredigion on Jan. 3rd and at Teifi Marshes, Pembroke on Jan. 13th, 1995; 1 – 2 birds at Malltraeth, Anglesey in December 1997, with one at the same site from Feb. 10th, 1998; two at Llangaffo, Anglesey on Feb. 3rd, 1998 and 1-2 birds again at Malltraeth, Anglesey up to December 31st 1998; in 2000 at Neath Abbey, Glamorgan Dec. 10th, at Bosherston, Pembroke Dec. 2nd and on Anglesey at Malltraeth Dec. 10th, Llanfaelog Dec. 12th – 16th with 2 at Mona Dec. 13th.

The only spring passages of 1000 or more were 1000+ at Kenfig Pool Glamorgan on April 26th 1995; 2500 at Eglwys Nunydd Res. Glamorgan on April 18th 1996, 1500 at Skokholm Pembroke on May 14th 1997, 2000 at Ramsey, Pembroke on Apr. 28th 1999.

and one at the same site and 1000 at Pen-y-Cae Res., Denbigh Apr. 22nd 2000.

Birds in Wales 1992-2000

Large autumn gatherings and movements of 1000 or more were:
1993   1000 at Eglwys Nunydd Res. on Sept. 7th; 4000+ at Wernffrwd, Glamorgan on Sept. 17th and 1840+ over Port Talbot, Glamorgan in 3 hours the next day
1994   2000 at Llanvaches, Gwent on August 30th, 1000 at Collister Pill, Gwent on Sept. 4th and similar numbers at Llandegfedd Res., Gwent on the 17th; 4000+ at Kenfig Pool on 1st Sept.; 5000 E at Lavernock Point, Glamorgan on Sept. 16th and 10,000 E here on the 24th, with similar numbers at Nash Point, Glamorgan and 6500 per hour E at Goldcliff, Gwent on the same date; 2000+ at the Teifi Marshes, Ceredigion/Pembroke during July and August, and similar numbers at the Dyfi Estuary, Ceredigion on August 5th
1995   1700 through Blackpill, Glamorgan in 1 1/2 hours on Sept. 20th; 3000 at Skokholm on the same day; up to 5000 at the Teifi Marshes in August and Sept.; 5000+ at Ynyslas/Cors Fochno Ceredigion on July 31st and August 2nd and 1000 W in 15 minutes at Glasbury, Brecon on Sept. 15th
1996   2000 roosting at Crymlyn Bog, Glamorgan on Sept. 4th and 2400 at Clyne Common, Glamorgan on the 23rd; 1000 per hour passing Strumble Head, Pembroke on Sept. 14th, with similar numbers here for 3 1/2 hours on the 26th; 3000 at Skokholm on Sept. 22nd and 1000 at Bardsey, Caernarfon on Sept. 13th
1997   1000+ at Peterstone/Wentloog, Gwent on Sept. 20th; 3000 at Kenfig on August 2nd; 3200+ at Clyne Common, Glamorgan on Sept. 22nd; 5000 at Lavernock on the 23rd, with 3500 there the next day; 1000 at Aberthaw Marsh, Glamorgan on Sept. 27th; 2000 at Skomer and 1000 at Strumble Head on Sept. 22nd; 6000 at Ramsey, Pembroke on the 23rd; 3-5000 at Gelly Wood, Pembroke on the 26th; 1500 SE in an hour at Cradoc, Brecon on Sept. 22nd and 1000 at Bardsey on the 25th
1998   A large passage during the period Sept. 19th – 23rd, with 2390+ past Blackpill in 15 minutes on the 19th; 6000+ over Clyne Common on Sept. 21st; 2030 E over Mynydd Margam, Glamorgan on the 22nd, with c.2700 E over Taibach and 10,000 in 3 hours at Lavernock Point, both Glamorgan the same day; 4000 at Ramsey on the 20th with 3500 past Strumble Head on the same date. A later wave of passage saw "thousands" at the Ogmore estuary, Glamorgan on Sept. 27th; 3000 on Skomer the same day and 2000 on Skokholm on Oct. 7th. There were roosts of 10,000 at Malltraeth RSPB, Anglesey on Sept. 1st and 3000 – 5000 at Slebech, Pembroke on Sept. 10th
1999   6000 at Kenfig on Aug. 13th, 1500 at Eglwys Nunydd Res. Aug. 22nd with 1100 there on the 26th, 1000 at Lavernock Point Sept. 26th. 3000 at Castle Martin Aug. 3rd, 5200 at Strumble Head Sept. 25th, 5000 at Porth Meudwy, Caernarfon Aug. 11th and 1008 at Bardsey Sept. 25th.
2000   2000 at GLWR Goldcliff, Gwent on Sept. 22nd and 2000 moving SE in 2 hrs. on Oct. 1st, 2000 moving SE over Lavernock, Glamorgan on Sept. 19th and 2500 on Oct. 1st, 2000 at Eglwys Nunydd Res., Glamorgan on Sept. 3rd, 2000 at Crymlyn Bog, Glamorgan Sept. 11th and in Pembroke, 3000 moving SE over Strumble Head on Sept. 25th, 10,000 in 2 hrs. at Skomer on Sept. 20th and 6000 moving E over Skokholm on Sept. 30th.

RED-RUMPED SWALLOW *Hirundo daurica*   *GWENNOL DINGOCH*
(B) A vagrant to Wales.

Eight records listed in *Birds in Wales*, since then a further two, both of singles in Pembroke. The first at Marloes Mere on May 7th 1995 and the second at Bosherston Pools on Feb. 15th-24th 1998 (earliest British record).

Of these 10 records, all have been of singles except for the record of at least 5 possibly 7 at Point of Air, Flint in October 1987.

HOUSE MARTIN *Delichon urbica*   *GWENNOL Y BONDO*
(C) Summer visitor, well distributed and locally common. Increasing.

Birds in Wales states that House Martins breed throughout Wales although colonies are generally small, often fewer than a dozen. During the period of this review there were few systematic breeding records are received, but there were over 250 pairs estimated breeding along sea cliffs around Castle Martin Pembroke in 1997.

Using data from the National Atlas produces a Welsh population estimate of 37,000 pairs. Results from BBS, 1994-1999 suggests a population change of 204%, producing a revised population estimate of 76,000 pairs.

Passage numbers of 500 or more were:
1993   1000 at Eglwys Nunydd Res., Glamorgan on August 4th and Sept. 7th
1994   There was a large movement on Sept. 16th, with 20,000 E. in 3 hours at Goldcliff, Gwent and 4000 E at Lavernock Point, Glamorgan 1000 at Eglwys Nunydd Res. on Aug. 4th and Sept. 7th 1993, on Sept. 16th 1994 and 5000 E at Nash Point, Glamorgan on Sept. 24th 1994
1995   2000 at at Eglwys Nunydd Res. on May 11th. In autumn 1200 here on August 24th; 600 at Llangorse Lake, Brecon on August 14th
1996   1500 at Eglwys Nunydd Res. on August 19th; 600 SE in 2 1/2 hours at Strumble Head, Pembroke on Sept. 20th
1997   690 E at Mynydd Margam, Glamorgan on Sept. 26th; 800 – 1000 at Tregoyd, Brecon on Sept. 11th
1998   Up to 5000 at Lavernock Point on Sept. 27th
1999   2000 at Eglwys Nunydd Res. Aug. 17th, 700 at Cwm Clydach Sept. 19th both Glamorgan and 1000/hr on Oct. 5th passing over the Epynt in Brecon
2000   700 moving SE over GLWR Goldcliff, Gwent on Oct. 1st

RICHARD'S PIPIT *Anthus novaeseelandiae*   CORHEDYDD RICHARD
(B) A scarce autumn passage migrant.

Birds in Wales quotes a total of 89 individuals recorded in Wales up to 1991. In the 9 years under review there were a total of 43 accepted individuals. The annual totals of individuals recorded in Wales during the period, 1992-2000, is shown in the chart.

All records were:
Glamorgan: singles Lavernock Point Oct. 19th 1997 and Rest Bay Nov. 3rd 2000.
Pembroke: singles Skomer Mar. 11th & Nov. 6th – 9th 1992, Skokholm Oct. 8th 1993, Oct. 14th – 15th 1994, Herbrandston Oct. 15th 1995, Ramsey Nov. 1st – 9th 1994, Strumble Head Oct. 22nd & 31st 1995 (probably two birds), Marloes 3 on Oct. 17th, only 2 the following day 1995 and Angle Bay Dec. 18th 1997.
Ceredigion: 2 Ynyslas December 1994 – January 1995 and single Mwnt Nov. 5th 1994.
Meirionnydd: single Fairbourne Oct. 2nd 1996, first county record.
Caernarfon: singles at Bardsey Sept. 27th 1992, Sept. 27th, Oct. 10th – 16th & 20th 1994 (3 birds), Sept. 14th & Oct. 8th 1996, Sept. 21st – 22nd & Oct. 18th 1999, Oct. 8th 2000 and 2 on the 21st; at Great Orme Oct. 7th – 8th, with 3 there 10th – 13th 1995 and Oct. 17th – 18th 1999, Llandudno Jan. 16th 1996 and Nov. 11th 2000, Conwy RSPB Oct. 4th – 14th & Nov. 15th – 16th 1996, Llanfairfechan Dec. 27th 1996, Carreg y Defaid Nov. 6th 1998 and Foryd Bay Nov. 14th 1999.
Anglesey: single at Braint Estuary Nov. 27th 2000.
Flint: singles at Oakenholt Nov. 23rd – Dec. 29th and at Greenfield Dock Dec. 10th – 31st 1999.

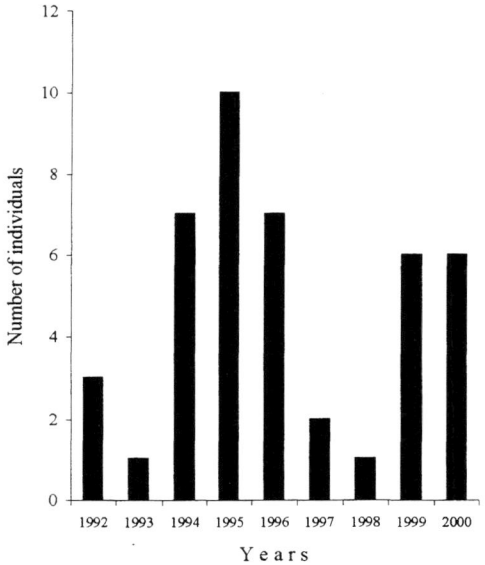

Breakdown of all Welsh records by month and by county, for Pembroke and Caernarfon the total in the county is given along with separate totals for the islands and mainland:

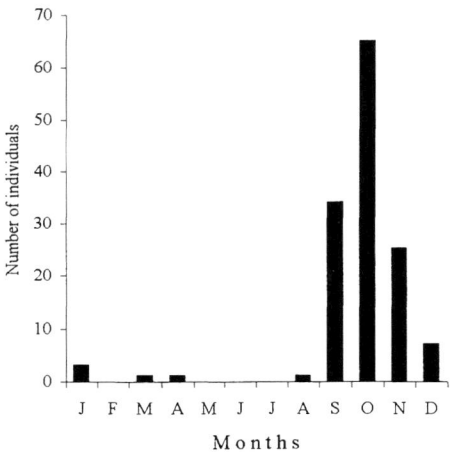

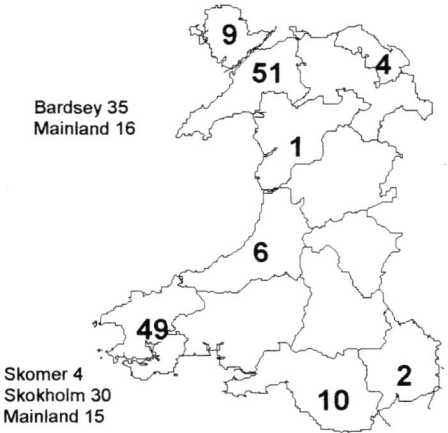

TAWNY PIPIT *Anthus campestris* *CORHEDYDD MELYN*
(B) A rare visitor.

Birds in Wales quotes 15 records of 16 individuals: Gwent: at Chepstow Apr. 27th 1988; Glamorgan: at Sker Farm Sept. 7th 1982, at Forway y Witch Apr. 28th 1990 and at Nash Point Aug. 31st 1991; Pembroke: at Skokholm Sept. 19th 1961, Sept. 18th 1968, Sept. 26th – 28th 1968, Sept. 13th 1970 and Oct. 11th – 12th 1991; at Skomer May 13th 1974, two Oct. 10th 1975, single Sept. 9th – 11th 1988; Ramsey Sept. 18th 1978; Caernarfon: at Aber Nov. 26th – 29th 1977, at Bardsey Sept. 1st 1980 and May 13th 1990.

An additional record of a single at Llandegfedd Res., Gwent Oct. 6th 1991. Since 1991 there has only been one further record of a single at Ramsey, Pembroke on May 7th 1994. This species remains one of the real gem finds in Wales, unfortunately usually on an island.

TREE PIPIT *Anthus trivialis* *CORHEDYDD Y COED*
(C) A widely distributed and locally numerous summer visitor, increasing. Passage migrant in fairly small numbers, best marked on the south coast in autumn.

Few systematic counts are available, but there were 38 territorial males in 3 Km^2 of Glasfynydd Forest, Brecon in 1993; 69 territories on Mynydd Epynt MoD, Brecon in 1997, 80 in 1999 & 52 in 2000 and 35 pairs at Lake Vyrnwy RSPB, Montgomery also in 1997.

Sample counts available from Dinas/Gwenffrwd, Carmarthen, 15-20 prs. in 1992, 24 in 1995, 14 in 1996 and 10 in 1999.

Using data from the National Atlas produces a Welsh population estimate of 21,500 pairs. Results from BBS, 1994-1999 suggests a population change of 121%, producing a revised population estimate of 26,000 pairs.

These figures are significantly higher than those quoted in *Birds in Wales* of between 8,000 and 10,000 pairs. The differences may be real, an large increase has been shown by BBS but some of the discrepancy probably lies at the extrapolation methods used in the Atlas.

MEADOW PIPIT *Anthus pratensis* *CORHEDYDD Y WAUN*
(C) A widespread and numerous breeding species, especially on moorland. Winter visitor and passage migrant. Most breeding birds are summer visitors.

One of the most widespread breeding species in Wales. Again few systematic counts are available, but in Brecon there were 520 pairs or territorial males on 39 km^2 of common in 1995; 40-50 territorial males in 5km^2 on the Black Mountains in 1996; 53 territorial males on 250 ha. of bog and old pasture at Bronffynnon, Llanwrtyd and 30 on 120 ha. of commons at Llanafan in 1997; 247 singing males or pairs on 829 ha. of bog, grassland and common in the north of the county and 130 on 9.5 km^2 of common south of Mynydd Epynt in 1998. In 1999 380 pairs were counted on Mynydd Epynt while areas around Llanwrtyd Wells showed a 28% decline compared to 1998. In 2000 there were 179 prs. On 703 ha of bogs in the Llanwrtyd Wells area and 15 prs/km^2 of ffridd near Llanwrtyd.

Large passage movements included c.1000 through Bardsey Caernarfon on March 22nd 1995, 723 on Sept. 25th 2000, 500 at Skomer, Pembroke Sept. 19th 2000 and c.2000 at Nash Point Glamorgan on Sept. 21st 1997.

Using data from the National Atlas produces a Welsh population estimate of 167,000 pairs. Results from BBS, 1994-1999 suggests a population change of 144%, producing a revised population estimate of 240,000 pairs. Analysis of CBC data however generates a population in between these figures, of 200,000 pairs.

RED-THROATED PIPIT *Anthus cervinus* *CORHEDYDD GYDDFGOCH*
(B) A vagrant to Wales.

There were singles in Glamorgan at Kenfig Pool on May 3rd 1992 and in Pembroke at Skomer on 6th-8th Sept. 1991 & Oct. 19th – 21st 2000 and at Skokholm on Sept. 17th and 27th 1992.

The only other Welsh records were from Skokholm Oct. 13th 1970 & Sept. 19th – 23rd 1989 and at Bardsey, Caernarfon Oct. 24th – 26th 1988.

ROCK PIPIT *Anthus petrosus* *CORHEDYDD Y GRAIG*
(C) A breeding resident of rocky coastlines in all coastal counties except Flint and Gwent. Winter visitor in small numbers.

Birds in Wales states that breeding is correlated with the presence of rocky shores and consequently this species is absent as a breeding species only from Gwent and Flint. In other counties it is a common inhabitant. Using data from the National Atlas produces a Welsh population estimate of 3,300 pairs.

Birds in Wales 1992-2000

The table below shows the number of breeding pairs recorded on the Pembroke islands and on Bardsey, Caernarfon:

	1992	1993	1994	1995	1996	1997	1998	1999	2000
Skokholm	53	49	44	57	57	33	20	14	29
Skomer		51			34	19	19	32	50
Ramsey	28	29	25	32	27	25		30	
Bardsey			34	35	33	34		24	23

Birds showing characteristics of the race *A. p. littoralis* were:
1993 One on Skokholm, Pembroke on April 9th
1995 Singles at Skokholm on May 9th - 30th and at Ynyslas, Ceredigion on December 27th
1996 Two at the Gann, Pembroke on April 13th, two on Bardsey, Caernarfon on April 10th - 19th and a breeding female here on May 15th, one at Red Wharf Bay, Anglesey on April 23rd
1997 Singles at the Ogmore estuary, Glamorgan on Nov. 2nd - 3rd, at Barafundle, Pembroke on Sept. 11th and on Bardsey on May 10th
1998 One at Stackpole, Pembroke on April 14th, two on Bardsey on March 30th
1999 Singles at Bardsey Apr. 15th & 26th
2000 2 at Bardsey on both Mar. 3rd and Apr. 25th.

WATER PIPIT *Anthus spinoletta*   CORHEDYDD Y DWR

(B) A winter visitor in small numbers, mainly recorded in coastal areas but also, less frequently in Wales, inland. Passage migrant in small numbers.

Recorded from all counties during the period under review, except Denbigh, with more of records in the south than in north Wales.

The total number of individuals recorded in each county during 1992-2000 is shown on the map opposite.

A large congregation built up at Ffwrd Fen, Carmarthen in 1999. A single present in January was joined by others, increasing to 14 by March.

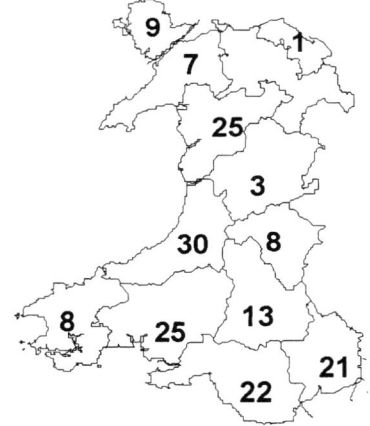

YELLOW WAGTAIL *Motacilla flava*   SIGLEN FELEN

(C) A summer visitor, now breeding regularly in only six counties.

Birds in Wales quotes a total of 180-190 probable breeding pairs in Wales. Although the above data are not complete, the figures suggest that there has been a marked decline with at most 100 prs. at the present time. It also quotes breeding along the Gwili Valley, Carmarthen as far as Conwyl Elfed – breeding this far north is now thought to be unlikely. The following table summarises the number of pairs or territorial males in each county during the period under review.

4-5 pairs bred along the River Severn, Montgomery, between Caersws and Newtown in 1995 and 30 breeding pairs were discovered at Shotwick Fields, Flint in 1999.

	1992	1993	1994	1995	1996	1997	1998	1999	2000
Gwent	1+	1-2	15		13+	5		7+	12-13
Montgomery	4+	7+	8-10+	8+	6	8	3	4-5	
Brecon	16-19	14	17-18	23-24	38	19-21	25	19-20	23
Radnor					5	4-7	4		
Denbigh						poss. 1			
Flint	bred	poss. 1	6			6+		33	18+

A large number at GLWR Goldcliff, Gwent in 2000, 93 on Aug. 30th and 30+ still there on Sept. 9th. 25 were also roosting at Rhymney, Glamorgan Sept. 12th 2000.

The map opposite shows the total number of individuals of the race *M.f. flava* (Blue-headed) recorded in each county during 1992-2000.

Records of the other races of Yellow Wagtail:
M.f. cinereocapilla (Ashy-headed): Single at Skomer, Pembroke on May 17th - 20th 1992.
Birds in Wales quotes 3 autumn birds on Bardsey, Caernarfon in 1913, another there September 1956 and April 1984. The only mainland record was from Glamorgan at Lisvane Res. May 3rd 1981.

M.f. thunbergi (Grey-headed): Singles at Peterstone/Wentloog, Gwent on May 30th 1996, at Skokholm, Pembroke on June 25th 1992, at Bardsey, Caernarfon on May 27th 1993, at Flimstone, Pembroke May 30th 1998 and at Conwy RSPB, Caernarfon May 20th 1999. *Birds in Wales* quotes 2 records on Bardsey, one Little Orme, 5 on Skokholm, and one at Strumble Head, all in May.

M.f. feldegg (Black-headed): A male at Conwy RSPB, Caernarfon on May 8th - 9th 1998 was the second Welsh record as a recent review of all records by BBRC led to the rejection of the record from Bardsey, Caernarfon on May 8th, 1976. The only other Welsh record was at Skomer, Pembroke May 7th 1986.

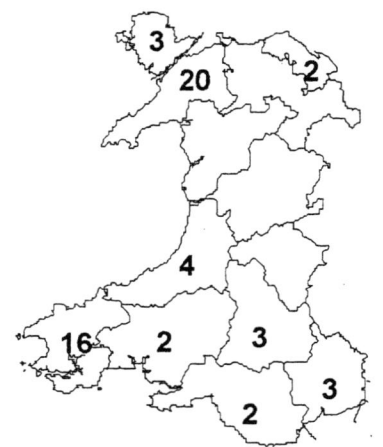

CITRINE WAGTAIL *Motacilla citreola*
(A) Vagrant.
The 1st Welsh record was of a 1st winter individual at Skomer, Pembroke Sept. 27th 2000.

SIGLEN SITRAIDD

GREY WAGTAIL *Motacilla cinerea* *SIGLEN LWYD*
(E) Locally common breeding resident, mainly along fast-flowing streams in all mainland counties of Wales. Increasing but very scarce on Anglesey. Most move away from hills in the autumn and either emigrate or frequent sites on lower ground.

Few systematic counts are available. In Gwent/Brecon, there were 21 pairs on the Grwyne Fawr/Fechan rivers in 1992 and 39 in 1995, 13 pairs on the Gwent section of the Grwyne Fawr in 1993 and 10 pairs on this section in 1994. In 2000 there were 5 pairs on the Afon Llwyd, 6 pairs on the River Ebbw Fach, 2 pairs on the River Ebbw and 2 on the Clydach, all Gwent.

The RSPB survey of 15,000 ha of Mynydd Du in Carmarthen/Brecon in 1996 found 15 territories. In Brecon, there were 11 pairs on streams within the MoD ranges on Mynydd Epynt in 1992, with 6 pairs here in 1996, 12 pairs in 1997, 16 pairs in 1999 and 17 in 2000. There were 15 pairs along the R. Irfon from Garth to the Devils's Staircase in 1992, but only 2 pairs along the Irfon above Llanwrtyd in 1999 and three pairs in 1997. There appears to have been a shift in the breeding population upstream in many rivers in this county (this explains the decline along the Irfon).

This species continues to be a scarce breeder on Anglesey. It is not uncommon on parts of the River Cefni and two pairs bred at Llangefni in most years.

Using data from the National Atlas produces a Welsh population estimate of 4,250 pairs. Results from BBS, 1994-1999 suggests a population change of 140%, producing a revised population estimate of 6,000 pairs.

PIED WAGTAIL *Motacilla alba* *SIGLEN FRAITH*
(C) A common and widespread breeding resident in many habitats. Autumn passage migrant and winter visitor. The continental race, White Wagtail, M.a. alba, is a regular passage migrant, occasionally breeding.

Using data from the National Atlas produces a Welsh population estimate of 30,600 pairs.

Roost counts of 200 or more were in Gwent 200+ at a factory in Pontypool at both ends of 1997 and 200 at Abergavenny on Dec. 21st 1997 & December 2000; in Glamorgan, 226+ at Oxwich Marsh on Nov. 3rd 1996, 300 at Cardiff City centre on Dec. 1st 1998 and 200 on the Christmas tree lights in Cardiff City in December 2000. There were 1000 at the Gulf Refinery, Pembroke on May 18th 1998, 250 at Llangorse Lake, Brecon in 1995, 600 at Ysbyty Gwynedd, Caernarfon on Oct. 9th 2000. 300-400 passed SE over Pen-y-Cae Res., Denbigh on Dec. 2nd 2000.

White Wagtails, the race *alba*, are regular spring and autumn migrants, occurring widely in Wales in late spring and autumn. Large counts included: 60 at Uskmouth, Gwent on Apr. 16th 2000, 50 there on the 20th, 68 at Ynys-hir, Ceredigion on Apr. 12th 1997, 106 at Llyn Traffwll, Anglesey on Apr. 19th, 1995, 50 at Llyn Alaw, Anglesey on May 1st, 1995, 100 at Shotwick, Flintshire, on Apr. 23rd 1995 and 100 at Cemlyn, Anglesey on Apr. 19th 1999.

In autumn, a significant passage was recorded in 1996, 1997 and 1999, large counts being 100 at the Gann, Pembroke on Sept. 1st, 1996, 80 on Bardsey, Caernarfon on Sept. 9th 1996, 100 here on August 30th 1997 & 81 on Aug. 30th 2000 and a total of 130 in September 1999 at Cemlyn, Anglesey.

WAXWING *Bombycilla garrulus* *CYNFFON SIDAN*
(B) An irruptive winter visitor from Northern Scandinavia and north-east Europe; occurs erratically in Wales in small numbers, most usually in years of major irruption in eastern England.

There were no records in the winters of 1991/2, 1992/3, 1993/4 and 1994/5. There was a major invasion of England in the winter of 1995/6, with good numbers of birds reaching Wales by late January 1996, when a total of 157 birds were recorded between Jan. 27th and Mar. 10th.

The map opposite shows the number of individuals recorded in each county in that winter.

In the subsequent winters:
1996/7 there were 3 individuals in Gwent, one in Carmarthen, 3 in Ceredigion, 4 in Radnor and one in Caernarfon.
1997/8 there were 5 in Glamorgan.
1998/9 there were singles in both Caernarfon and Flint.
1999/00 there was a single in Brecon and 6 in Radnor.
2000/01 there was a single in Brecon.

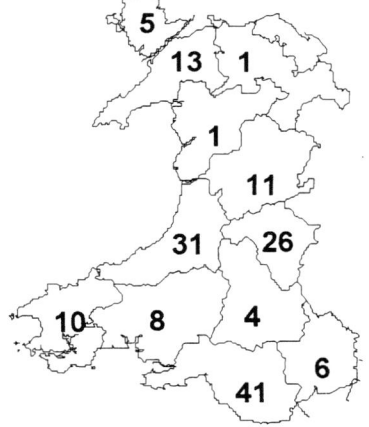

Birds in Wales 1992-2000

DIPPER *Cinclus cinclus* *BRONWEN Y DWR*
(C) Common resident of fast flowing streams, particularly in hill areas. In winter some individuals move into the lowlands and also make use of estuaries, rocky coasts and lake edges.

Birds in Wales states that Dippers are evenly distributed throughout the hill areas of Wales, although its density varies dependent on such factors as gradient, availability and more recently acidification. Using data from the National Atlas produces a Welsh population estimate of 1,750 pairs compared to the population estimate of 2,000 – 2,500 reported in *Birds in Wales*.
Other surveys showed 65 birds along the River Tanat and tributaries, Montgomery, in February 1993 and 27 along the River Wye from Newbridge to Hay, Radnor, on Feb. 4th 1996. Streams within the MoD ranges on Mynydd Epynt, Brecon held 5 pairs in 1992 and 4 pairs in 1996.

There was very little breeding data reported during the period 1992-2000, with only one river counted annually, the Grwyne Fawr and Fechan on the Gwent / Brecon border. Annual counts were 26 in 1992, 26 in 1993, 10 in 1994, 33 in 1995, 11 in 1997, & 1999 and 13 in 2000 (the counts for 1994, 1997, 1999 and 2000 were for the River Grwyne Fawr only).
In 1999 there were 5 pairs on the Afon Lwyd, Abersychan – Cwm Avon (3 in 2000) and 4 pairs on the River Clydach (5-9 in 2000), both Gwent; 4 pairs Cwm Clydach in Glamorgan; 7 pairs on the Epynt, Brecon; 6 pairs at Lake Vyrnwy (8 in 2000); Montgomery; 12 pairs along the Afon Ogwen, Caernarfon from Halfway Bridge to Aber Ogwen.

WREN *Troglodytes troglodytes* *DRYW*
(E) Abundant resident, found from sea-level to 3,000ft.

The Wren is one of the most numerous of Welsh birds, breeding in a wider variety of habitats than any other species. Using data from the National Atlas produces a Welsh population estimate of 680,000 pairs. Results from BBS, 1994-1999 suggests a population change of 115%, producing a revised population estimate of 783,000 pairs. A pair nested within 50m of the summit of Snowdon, Caernarfon in 1999.

DUNNOCK *Prunella modularis* *LLWYD Y GWRYCH*
(C) Abundant breeding resident. Passage migrant and possible winter visitor in small numbers.

One of the most widespread and abundant birds in Wales, breeding throughout the country in a variety of habitats. Using data from the National Atlas produces a Welsh population estimate of 200,000 pairs. Results from BBS, 1994-1999 suggests a population change of 116%, producing a revised population estimate of 213,000 pairs. At the end of September 2000 there was a large influx onto Skomer, Pembroke, with 80 by the end of the month and 100 on Oct. 10th – 13th.

ALPINE ACCENTOR *Prunella collaris* *LLWYD Y MYNYDD*
(B) A vagrant.
A bird found on the lighthouse steps, Strumble Head, Pembroke on Oct. 30th 1997 was only the second record for Wales. The first Welsh record was of one on the Llanberis side of Snowdon, Caernarfon Aug. 20th 1870.

ROBIN *Erithacus rubecula* *ROBIN GOCH*
(C) A common and widespread resident in all counties and a passage migrant.

One of the most abundant and widely distributed birds in Wales, breeding in every 10-km square throughout the country. Using data from the National Atlas produces a Welsh population estimate of 437,000 pairs. Results from BBS, 1994-1999 suggests a population change of 111%, producing a revised population estimate of 485,000 pairs.

This is consistent with estimates generated from CBC data.
A completely white individual was at Llanberis, Caernarfon Aug. 1998 until 1999. Influx onto the Pembrokeshire islands at the end of September 2000, with 150 on Skomer on Sept. 30th, 102 on Oct. 15th and 70 on Nov. 13th, 100 counted on Ramsey at the end of October.

NIGHTINGALE *Luscinia megarhynchos* *EOS*
(B) Scarce passage migrant; formerly bred locally in the east. No breeding since 1981.

Since *Birds in Wales* there have been 20 records of this species in Wales, most were only birds of passage but at least 5 remained for a short period of time. No definite breeding records although the series of reports from Ynyshir was possibly a breeding bird. A breakdown of these by county:

Gwent: at Mamhilad May 22nd – June 1992, Wyesham May 1997 & May 8th – 24th 1998.
Glamorgan: at Kenfig Nov. 6th 1993 and in Cardiff Apr. 24th – June 5th 1995.
Pembroke: at Strumble Head May 14th 1992, Goodwick May 4th 1995, Skomer June 14th 1995 and at Skokholm

May 27th 1992, May 10th 1993 & Apr. 16th 1995.
Ceredigion: at Ynyshir May 5th & June 28th - July 1st 1995 (presumably the same bird) and May 20th 1997.
Montgomery: at Penstrowed May 14th 1995.
Caernarfon: at Bardsey May 25th 1992, May 13th & 14th 1994 (both trapped), May 5th 1995, May 15th & Aug. 7th 1997. Single at the Great Orme Apr. 23rd 1998.

BLUETHROAT *Luscinia svecica*

BRONLAS

(B) A rare and irregular passage migrant in spring and autumn.
Birds in Wales quotes 30 individuals recorded in Wales, since then there have been a further 8. All records were: Gwent: at Uskmouth, Oct. 1st 1999; Pembroke: at Skokholm Sept. 27th 1992 & a red-spotted male May 29th 1995 and at Bosherston June 8th 1993; Caernarfon: at Bardsey Oct. 10th 1994 and at Conwy RSPB a 1st winter male Nov. 1st – 11th and an adult male 9th – 21st 1997; Anglesey: a 1st winter female at South Stack Sept. 27th 1997.

Breakdown of all Welsh records by arrival date and by county:

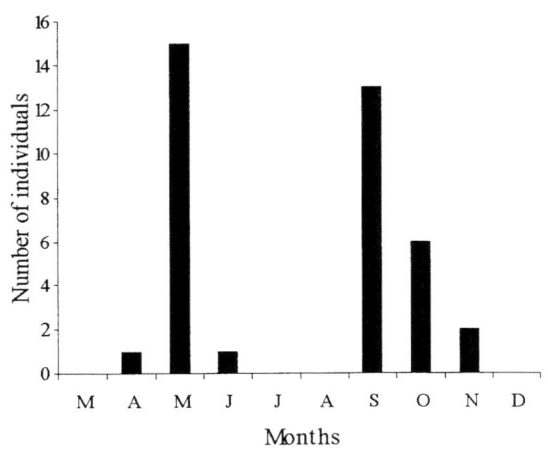

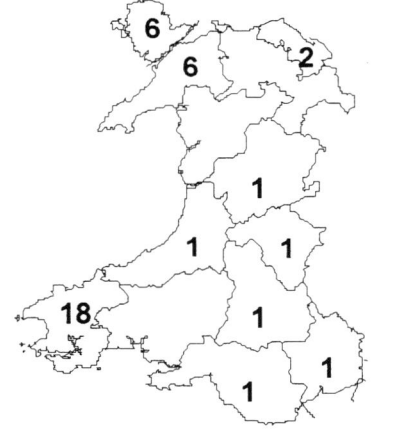

BLACK REDSTART *Phoenicurus ochruros*

TINGOCH DU

(C) Spring and autumn passage migrant. Winter visitor in small numbers. Has bred at least twice.

	1992	1993	1994	1995	1996	1997	1998	1999	2000
Gwent					2 & 0	2		2	1 & 0
Glamorgan	3 & 12	7 & 5	2 & 11	10 & 12	3 & 0	5	4	6	3 & 0
Carmarthen	1 & 4	5 & 0	1 & 4	8 & 0	1 & 0		1		3 & 0
Pembroke	7 & 3-6	3 & 2	4 & 1	7 & many	10+ & 6	4-5 & 2	3 & 7	14 & 4	7 & 15
Ceredigion	1 & 3	4 & 0	3 & 6	2 & 1	6 & 0	3		2	7 & 0
Brecon	1 & 0	1 & 0	0 & 1						
Montgomery			0 & 1	1 & 0					
Meirionnydd				0 & 1				2	0 & 1
Caernarfon	0 & 6	6 & 4	0 & 6				1	2	5 & 1
Anglesey	1 & 4	1 & 1	0 & 4	4 & 2	2+ & 5	1 & 0	2 & 5	2	1 & 3
Flint				2 & 0		1			1 & 0
Approx. Totals	14 & 32	27 & 12	10 & 34	34 & 16+	24 & 11	17 & 6	9 & 13	31 & 9	28 & 20

Small numbers over-winter in coastal areas each year. The table above is a breakdown of wintering individuals by county, the first number is the approx. number of birds present in the early months of each year, the second the approx. number of birds present in November / December.

Spring passage, again mainly in coastal areas, beginning the 1st week of March, lasting into early June. Below is a summary of passage periods and the numbers of individuals involved.

Birds in Wales 1992-2000

		1992	1993	1994	1995	1996	1997	1998	1999	2000
Number		-	-	20	150	40	30 - 40	13	20	12
Period	from	Mar. 4th	-	Mar. 5th	Mar. 8th	Mar. 14th	Mar. 3rd	Mar. 9th	Mar. 4th	Mar. 3rd
	to	May 29th	-	May 28th	June 5th	June 16th	May 30th	May 3rd	May 17th	May 20th

Inland individuals at Llandewi Rhyddech, Gwent May 3rd and Cradoc, Brecon Apr. 15th, both 2000.

Late birds include one or 2 on Bardsey into July 1993 a single there July 18th 1994, a male singing in Cardiff May 18th 1995, at Skokholm, Pembroke July 6th 1997 and a male at Machen, Glamorgan June 22nd 1998.

Black Redstarts are fairly late autumn migrants, with passage not usually until mid-September, continuing into November and occasionally December. Below is a summary of passage periods and the numbers of individuals involved.

Large counts in Pembroke include 13 at Skokholm Oct. 7th 1992, 10 at Castle Martin, Ramsey & Skokholm in October 1995, 7 at St. Brides in November 1996 and 9 on Skomer in 2000.

		1992	1993	1994	1995	1996	1997	1998	1999	2000
Number		c30	45	73+	135	over 100	c25	30	32	39
Period	from	Oct. 13th	Sept. 2nd	Oct. 7th	Oct. 11th	Sept. 18th	Aug. 5th	Aug. 26th	Sept. 17th	Sept. 22nd
	to	late Nov.	Nov. 11th	Nov. 15th	Nov. 24th	Nov. 5th	Dec. 4th	Nov. 25th	Nov. 21st	Nov. 28th

REDSTART *Phoenicurus phoenicurus*   *TINGOCH*
(C) A numerous breeding species in deciduous woodland and ffridd in many parts of upland Wales. Locally common elsewhere; scarce on Anglesey and Pembroke.

The Redstart is one of the most attractive constituents of the ancient sessile oak woods of upland Wales, where it is quite numerous. The Principality is one of the main strongholds for this species in Britain. Using data from the National Atlas produces a Welsh population estimate of 28,260 pairs.

Few sites were counted on an annual basis. Those that were include (numbers refer to the number of breeding pairs / territorial males): Glamorgan: 23 Cwm Clydach RSPB in 1995, 15 there in 1999; Carmarthen: 30 Dinas / Gwenffrwd in 1992, 70 in 1994 and 23 in a 45 ha sample in 1997; Brecon: 53 Epynt in 1992 & 25 in 1996, 60 males counted on MOD Sennybridge in May 2000; 40 Cradoc / Aberyscir / Pontfaen, Brecon / Carmarthen in 1996, 32 in 15,000 ha Mynydd Du, Carmarthen / Brecon 1996; Ceredigion: 35 pairs Ynyshir in 1993;
Montgomery: 70 Lake Vyrnwy 1993, 24 in 1995, 13 in 1996, 19 in 1997, 10 prs. with a total of 13 young fledged in 1999; Meirionnydd: 8 prs. on 40 ha. Coedydd Afon Gwynant, 6 prs. on 50 ha Coed Garth Cell, Mawddach 1999.

The main spring arrival is concentrated around the 3rd week of April and autumn passage peaks in September with few individuals to be found on the mainland by the end of that month. Unseasonal / late records were: an early bird in at Skomer, Pembroke Mar. 23rd 1995; late records were: in Glamorgan at Bacon Hole Nov. 9th 1995, Pennard Oct. 31st – Dec. 26th 1995, Sketty Park Oct. 26th and Killay Oct. 30th 1997; at Skomer Nov. 6th 1992; at Cwm Ystwyth, Ceredigion Oct. 25th – 31st 1995; at Trefeddian, Meirionnydd Nov. 1st 1995. Abercastle, Pembroke Nov. 20th 1999 – latest ever in that county.

WHINCHAT *Saxicola rubetra*   *CREC YR EITHIN*
(C) Summer breeding visitor; now predominately a species of upland areas. Declining.

Birds in Wales states that Whinchats are among the latest summer visitors to arrive, usually not until late April. On the uplands it is a fairly common breeding bird throughout Wales, although less so on Anglesey and in the south west. The Welsh strongholds are the extensive areas of bracken dominated ffridd of the valleys and hillsides. Few areas were counted, during the period under review, on a regular basis. Those that were are included in the table below, of breeding counts (no data available for 2000.

Numbers fluctuated annually but data from Brecon of common & boggy land, 1997, similar to that of 2.5 prs / 10ha, as quoted in *Birds in Wales* (from Carngafallt in Brecon). Figures do however suggest that this species is declining. CBC data suggest a decline of 16-20% since 1991 and that the Welsh population is estimated at 2,400 pairs, much less than the 5-6,000 prs. as quoted in *Birds in Wales*.

		1992	1993	1994	1995	1996	1997	1998	1999
Gwent	Garnclochdy		30	30			21	29	32
	Blorenge		35	40			19	35	40
	Mynydd-y-Garn Fawr		17	19			4	21	14
Carmarthen	Dinas/Gwenffrwd	7	5	10	12	7	11	16	

Other breeding data:
1992   30 males on 250 ha of boggy pasture Llanwrtyd Wells, 33 on the Epynt, both Brecon
1993   20 on Coty Mountain, Gwent and 30 in 3 km^2 of Glasfynydd Forest, Brecon
1994   50% decline in the Llanwrtyd Wells area and 27 % decline on the Epynt
1995   46 prs. located in 39 km^2 in Brecon, 34 prs. on 150 ha Elenydd, Brecon/Radnor/Ceredigion.
1996   RSPB survey of 15,000 ha Mynydd Du found 31 territories, in Brecon 13 prs. in 4000 ha Cnewr estate and 22 on the Glasfynydd forest
1997   48 prs. on 220 ha of boggy pasture Llanwrtyd Wells, 16 on 120 ha of common at Llanafan and 14 at Carngafallt, all Brecon
1998   North Brecon: 70 prs. on 442 ha of boggy pasture, 17 prs. on 139 ha common (15 prs. / km^2 – a slight decline)
1999   38 prs. on 200 ha Cwm Onnen, Brecon, 10 prs. on 5km^2 of moorland at Llyn Dwr. 21% decline at Llanwrtyd Wells but increasing on Mynydd Epynt
2000   19 males on Cors Caron, Ceredigion

Early records were from Skomer, Pembroke Mar. 11th 1995 & Mar. 27th 1996. Late records at Peterstone, Gwent Dec. 27th 1993 and at Ystumtuen, Ceredigion Oct. 16th 1995.

STONECHAT *Saxicola torquata*   *CLOCHDAR Y CERRIG*
(C) Breeding resident, locally common in coastal areas and thinly distributed inland. Numbers regularly reduced after hard winters. Increasing.

Stonechats are familiar birds of undeveloped areas almost all round the coasts of Wales, especially on gorse and brambles mixed with bracken. The western coasts of Wales are a stronghold for his species in Britain. Generally numbers increasing due to milder winter weather. The table below summarises the approximate number of breeding pairs in each county, where data is available, compared to the results of the last detailed survey in 1968 (the blanks refer to no data). The 1968 census estimated the total of Welsh breeding population as 420 pairs. There has been a significant increase since then and using data from the National Atlas puts the current Welsh population at 3,000 – 3,500 pairs.

Birds in Wales 1992-2000

	1968	2000		1968	2000
Gwent	-	25+	Montgomery	1	8+
Glamorgan	59	120+	Meirionnydd	32	-
Carmarthen	2	42+	Caernarfon	73	widespread
Pembroke	97	widespread	Anglesey	143	widespread
Ceredigion	-	widespread	Denbigh	1	14
Brecon	5	22+	Flint	7	-
Radnor	-	55+			

Few sites were censused regularly enough to gain any meaningful data. Results of surveys, grouped by County are shown below.

Gwent: 18 – 19 prs. in the Valleys & south Black Mountains in 1993.

Glamorgan: 20 prs. along 8km sea cliff, Caswell to Little Tor in 1998. In 1999 58 prs. found by GOS/RSPB survey of unenclosed cliff, Mumbles to Rhosilli, 24 prs. at other sites.

Carmarthen: significant increase noted, particularly inland. County total estimated at 6 prs. in 1992, 10 in 1993, 28 in 1994, 36-40 in 1995, 42 in 1996 (20 of which were on Mynydd Du) and 20+ in 1997. 17 prs. were found at Pembrey/Burry Port in 1998.

Brecon: 7 prs. on 4,000 ha of Cnewr estate and 3 prs. Garnclochdy forest in 1996 (highest ever recorded in Brecon).

Radnor: 10 prs. located on 150 km^2 Elenydd, nearly all to the east of the reservoirs in 1995, 21 – 22 prs. located in 19 1-km squares in 1997.

Birds in Wales quote 5 records of Siberian Stonechat *maura/stejnegeri*, from Bardsey October 1983, Llanfairfechan November 1983, Strumble Head October 1986 and 2 at South Stack in October 1987. Since then there have been a further 5 records, in Pembroke: on Skokholm Oct. 11th – 15th 1991 & Sept. 30th – Oct. 2nd 1992 and Strumble Head Oct. 6th 1997 (1st w / female); Bardsey, Caernarfon a 1st w / female Oct. 29th 1997 and Oct. 14th 2000.

WHEATEAR *Oenanthe oenanthe* — TINWEN Y GARN

(C) Summer visitor and passage migrant. Locally common, favouring upland areas with sheep-grazed turf.

The Wheatear is a numerous visitor to Wales between March and October, breeding throughout all the counties of Wales. The Welsh population is estimated at between 5,000 and 10,000 pairs based on the National Atlas. There were few breeding surveys during the period:

- 1992 345 prs. on 126.5 km^2 of sheep-walk, Mynydd Du, Carmarthen, an average of 13 prs. / km^2 (281 prs. in 1979, although there was less coverage); 34 prs. Epynt MOD, Brecon
- 1994 191 prs. on 150 km^2 Elenydd, 33 prs. Epynt MOD and 60 on 39 km^2 of commons in the centre of the county
- 1996 648 territories found by RSPB on 15,000 ha of Mynydd Du (4.4 prs. / km^2), 80 on 4000 ha Cnewr estate Brecon (2 prs. / km^2). 25 territories on the Epynt, a 24% decline on 1995 figures
- 1997 Decrease noted at Blorenge, Gwent. 36 prs. on the Epynt
- 2000 45 prs. Epynt

The numbers of breeding pairs on the islands, are tabulated below:

	1992	1993	1994	1995	1996	1997	1998	1999	2000
Skokholm	11		12	9	11	17	10	5	9
Skomer	12		32	38	31	20	29	38	11
Ramsey	25		67	68	78	88	57	79	42
Bardsey		21	19	18	20	18	7	8	4

Large counts of migrants were:
- 1993 107 at Port Talbot, Glamorgan Aug. 23rd
- 1994 at Bardsey, Caernarfon 7 days of 100+ in April / May, 200 on Apr. 26th
- 1996 at Bardsey, 200 on Mar. 24th, 100 on the 26th and 150 on Apr. 20th; 102 at South Stack, Anglesey Apr. 21st
- 1997 200 at Bardsey Apr. 26th; 60 at Ramsey, Pembroke Sept. 21st. 40 Greenland Wheatears *leuchorhoa* at Skomer, Pembroke Apr. 17th
- 1998 561 at Bardsey on Mar. 30th, 85 of which were of the Greenland race, 150 at Cemlyn, Anglesey Mar. 29th
- 1999 500 at Ramsey Apr. 27th, 165 at Skokholm on the 29th and 100 at Skomer on the 30th, all Pembroke
- 2000 152 at Bardsey Apr. 25th – 26th (including many of the Greenland race)

Exceptionally early individuals were at Skomer, Pembroke Feb. 25th 1998 and at Bangor, Caernarfon Jan. 22nd 2000. Late individuals at Port Talbot, Glamorgan Nov. 1st 1998, Ginst Point, Carmarthen Dec. 8th 1996, Borth Ceredigion Nov. 28th 1993, at Oakenholt, Flint Nov. 7th 1996 and at Llanrhidian, Glamorgan on Nov. 25th 1999.

PIED WHEATEAR *Oenanthe pleschanka* *TINWEN FRAITH*
(B) A vagrant.
A male at Ramsey, Pembroke Oct. 25th 1993 was the second Welsh record. The first, also from Pembroke but from Skokholm was a female Oct. 27th – 29th 1968.

BLACK-EARED WHEATEAR *Oenanthe hispanica* *TINWEN CLUSTIOG DU*
(B) A vagrant.
The only definite record of this species in Wales, as published in *Birds in Wales*, was of a male at Bardsey, Caernarfon Apr. 18th 1970. There is an additional record of a male at Skomer, Pembroke May 4th 1990. Since then there has been one further record, of a male in song at Bardsey May 6th 1992.

ISABELLINE WHEATEAR *Oenanthe isabellina* *TINWEN ISABELLA*
(A) A vagrant.
The first record for Wales was of a first winter bird at Bardsey, Caernarfon Sept. 20th – 21st 1997. Remarkably it was immediately followed by another, established by moult differences as definitely a different bird, at Skokholm, Pembroke from Sept. 24th – 26th 1997.

DESERT WHEATEAR *Oenanthe deserti* *TINWEN Y DIFFAETHWCH*
(B) A vagrant.
The only record in *Birds in Wales* was of a first summer male at Penclawdd, Glamorgan Nov. 21st – 22nd 1989. Since then there have been two further records; at Peterstone Church, Gwent Dec. 16th – 20th 1996 and a female at Skokholm, Pembroke Dec. 12th 1997.

BLUE ROCK THRUSH *Monticola solitarius* *BRONFRAITH LAS Y GRAIG*
(A) A vagrant.
A male at Moel-y-Gest, Caernarfon on June 4th 1987 which was originally accepted onto Category D of the British list, has now been transferred onto Category A as a genuine vagrant. This remains the only Welsh record.

RING OUZEL *Turdus torquatus*   MWYALCHEN Y MYNYDD
(E) Summer visitor breeding in the uplands in decreasing numbers; passage migrant.

Tyler & Green (1988) suggest a Welsh population of 400 – 500 prs, Birds in Wales 365 – 425 prs. Present figures suggest that this species continues to decline. Few detailed breeding surveys. In 1995 a CCW survey of north Wales (repeat of 1988 – 91 survey) recorded this species in only 26 out of 52 tetrads – a real decline. As part of a UK survey of this species in 1999 a total of 35 tetrads were surveyed. This gives a population estimate of 293-392 pairs in Wales.

Breakdown of breeding records by county:

Gwent: The only recorded breeding at Trefil quarries, a long established site was in 1997.

Glamorgan: Single pairs bred in Glamorgan at 4 sites in 1994. None in 1999.

Carmarthen: A pair located plus 2 other males in 1999.

Pembroke: A pair bred in Pembroke in 1995, the first time since 1971.

Brecon: 8 prs. Mynydd Du in 1992 on 126.5 ha (cf. 17 in 1978). 14 prs. bred on the Beacons/Black Mountains in 1994, 6 of which were successful. An RSPB survey of 15,000 ha of Mynydd Du in 1996 found 12 territories, a 30% decline since 1978. Elsewhere, the total numbers of pairs / sites for other years was 8 – 10 prs. at 6 sites in 1995, 9 prs. at 8 sites 1996, 8 prs. at 6 sites 1997, 9 prs. at 7 sites in 1998 and 11-16 prs. in 1999.

Radnor: RSPB / CCW survey of 150 km^2 of Elenydd (also in Brecon / Ceredigion) in 1995 found only 6 pairs (cf. 13 in 1975). The only record for 1997 was of a single breeding pair, while in 1999 there were 13 prs. at 8 sites.

Montgomery: Single pair Lake Vyrnwy 1994 & 1997, 4 prs. there in 1998 and 2 prs. in 1999 & 2000. Bred at 2 other sites in 1999.

Meirionnydd: 6 territories at Blaenau Ffestiniog in 1995. Single pair 1997. Survey of the Rhinog Mountains, in 1998, a comparable study to the survey 1966 – 1975, using the same criteria of site occupancy found 12 territories – 44% of earlier figures (cf. 27 territories). No apparent reason for decline, habitat still all right except in a few cases where there had been loss of heath. In 1999 an RSPB survey found 49-67 territories in 11 locations (2 further sites did not hold any birds), of which 21-31 territories on Cadair Idris, 7 at Moelwyn, 5-8 at Diffwys Rhinogau, 4 at Arenig Fawr.

Caernarfon: Bred at 8 sites in 1994, a single pair bred in 1997. In 1998 there were 2 prs. at Cwm Ystradllyn & Nant Peris and present at 6 other sites. In 1999 there were 15 prs. at Llanberis. Birds were reported to be more widespread than in recent years at Cwm Idwal, where there had been a reduction in sheep grazing. 1 pair at Cwm Pennant and birds present at 4 other sites. In 2000 breeding was reported at 4 sites.

Denbigh: Extinct as a breeding species at Mynydd Hiraethog (cf. 5 in 1977) in 1994. Decline reported in the Horse Shoe Pass and Eglwyseg rocks in 1997, a pair at the Horseshoe Pass in 1998 & 1999 and 1-2 prs. at 4 other sites in 1999.

This species remains an uncommon passage migrant, with several records annually in most coastal counties.

EYE-BROWED THRUSH *Turdus obscurus*   BRYCH AELIOG
(A) A vagrant.
A first winter bird ringed on Bardsey, Caernarfon on Oct. 12th 1999 was the first Welsh record.

BLACKBIRD *Turdus merula*   MWYALCHEN
(D) Abundant breeding resident, passage migrant and winter visitor.

Birds in Wales states that Blackbirds are inhabitants of a wide spectrum of habitats from gardens to farmland, woodland, cliffs and moorland. Recorded as a breeding bird in every 10-km square in Wales. Using data from the National Atlas produces a Welsh population estimate of 480,000 pairs. This is consistent with estimates generated from CBC data.

FIELDFARE *Turdus pilaris*   SOCAN EIRA
(C) A winter visitor and passage migrant in variable numbers; usually abundant.

Numbers of wintering Fieldfares in Wales varies from one year to the next depending on the weather and the availability of fruit crops. Large flocks, over 500 recorded at:

1993 600 Collister Pill, Gwent Nov. 21st, 400-500 Epynt, Brecon Dec. 23rd, 2000 / hr. moving south at Porth Meudwy, Caernarfon Oct. 14th and c3,000 in a mixed Redwing roost at Gwysaney, Clwyd

1994 500 Olway Meadow, Gwent Mar. 5th, 650 Merthyr Mawr, Glamorgan Nov. 21st, 500 Llandeilo, Carmarthen Nov. 29th, 1200 passing SE over Llanwrtyd Wells, Brecon Nov. 2nd with 750 at Upper Chapel / Cwm Owen on the 14th

1995 850 Llyn Clwyd Jan. 7th, 500 Llanwrthwl, Brecon, Nov. 6th – 10th, 700 Lake Vyrnwy, Montgomery in November

1996 800 Pitton Cross, Gower Feb. 17th; 500 in Carmarthen at Cilsarn Bridge, Feb. 27th & Cynghordy Nov. 21st; in Brecon 500 at Cefn-gorwydd Jan. 14th, and Llanafan Apr. 4th, with 900 at Blaengwdi Jan. 22nd; 1100 flew W over Presteigne, Radnor Feb. 5th

1997	few in second part of the year, possibly due the failure in the Hawthorn berry crop. Only large count was of 10,000 going SW over Cwm Ystywth, Ceredigion Nov. 11th
1998	1,500 Ogmore Est., Gwent Jan. 11th, 600 Cynghordy, Carmarthen Nov. 11th, 800 Cors Caron, Ceredigion Nov. 6th, 1,800 Epynt, Brecon Nov. 5th and 1000 over Ruabon, Denbigh
1999	500 Glasbury, Brecon Jan. 2nd, 1,000 at Goldcliff, Gwent Nov. 4th with 750 on the 7th, 2,000N over the Epynt, Brecon on Nov. 5th, 2,000 at Merthyr Mawr, Glamorgan on Dec. 10th
2000	750 at Redwick and 600 at Uskmouth, Gwent Feb. 6th, 1000+ at Candleston, Glamorgan Jan. 21st and 500 at Kenfig Feb. 23rd. 1500 at Uskmouth Nov. 19th, 600 at Cynnant, Carmarthen Nov. 11th, 600 at Dolmenyn, Brecon Nov. 8th, 1500 Pantyllyn Nov. 12th and 500+ Llanfwrog, Anglesey Nov. 13th

The only unseasonal records were of singles at Llanberis, Caernarfon on Aug. 8th 1994 and at Tyle Garn, Brecon on May 15th 1997.

Birds in Wales 1992-2000

SONG THRUSH *Turdus philomelos*   *BRONFRAITH*
(E) A widely distributed and fairly common resident, formerly abundant. Increasing. Winter visitor and passage migrant.

Birds in Wales states that Song Thrushes can be found as a breeding bird in every 10-km square in Wales. Recently however it appears to be declining in lowland Wales but this is more than offset by increases in the uplands due to colonisation of new forestry plantations. 130 singing males counted at MOD Sennybridge in May 2000.

Using data from the National Atlas produces a Welsh population estimate of 83,000 pairs. Results from BBS, 1994-1999 suggests a population change of 112%, producing a revised population estimate of 93,000 pairs. This is consistent with estimates generated from CBC data.

REDWING *Turdus iliacus*   *COCH DAN-ADEN*
(C) An abundant winter visitor and passage migrant; particularly numerous in severe winters.

Numbers of wintering Redwings in Wales varies from one year to the next depending on the weather and the availability of fruit crops. Large flocks over 1,000 strong were counted at:

1992	roost at Gwsaney, Clwyd 10,000 in December
1993	3,000 attracted to the Bardsey lighthouse, Caernarfon Mar. 16th, 500 there Oct. 13th
1994	1,400 over Port Talbot, Glamorgan Nov. 1st, 1,800 on the 2nd, 1,650 on the 16th, five flocks totalling 1,700 Llanwrtyd Wells, Brecon Nov. 15th
1995	5,000+ Llysdinas, Brecon in the early months, an attraction of 7,000+ to Bardsey light Mar. 29th / 30th, 2,800 over Port Talbot Oct. 22nd, 600 / hr. over Holyhead, Anglesey for 8.5 hrs. Oct. 22nd / 23rd, 2,000+ on Rowans at Llanwrthwl, Brecon Oct. 23rd – Nov. 10th, 1,000 at Llanwrtyd Wells, Brecon Oct. 29th – 30th and a similar number on Usk Res., Brecon on the 30th, 2,500 Llandinam, Montgomery in 30 minutes Nov. 12th
1996	1,250 at Taibach, Gower Nov. 9th was the only flock reported over 400
1997	1,630 S in 40 minutes Port Talbot, Glamorgan Jan. 1st, 3-5,000 roosting at Stackpole, Pembroke in January. 2,500 Brecon on Oct. 22nd, 1,400 over the Dee valley, Denbigh Oct. 28th and 1,500 Llanmadoc, Glamorgan Dec. 20th
1998	relatively scarce; 1,450 Taibach, Glamorgan Oct. 27th, c2,100 Nov. 4th and 1,600 Nov. 11th. Roost at Aberthaw, Glamorgan peaked at 1,000+ Nov. 14th
1999	5,500 attracted to the light at Bardsey on the night of Oct.15th/16th, 2900 the following night, 1,000+ at Glan y Mor Elias, Caernarfon on Nov. 7th, 2,890 over Taibach, Glamorgan on 5 dates in the period Dec. 9th – 20th

In 1994 two individuals summered at Aber, Caernarfon.

MISTLE THRUSH *Turdus viscivorus*   *BRYCH Y COED*
(D) A fairly common resident; many young birds emigrate in their first autumn.

Mistle Thrushes are relatively common birds throughout Wales, widely distributed but nowhere numerous. Using data from the National Atlas produces a Welsh population estimate of 19,000 pairs. Results from BBS, 1994-1999 suggests a population change of 114%, producing a revised population estimate of 22,000 pairs.

CETTI'S WARBLER *Cettia cetti*   *TELOR CETTI*
(E) Unusual breeding resident in south Wales, probably increasing; occasionally recorded on passage.

Birds in Wales quotes that the first Welsh record was at Bardsey, Caernarfon in 1973 and the first confirmed breeding in Wales took place at Oxwich, Glamorgan in 1985. Since then this species' range has increased steadily, finally reaching Anglesey in 1997, although there remains a large gap (probably related to the limited amount of suitable habitat in mid Wales). The 1996 UK survey located at total of 40-42 singing males at 19 sites in 5 counties. In mind of the mild winters since 1996 it is likely that the Welsh population has at the very least been maintained and has probably increased further. Total Welsh population is probably of the order of 60-80 pairs. The table below is a summary of all the singing males at the main sites and for individual counties.

Population crashed on the Teifi in 1996 due to cold winter periods but numbers are starting to increase slowly. A similar situation in Pembroke although a secondary influx may be on the way. Winter records in 2000 include individuals at Westfield Pill & Bosherston in Pembroke, Llangorse, Brecon and Conwy RSPB in Caernarfon.

County	Location	1992	1993	1994	1995	1996	1997	1998	1999	2000
Gwent	County total					7	3-4	8		7
Glamorgan	Kenfig			1					3+	
Gower	Oxwich	18	6*	4*	bred	bred	10	10	bred	
	Pant y Sais						3	3		
	Crymyln				4+	bred		2		12
	elsewhere					2	4	1	1	8+
Carmarthen	Witchett	3	3	4	3	3			16	5
	Dyfalty / Ashpits	1	5	6	8	5				
	Ffrwd Fen	1	1	2	3-5	2				4
	County total	5+	8	16+	16	21-23	16	16	16	20-25
Cere/Pemb	Teifi Marshes	5	8	10+	23	4-5	6	2	3	3
Pembroke	elsewhere	2	2	2	0	4	1	2	5	1
Anglesey									6+	
		30	28+	33+	46+	42+	44+	43+	25+	57+

* data for Oxwich 1993 & 1994 incomplete.

LANCEOLATED WARBLER *Locustella lanceolata* *TELOR RHESOG*
(B) A vagrant.

The first Welsh record was of one caught at Bardsey lighthouse, Caernarfon Oct. 18th 1990. Since then there have been two more records, both at Bardsey. One was picked up moribund on Oct. 8th 1995 and a first winter was trapped on Sept. 27th 1997.

GRASSHOPPER WARBLER *Locustella naevia* *TROELLWR BACH*
(C) Summer visitor, widely but thinly distributed in all counties. Passage migrant.

Birds in Wales states that Grasshopper Warblers breed throughout Wales although the New Atlas demonstrates a markedly western bias to the breeding population, with the largest numbers in Carmarthen, Anglesey and around the coast of Cardigan Bay, Pembroke to the Llyn. Using data from the National Atlas produces a Welsh population estimate of 1,700 pairs. Few annual breeding censuses.

Declines were noted in Brecon, where populations in plantations have decreased as the age of the forestry increased. However the birds appear to have moved back into more traditional rhos type habitats. In Carmarthen 18 males located at 12 sites in 1997 & 1998.

Main surveys were:

1997   present in 16 out of 73 1km^2 on Anglesey (cf. 5 in 1986) an increase of 220%.
1998   12-14 territories on 442 ha of Cors Caron, Ceredigion
1999   28 territories on Cors Caron, 16 on Cors Fochno, both Ceredigion
2000   30+ territories located in Brecon

Large counts of passage birds at Bardsey, Caernarfon 50 Apr. 30th 1995, 20 on Apr. 27th 1996 and 25 on Apr. 26th 1997. An unprecedented 65 there on the late date of Oct. 20th 1999.

SAVI'S WARBLER *Locustella luscinioides* *TELOR SAVI*
(B) A vagrant to Wales.

Three records quoted in *Birds in Wales*, at Skokholm, Pembroke Oct. 31st 1968, at Dowrog, Pembroke June 18th 1983 and at Oxwich, Glamorgan May 13th – 20th 1987. Since then there have only been two acceptable records. One was caught at Llangorse, Brecon on July 4th and heard again on the 7th 1994. The other record was of a singing bird at Malltraeth Marsh, Anglesey June 8th – 11th 1999. A record of one at the Teifi Marshes, Pembroke/Ceredigion in April 1993, although previously accepted and published in Welsh Birds, was re-assessed and deemed unproven. This species reverted back to being a BBRC description species in 1998.

AQUATIC WARBLER *Acrocephalus paludicola*
(B) A scarce autumn passage migrant.

TELOR Y DWR

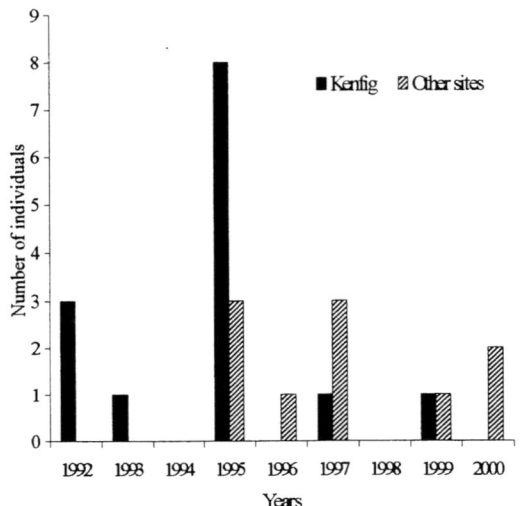

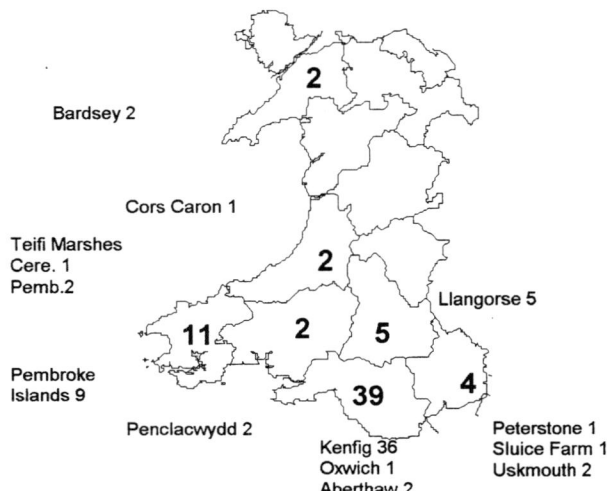

Birds in Wales quotes 35 records of a total of 41 individuals up to and including 1991. Since then there has been a further 24, 14 of which were at Kenfig. The annual totals are graphed below, with a best year being 1995 when there were 11 individuals recorded in Wales.

The top site for this species in Wales is Kenfig, with a total of 36 individuals recorded. The non-Kenfig records 1992-2000 came from Uskmouth in Gwent (2), Oxwich in Glamorgan, Penclacwydd in Carmarthen, Skomer in Pembroke, Teifi Marshes in Ceredigion / Pembroke (3), Cors Caron in Ceredigion and Llangorse in Brecon.

A breakdown of all Welsh individuals, by county by site on the map. Totals are given for each county and for each of the sites e.g. 39 recorded in Glamorgan, of which 36 were at Kenfig, 2 at Aberthaw and one at Oxwich. The above chart of arrival dates of all Welsh individuals clearly shows peak passage periods of mid to late August and a small peak mid / late September.

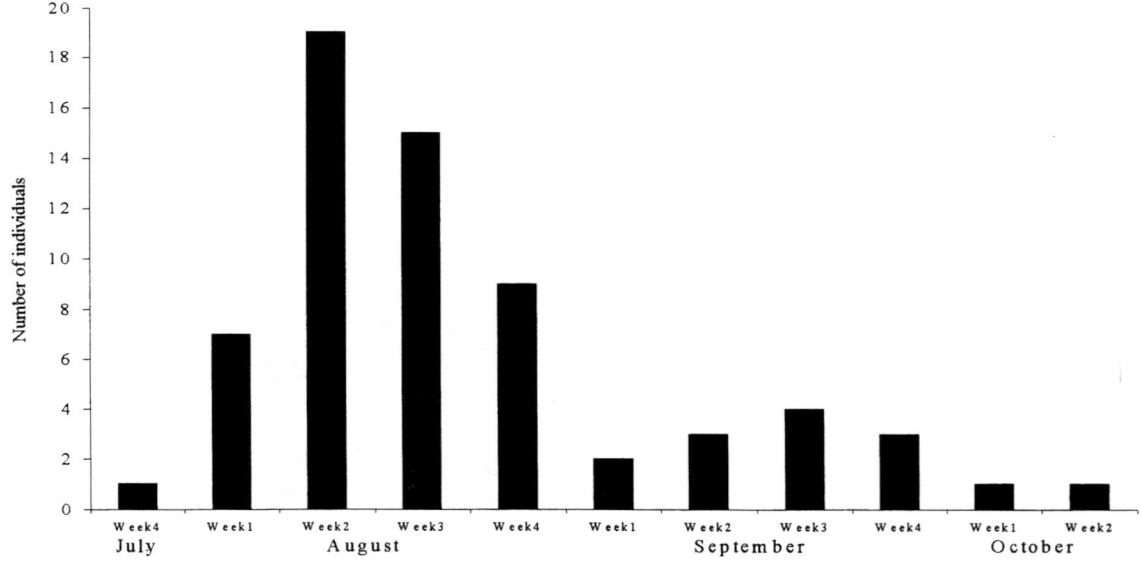

SEDGE WARBLER *Acrocephalus schoenobaenus*   *TELOR YR HESG*
(E) A locally distributed summer visitor, numerous in some sites, increasing; plentiful passage migrant.

Birds in Wales states that Sedge Warblers are more widespread and generally more numerous than Reed Warblers, breeding in all Welsh Counties, although scare in Radnor, Montgomery and Brecon. Using data from the National Atlas produces a Welsh population estimate of 15,500 pairs. Results from BBS, 1994-1999 suggests a population change of 114%, producing a revised population estimate of 17,700 pairs.

During the period under review there were few annual breeding censuses, the results of which are tabulated below.

1994   125 in Carmarthen between Cydweli & Hendy and 60 on the Tywi, Llanwrda to Carmarthen
1997   present in 30 out of 73 1-km squares on Anglesey (cf. 19 in 1986) an increase of 73%.
1999   110 singing at Cors Caron and 48 at Cors Fochno, both Ceredigion

Counts of breeding pairs on sites with over 30 pairs:
1996   36+ Eglwys Nunydd Res., 30+ Crymlyn Bog both Glamorgan and 50+ Pant y Sais, Carmarthen
1997   56 at Valley Lakes, Anglesey
1998   30 Eglwys Nunydd Res., Glamorgan, 74 on part of Cors Caron, Ceredigion
1999   46 at Eglwys Nunydd, 50 at Pant-y-Sais and 31 at Cors Goch NWWT, Anglesey
2000   51 at GLWR Goldcliff, Gwent, 40 at Oxwich and 64 at Eglwys Nunydd, 35 on Skomer, Pembroke and 33 at Malltraeth RSPB, Anglesey

Although numbers are increasing they still haven't recovered to the pre-crash figures (1968). The Brecon population is now estimated at 40-45 pairs compared to 25 pairs in 1990.

Large counts of over 100 birds on passage were noted at:
1992   150 at Kenfig, Glamorgan Aug. 6th
1993   100 at Bardsey, Caernarfon Apr. 28th, 300 on May 12th
1994   100 at Bardsey on four days in May between the 9th and the 15th with 250 on the 13th
1995   100 at Bardsey May 2nd. CES ringing at Llangloffan, Pembroke produced a total of 500 birds, two of which originated in Scotland, 278 trapped Teifi Marshes, Pembroke (217 of which were juvs.).
1996   350 at Bardsey May 12th, 100 on the 14th and 120 on Aug. 20th
1999   219 ringed at Uskmouth NR, Gwent during August.

MARSH WARBLER *Acrocephalus palustris*   *TELOR Y GWERNI*
(B) An extremely rare summer visitor and passage migrant. Bred in 1972.

Birds in Wales quotes 10 records: in Gwent bred at Llanwern in 1972, noted at Caerleon in 1983, possibly 2 at Magor in 1964, singles at Wentloog 1981, at Caerleon in 1983 and at Magor in 1991; in Glamorgan at Oxwich in 1963; at Bardsey, Caernarfon in 1959 (2) & 1988; at Llanfachraeth, Anglesey in 1986.

The Oxwich records is now deemed unacceptable.
Since then there have been only 3 accepted records at: Bardsey on June 12th 1992 and Llyn Cwellyn June 30th 1995, both Caernarfon and at Peterstone, Gwent June 22nd – July 10th 1996.

REED WARBLER *Acrocephalus scirpaceus*   *TELOR Y CYRS*
(E) Local summer visitor, breeding in reedbeds in many counties, increasing; scarce passage migrant.

According to Birds in Wales Reed Warbler's range has been slowly expanding northwards in Wales since the 1930's. This has continued during the period of this review, with recent colonisation of Morfa Dyfi, Montgomery and Anglesey. 17 prs. at 3 sites (Teifi / Dyfi) in 1984 had increased to 26 singing males at Ynyshir in 1994 and 51 at Teifi Marshes (Ceredigion / Pembroke) in 1999. The Anglesey population is rapidly expanding, 95 prs. at Valley and 20 prs. at Malltraeth in 2000.

Using data from the National Atlas produces a Welsh population estimate of 3,100 pairs. Results from BBS, 1994-1999 suggests a population change of 115%, producing a revised population estimate of 3,600 pairs.

The only county wide census was in the SE section of Carmarthen in 1994, where a minimum of 333 singing males counted between Penallt and Hendy – the county's population is estimated at 400-500 pairs.

Birds in Wales 1992-2000

Data of breeding pairs at the main sites (over 30 pairs) or where extension of the range:
1992   89 territorial males Llangorse, Brecon
1993   3 prs. Morfa Dyfi, Montgomery – a new site
1994   in Carmarthen, 30 at Ffrwd Fen, 31 Ashpit ponds, 54 Llangennech; 26 at Ynyshir, Ceredigion (cf. 4 in 1992), 3 prs. Morfa Dyfi, Montgomery, 6 sites on Anglesey
1998   8 singing males at Dolydd Hafren, Montgomery – a new site
1999   12 at Aberleri, Ceredigion and 51 at Teifi Marshes, Ceredigion/Pembroke
2000   122 at GLWR Golcliff, Gwent, 220 at Witchett, Carmarthen

200 were ringed at Uskmouth NR, Gwent in August 1999. An early bird at Broad Haven, Pembroke Apr. 3rd 1999 was the earliest for that county and a late bird was at Porth Clais, Pembroke Nov. 12th 1994.

GREAT REED WARBLER *Acrocephalus arundinaceus*   *TELOR MAWR Y CYRS*
(B) A vagrant.
Seven records quoted in *Birds in Wales*, only one since, at Skomer, Pembroke May 21st 1998.
Seven of the eight Welsh records have been spring birds, May 11th being the earliest, June 15th the latest. The only non-spring bird was caught at Penmaenpool, Meirionnydd July 30th 1978.

Two have been on Skokholm, Pembroke May 1967 & 1970 and three on Bardsey, Caernarfon May 1976 & 1991 and June 1979. The others at Penmaenpool and at Llangynog, Montgomery June 15th 1988.

OLIVACEOUS WARBLER *Hippolais pallida*   *TELOR LLWYD*
The record at Skokholm Sept. 23rd – Oct. 3rd 1951 has been removed from the British List. There have therefore been no acceptable records of this species in Wales.

BOOTED WARBLER *Hippolais caligata*   *TELOR BACSIOG*
(A) A vagrant.
The first Welsh record was at Skokholm, Pembroke Sept. 25th – 28th 1993. The second was a first winter at Bardsey, Caernarfon Sept. 25th – 26th 1998. The third was at Skomer, Pembroke Sept. 14th – 15th 2000.

ICTERINE WARBLER *Hippolais icterina*   *TELOR AUR*
(B) *A scarce spring and summer passage migrant to coastal areas of Wales, half as numerous as the very similar Melodious Warbler.*
Correction *Birds in Wales*: 68 individuals up to and including 1991, since then there has been a further 21 recorded in Wales.

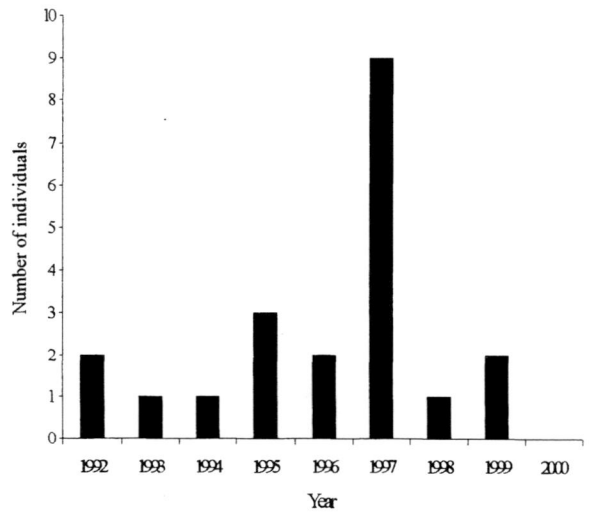

The annual totals are shown in the graph.
All records were:
Glamorgan: at Flat Holm Aug. 19th 1995.
Carmarthen: at Abergwili Dec. 2nd & 6th 1996 – latest Welsh record.
Pembroke: at St. Ann's Head Sept. 6th 1997, on Skomer May 20th & Sept. 25th 1992, Sept. 6th – 12th 1993, May 30th, Aug. 19th – 20th, Aug. 24th & Sept. 17th 1997. At Skokholm Oct. 19th – 20th & Oct. 31st – Nov. 1st 1995, July 20th 1996, Sept. 11th, Sept. 15th – 18th (possibly the same bird as earlier) and Sept. 28th 1997.
Caernarfon: Bardsey May 30th 1994, Aug. 10th 1995, Aug. 23rd 1997, Aug. 29th & Sept. 20th 1999.
Anglesey: single at the Skerries June 1st 1998.

Breakdown of all Welsh records by month in the graph and by county on the map. Totals for each county and the key sites are shown, e.g. a total of 31 recorded in Pembroke, of which 2 on mainland, 7 on Skomer and 22 on Skokholm).

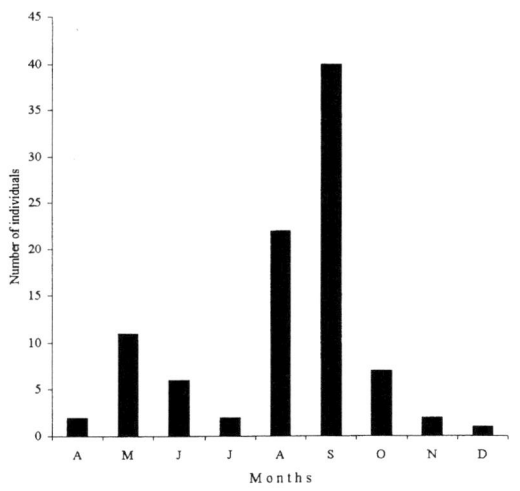
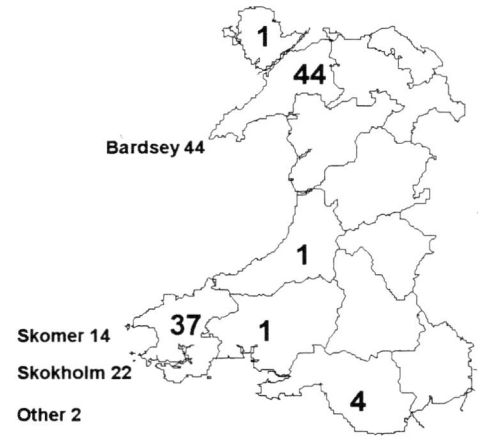

MELODIOUS WARBLER *Hippolais polyglotta* — *TELOR PER*

(B) *A scarce but regular migrant to the Welsh coast, principally in the autumn but with a scattering of spring occurrences.*

This species shows a marked south-westerly bias in its occurrence pattern in Britain, which doubtless explains the greater number of records in comparison with Icterine Warbler in Wales.

Birds in Wales quotes a total of 142 individuals up to the end of 1991. Since then there have been a further 28 records making a grand total of 170 birds recorded in Wales. Top site for this species is Bardsey, with 90 individuals, nearly twice as many as the second top site, Skokholm, Pembroke which has recorded 51 individuals. The only other sites that have recorded more than one individual are Skomer, Pembroke (5), Strumble Head, Pembroke (3) and Porth Meudwy, Caernarfon (4).

All records 1992-2000 were:
Glamorgan: Nash Point Oct. 7th 1993, Llanmadoc Oct. 31st 1995 and at an undisclosed coastal site Oct. 26th 1998.
Pembroke: at Skomer Sept. 17th 1993, Sept. 11th 1994 & Sept. 25th – Oct. 6th 2000, at Skokholm Aug. 7th & 30th 1994, Sept. 5th 1995 and Aug. 8th – 17th 1996 and at Strumble Head Oct. 4th – 5th 1998 & Aug. 7th – 8th 2000.
Ceredigion: at Cors Caron Sept. 28th 1992.
Caernarfon: at Bardsey Aug. 7th & 10th 1992, Aug. 20th 1993, June 13th & Sept. 3rd 1994, Aug. 5th & 13th, Oct. 28th 1995, June 6th – 7th & 14th 1996, June 11th 1997, May 24th 1998, Aug. 29th 1999 and June 6th 2000. Porth Meudwy Sept. 21st – 22nd 1998 was the only mainland record.

Breakdown of all Welsh records by month in the graph and by county on the map (totals for the key sites are shown, e.g. a total of 62 recorded in Pembroke, 51 of which were on Skokholm, 5 on Skomer and 6 on the mainland).

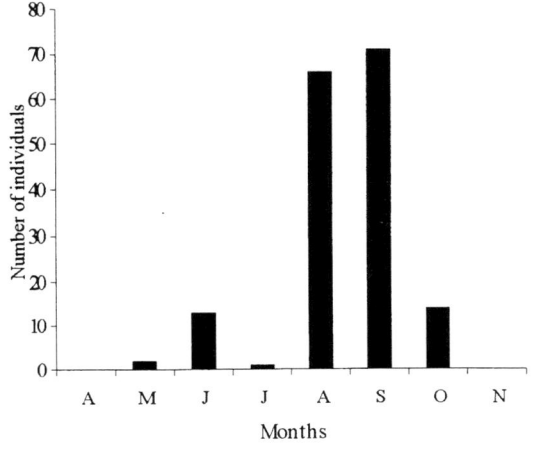
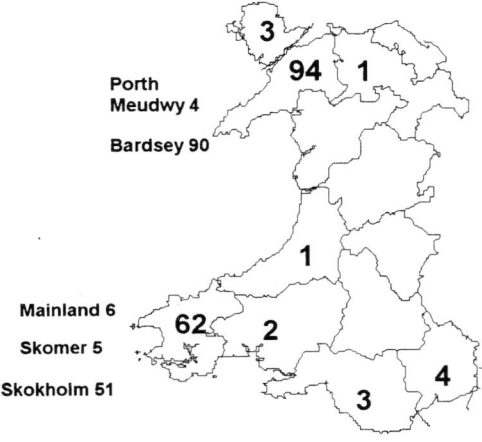

DARTFORD WARBLER *Sylvia undata*
(E) A rare migrant to coastal areas of south Wales. Bred in 1998.

TELOR DARTFORD

Birds in Wales quotes only 6 records up to 1991, since then there have been a remarkable 22 individuals and the first breeding of this species in Wales in 1998. With the English breeding population having increased and spread since the mid-1980's, the first Welsh breeding record was perhaps not totally unexpected but most welcome nevertheless. Breeding was confirmed in 2000 at 3 sites in Glamorgan and a pair was also noted at another.

Due to the nature of habitat selection it is possible that there are even more breeding in south Wales.

All records were:

Gwent: single at Uskmouth Nov. 28th 1992 – Mar. 13th 1993, pair bred in 1998, fledging 2 broods, individuals still present on Mar. 15th 1999. Single at Abertillery May 28th 2000.

Glamorgan: singles at Oxwich Mar. 20th – 21st 1994, Tutt Head Mar. 21st 1995, at Kenfig Apr. 6th 1996, Feb. 13th & 15th and Nov. 9th 1999, at Rhosilli / Worms Head Aug. 7th & 16th 1997, at Nash Point Sept. 27th 1998 & Oct. 14th 1999, at Port Eynon, Apr. 24th 1999. 3 breeding and one non-breeding pairs in 2000.

Pembroke: a male at St. David's Head Mar. 18th 2000.

SUBALPINE WARBLER *Sylvia cantillans*
(B) A rare visitor to Wales.

TELOR BRONGOCH

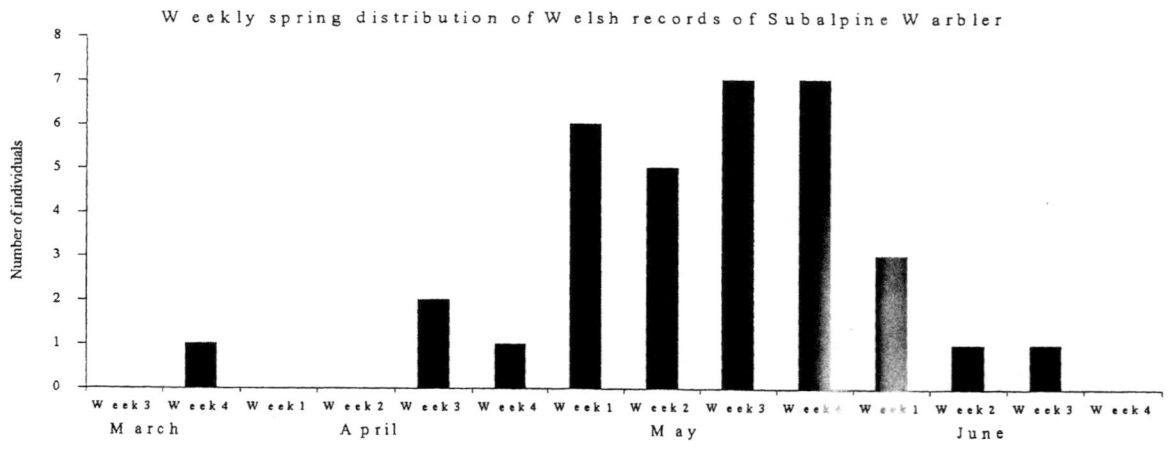

18 individuals recorded in *Birds in Wales*, 4 on Skokholm, 6 on Skomer and 7 on Bardsey. The only mainland record was from Porth Meudwy, Caernarfon May 30th 1985. Since then there have been a further 19 accepted records, 16 from the islands of Skokholm, Skomer and Ramsey in Pembroke and Bardsey in Caernarfon. The chart above shows the spring arrival dates of all the Welsh Subalpine Warblers.

All records during the period under review were from May except at Bardsey June 16th 1996 and June 11th – 16th 2000. In spring 2000 there were 3 mainland records, all males, 2 in Caernarfon at Penmaenmawr May 2nd and at Llandudno May 14th – 17th and one in Meirionnydd at Tal-y-bont Apr. 22nd.

Island records were:

Skokholm	male May 15th 1992, 1st summer May 29th 1994, females May 7th & 29th 1995
Skomer	males 14th & 29th May 1992 and May 14th 1994
Ramsey	male *albistriatal* May 12th 1993
Bardsey	1st year male May 20th 1992, male May 16th – 25th 1994, 1st summer female May 1st & 15th – 17th 1995, male June 16th 1996, 1st summer female May 14th – 18th 1997, 1st summer male May 15th – 29th 1997, a male May 24th 1998 and a male June 11th – 16th 2000.

RUPPELL'S WARBLER *Sylvia rueppelli*   TELOR RUPPELL
(A) A vagrant.
The first record, and only the 5th for Britain, was at Porth Meudwy, Caernarfon June 21st 1995.

SARDINIAN WARBLER *Sylvia melanocephala*   TELOR SARDINIA
(B) A vagrant.
The second Welsh record was of a male at Bardsey, Caernarfon June 2nd – 7th 1994. The first record was of a male at Skokholm, Pembroke on Oct. 28th 1968.

BARRED WARBLER *Sylvia nisoria*   TELOR RHESOG
(C) A rare autumn passage migrant.
41 accepted records of 43 individuals quoted in *Birds in Wales* with an additional record of a single at Bardsey, Caernarfon on Oct. 8th 1991. Since then there have been a further 17 accepted individuals all which all were of singles except for two on Skokholm Sept. 1992. All records were:

Gwent: Uskmouth Aug. 29th – Sept. 1st 1994; Glamorgan: Nash Point Sept. 26th 1993 and Kenfig Sept. 4th 1994; Pembroke: at Skokholm 2 on Sept. 30th, single Oct. 4th 1992, Oct. 12th – 15th 1993 and Oct. 2nd & 6th 1995, at Skomer Oct. 10th –15th 1995, at Strumble Head Oct. 8th 1993 and at Angle Nov. 26th 1999; Caernarfon: at Bardsey on Aug. 19th – 30th 1995, in 1998 one was found moribund on Sept. 18th & a live bird was seen the following day, Aug. 30th 1999 and Sept. 15th 2000; Anglesey: at South Stack Oct. 12th 1996.

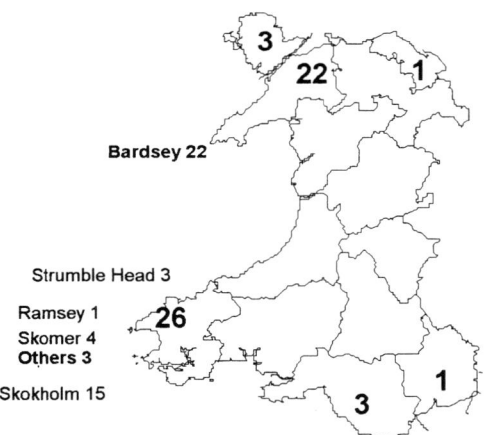

Weekly distribution of all Welsh Barred Warbler records shown in the table and county & site totals shown on the map:

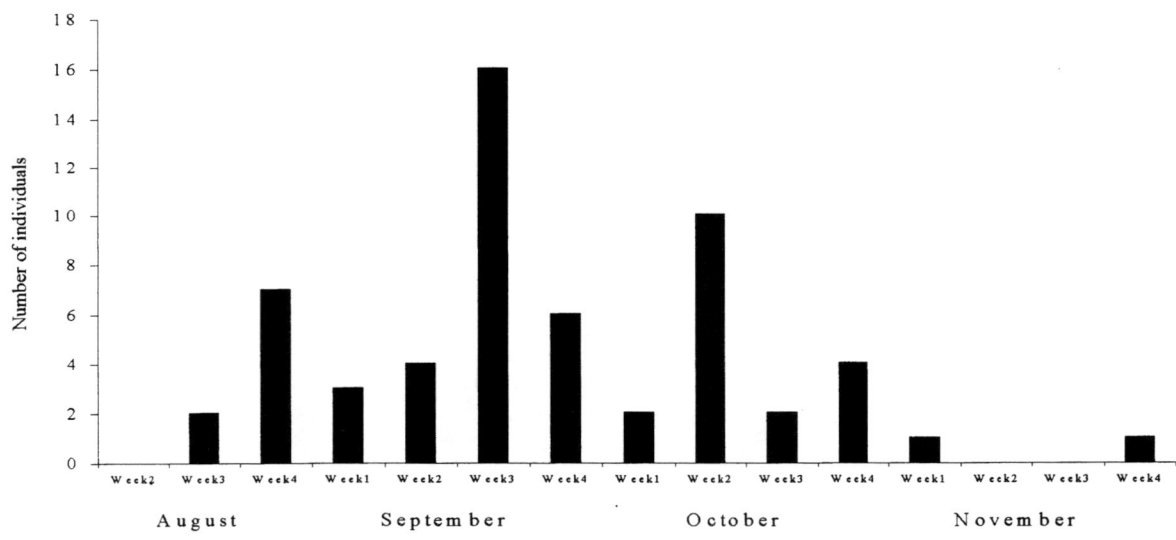

LESSER WHITETHROAT *Sylvia curruca* — *LLWYDFRON FACH*
(E) A summer visitor, declining. Passage migrant in small numbers.

Birds in Wales states that the distribution of Lesser Whitethroat in Wales is very localised and that the New Atlas suggests that this species' strongholds are in Gwent, Radnor, Carmarthen and in the Welsh Marches. Using data from the National Atlas produces a Welsh population estimate of 5,500 pairs. Results from BBS, 1994-1999 suggests a population change of 69%, producing a revised population estimate of 3,800 pairs.
An unseasonal record was of an individual at Cathays, Cardiff on Dec. 24th 1992.

WHITETHROAT *Sylvia communis* — *LLWYDFRON*
(C) A widespread summer visitor; formerly abundant. Common passage migrant.

The Whitethroat is an abundant summer visitor and in some parts more numerous as a breeding migrant than Willow Warbler. Using data from the National Atlas produces a Welsh population estimate of 69,000 pairs. There was no real change since. The only censuses were:

1997 44 males along the roadside hedges in a 10-km square in Gower. Survey of 73 1-km squares by the RSPB on Anglesey, presence noted in 62 (cf. 39 in 1986) – a 59% increase
1998 thought to be more widely distributed on Gower
1999 GOS/RSPB survey of unenclosed cliff on Gower, Mumbles to Rhosili, found 28 prs.
2000 33 males at GLWR Goldcliff, Gwent and 45 males at Kenfig, Glamorgan

Unseasonal records were: at RAF St. Athan, Glamorgan Feb. 24th 1992 and at Llansantffraid, Ceredigion Oct. 24th 1997.

GARDEN WARBLER *Sylvia borin* — *TELOR YR ARDD*
(C) A common and widely distributed summer visitor. Regular passage migrant.

Using data from the National Atlas produces a Welsh population estimate of 33,600 pairs. Results from BBS, 1994-1999 suggests a population change of 92%, producing a revised population estimate of 31,00 pairs. This corresponds to estimates from CBC data.

Unseasonal records came from Pembroke, where one was trapped at Bosherston in February 1992 and birds at Porth Clais Nov. 1st 1996 & Nov. 15th 1998.

BLACKCAP *Sylvia atricapilla*   *TELOR PENDDU*
(E) A plentiful summer visitor throughout Wales, increasing; passage migrant and small-scale winter visitor.

A very common summer visitor throughout all the Welsh Counties. Using data from the National Atlas produces a Welsh population estimate of 66,700 pairs, this corresponds to estimates from CBC data. Results from BBS, 1994-1999 however suggest a population change of 159%, producing a revised population estimate of 106,000 pairs. Further evidence for this increase comes from Carmarthen, where territorial males now occupying young (15 yr. old) planted woodland on the south coast.

The number of wintering individuals fluctuates greatly one year to the next. The following table gives county totals of wintering birds in each of the winter periods:

Birds in Wales 1992-2000

	1992	1993	1994	1995	1996	1997	1998	1999	2000
Gwent	24	14 & 16						6 & 7	10 & 13
Glamorgan		14 & c60	45 & 22	26 & 12+	18 & 10	30 & 14	8 & 13	8 & 16	18 & 27
Gower					18 & 6		0 & 5		
Carmarthen	8 & 0	0 & 9	10 & 6	4 & 4	7 & 3	10 & 5	1 & 3	1 & 5	5 & 12
Pembroke	24-34 & 2	2 & 2	7 & 12	3 & 12	15 & 10	1 & 12		3 & 9	
Ceredigion	5+	1 & 7	4-5 & 1	0 & 2	15 & 2	8 & 2	2 & 5	5 & 5	7 & 3
Brecon	0 & 0	2 & 9	1 & 2	1 & 2	3 & 1	1 & 3	0 & 2		
Radnor				2 & 1					
Montgomery	2			5 & 0		0 & 1			
Meirionnydd		0 & 2			1 & 1		0 & 0		0 & 1
Caernarfon	4			0 & 17	0 & 1	1 & 0	5 & 2	2 & 7	10 & 11
Anglesey		2 & 3	1 & 3		1 & 0		3 & 1	2 & 2	5 & 15
Denbigh	2	1 & 1	1 & 0	0 & 3		4 & 2			0 & 1
Flint					1 & 0	0 & 2		0 & 1	

GREENISH WARBLER *Phylloscopus trochiloides*
(B) A vagrant.

TELOR GWYRDD

Birds in Wales quote eight accepted records, 4 from Bardsey, Caernarfon, three from Skokholm and one from Skomer, both Pembroke. There have been a further five records in Wales: at Bardsey on Aug. 24th 1995, June 14th – 15th & Aug. 31st 1997, at Skomer June 27th 1996 and at Skokholm June 23rd 1997.

PALLAS'S WARBLER *Phylloscopus proregulus*
(B) A vagrant.

TELOR PALLAS

Birds in Wales quote 9 records, since then there have been a further 16 individuals:
Glamorgan: at Whitchurch Nov. 21st 1994, Aberdare Dec. 12th 1994 and Kenfig Dec. 27th 2000; Pembroke: at Strumble Head Oct. 9th 1994, St. David's Head Oct. 25th 1997, Bosherston Jan. 19th – 20th 2000 and 2 on Skomer (the first island records) Nov. 15th 2000; Ceredigion: at Pennant Oct. 22nd – 29th 1993; Caernarfon: on Bardsey Oct. 8th 1992, Oct. 27th 1993, two Oct. 30th – 31st 1997, Nov. 6th 1997 and on the Little Orme Nov. 15th 1997.

Breakdown of all Welsh records, autumn records by week in the graph and by county on the map (sites with multiple records are given, e.g. 8 individuals recorded in Pembroke, 2 of which were at Strumble Head, 2 at Skomer and 4 others).

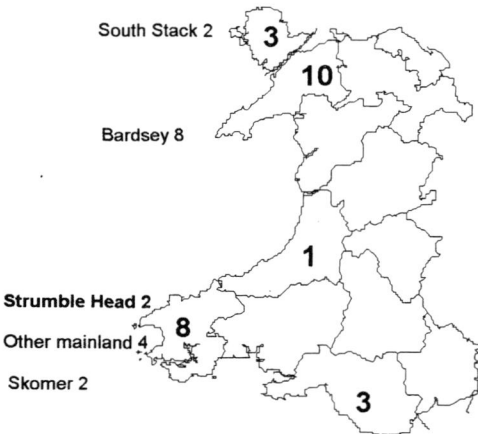

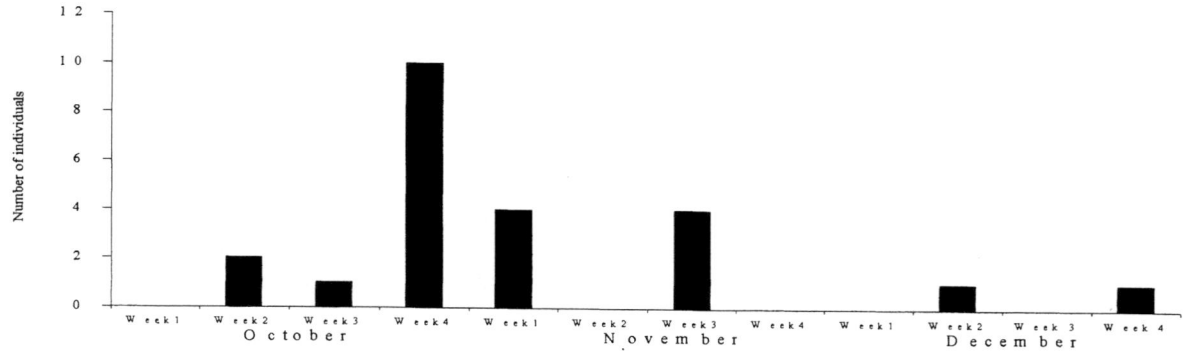

YELLOW-BROWED WARBLER *Phylloscopus inornatus* — *TELOR AELFELYN*

(D) A scarce to uncommon but rapidly increasing, autumn passage migrant.
Birds in Wales record 113 individuals as occurring in Wales, since then there has been a further 51 individuals in Wales.

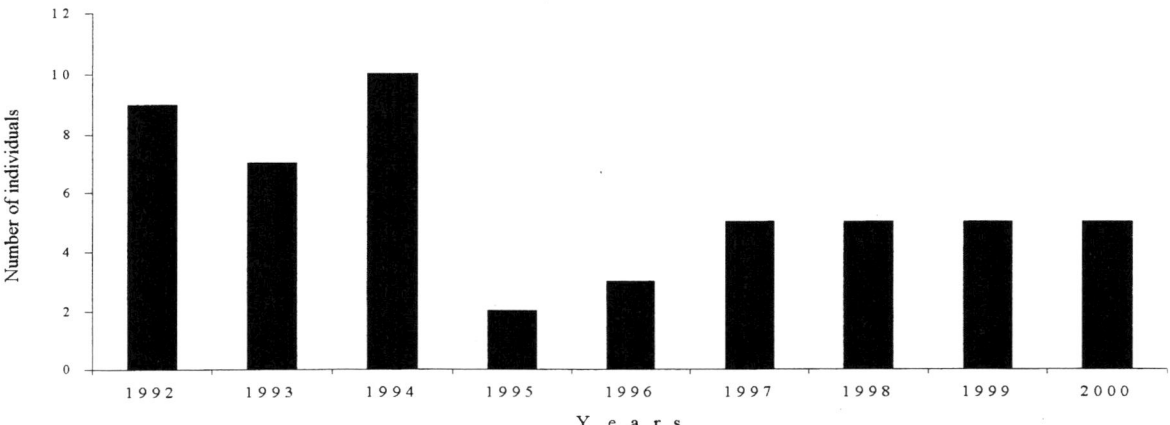

All records were:
Gwent (1), Glamorgan (5), Carmarthen (1) first and only county record at Pembrey Country Park, Oct. 4th – 5th 1997, Pembroke (17), Meirionnydd (1) first and only county record at Dolgellau Oct. 14th 1996, Caernarfon (19) Anglesey (4), Denbigh (1) first and only county record at Clocaenog forest Oct. 22nd 1994 and Flint (2).

Breakdown of all Welsh records by county is shown on the map. Sites with multiple records are given, e.g. in Pembroke, a total of 41 individuals recorded of which 15 were on Skokholm, 3 on Skomer, 4 on other islands, 4 at Porth Clais, 8 at Strumble Head and 7 elsewhere on the mainland; in Caernarfon a total of 86 individuals recorded, of which 68 were on Bardsey and 18 on the mainland.

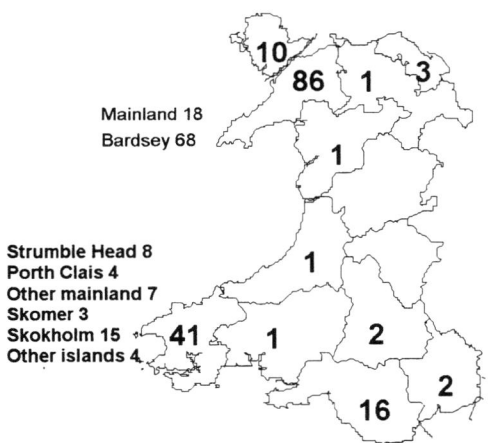

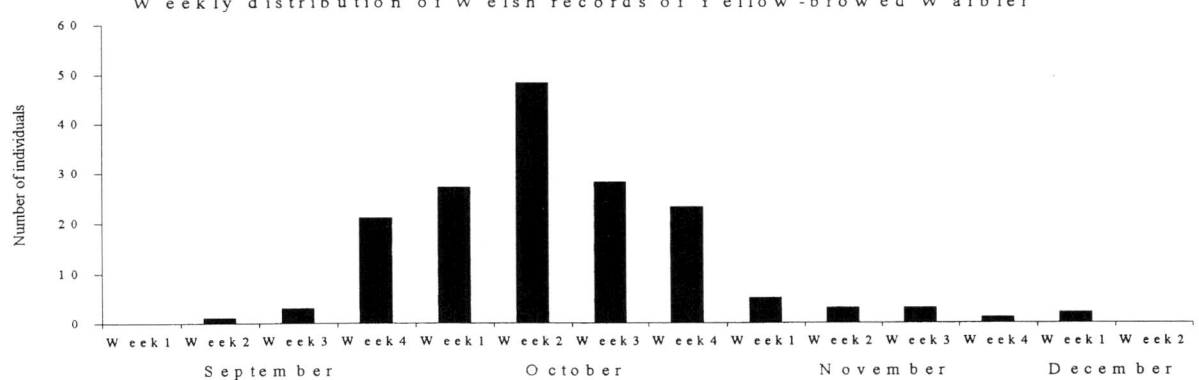

HUME'S LEAF WARBLER *Phylloscopus humei* *TELOR HUME*
(A) A vagrant.
The first and only Welsh record of this recently separated species was at Strumble Head, Pembroke Nov. 20th 1993.

DUSKY WARBLER *Phylloscopus fuscatus* *TELOR TYWYLL*
(B) A vagrant.
The fourth Welsh record was from Point of Air, Flint Nov. 11th – 12th 1997. The other three records, as quoted in *Birds in Wales*, were from Bardsey, Caernarfon Oct. 30th 1982, Nov. 7th 1987 and at Strumble Head, Pembroke Oct. 15th 1988.

WESTERN BONELLI'S WARBLER *Phylloscopus bonelli* *TELOR BONELLI*
(B) A vagrant.
With BOU splitting Bonelli's Warbler into two species, all records were re-assessed by BBRC. The following were all deemed to be of this "new" species: Skokholm, Pembroke on Aug. 31st 1948 was the first for Wales. Since then 5 individuals all at Bardsey, Caernarfon: Aug. 18th – Sept. 5th 1959, Sept. 10th 1959, Sept. 1st – 2nd 1962, Sept. 15th – 16th 1962 and Aug. 20th 1984.

The records from Lavernock Point, Glamorgan on Aug. 30th 1963, Llaniestyn, Caernarfon on Sept. 17th 1968 and Skokholm Aug. 31st 1991 were accepted by BBRC as being individuals of Bonelli's Warbler sp.

WOOD WARBLER *Phylloscopus sibilatrix* *TELOR Y COED*
(C) A summer visitor to oak woods throughout most of upland Wales, declining; passage migrant in small numbers at coastal sites.

The Wood Warbler is one of the characteristic summer visitors to the oak dominated woods of the hills and valleys in Wales, most notably in mid-Wales. The Welsh population is estimated at 5,000 pairs using data from the National Atlas. Results from BBS 1994-1999 suggest a population change of 55% producing a revised estimate of 2,800 pairs.
There is little breeding data to allow further comparisons.
1995: 25 territorial males Cwm Clydach RSPB reserve in Glamorgan and 60 at Lake Vyrnwy, Montgomery.
1999: 15 territorial males at Darren Woods (18 in 2000), Maesteg, 24 at Cwm Clydach, 7 in 40ha Coedydd Afon Gwynant and 10 in 50ha Coed Garth Gell both Mawddach, Meirionnydd.
At Dinas/Gwenffrwd, Carmarthen 80 territorial males in 1994.
Regular breeding bird in larch plantations of Brecon, 26 males on the Epynt in 1992, 31 in 1997, 17 in 1998 and 36 in 1999.

Two late birds on Ramsey, Pembroke Oct. 21st – 25th 1997 and one Bardsey, Caernarfon Sept. 28th 1999 (the latest record for the island).

CHIFFCHAFF *Phylloscopus collybita* *SIFF-SAFF*
(C) A locally common breeding visitor in most counties, declining, wintering in small numbers. Rare on Anglesey.

	1992	1993	1994	1995	1996	1997	1998	1999	2000
Gwent		1 & 7				2 & 2		4 & 0	1 & 0
Glamorgan	6 & 26	10+ & 12+	27 & 17	11 & 4	10 & 16	10 & 8	11 & 14	19 & 0	7 & 0
Carmarthen	2 & 8	16 & 8#	8 & 20#	23 & 2	7 & 3	2 & 4	4 & 6#	6 & 2	1+ & 1+
Pembroke	7-10 & 2	6 & 5	6 & 10	9 & 8	10 & 10	3 & 9	4+ & 17	11 & 15	10 & 19
Ceredigion		1 & 0	0 & 1	1 & 3	2 & 1	0 & 2	2 & 5	0 & 1	
Brecon							2 & 2	0 & 2	
Meirionnydd		0 & 1		1 & 0					
Caernarfon							2 & 12	3 & 4	4 & 7
Anglesey			5 & 1	2 & 2	0 & 1	4 & 1	1 & 0		
Denbigh									
Flint	2 & 0			2 & 12	0 & 2	1 & 0	0 & 1	1 & 0	

birds caught at Cydweli sewage works.

The Breeding Bird Atlas shows that Chiffchaffs breed in almost all 10-km squares in Wales, with the exceptions of the mountains of mid and north Wales. Using data from the National Atlas produces a Welsh population estimate of 88,000 pairs. Results from BBS, 1994-1999 however suggest a population change of 85%, producing a revised population estimate of 75,000 pairs.
Wintering recorded in most counties, data tabulated above. The first figure the number of individuals in the 1st winter period, the second the number of individuals in the 2nd.

The 1999 figures for Pembroke include 11 individuals at St. Ishmael's sewage works in January & 4 there in November – December. Figures clearly show a southern and coastal bias, with few records from inland counties or in the north, the exceptions being the 12 at Point of Air, Flint 1995 and in Caernarfon 1998.

Max. of passage birds recorded at Bardsey, Caernarfon, 100 Apr. 20th & 400 Sept. 10th 1996 and 30 Sept. 27th 1997.

Individuals showing characteristics of the eastern races *abietinus* or *tristis* recorded: in Glamorgan at Kenfig in one in January, 3 in November and one in December all 1994, Jan. 3rd 1997, Oct. 31st – Nov. 5th 1997, November 1998 – March 1999 and Nov. 11th 1999, at Roath Dec. 23rd 1998, at Magor Jan. 1st 1997 and at St. Mellons Nov. 13th 1999; Cydweli sewage works, Carmarthen, 5 during January and February 1993, 2 in both winter periods 1994 and 2 in December 1995; Caernarfon at Bardsey Oct. 30th – Nov. 1st 1997, at Bwlchtocyn in November 1998 and at Llandudno Dec. 26th 1999.

Individuals showing characteristics of *abietinus*: 2 at Cydweli sewage works Jan. 2nd 1995, at Porth Clais, Pembroke Nov. 8th 1998, at Kenfig Nov. 9th – Dec. 25th 2000 and Red Wharf Bay, Anglesey Dec. 21st.

Individuals showing characteristics of *tristis*: at Kenfig Apr. 8th 1995 and Feb. 27th 2000, in Pembroke at Skomer Nov. 9th 1996, at Skokholm Nov. 23rd 1998 and Porth Clais Dec. 1st 1998 and at Bardsey singles Apr. 24th & Oct. 18th with 2 on Apr. 10th all 1995 and 2 on Apr. 29th a single on the 30th 1999. Single at Conwy RSPB, Caernarfon Nov. 21st – 30th 2000.

WILLOW WARBLER *Phylloscopus trochilus*   TELOR YR HELYG
(C) The most abundant summer visitor. Numerous in suitable habitats up to around 670 m. Also numerous passage migrant.

The Willow Warbler is a very numerous species in Wales, widely distributed throughout a range of habitats, in all counties. Using data from the National Atlas produces a Welsh population estimate of 240,000 pairs. Estimates from CBC data however, suggest a Welsh population of between 160,000 and 200,000.

The only data of breeding numbers was:
In 1997: 46 – 50 pairs Llynfi Valley, 35 territories on 190 ha of industrial land Llansamlet and 22 Blaengwrach nature reserve, all Glamorgan.
In 1999 a max. of 72+ pairs at Cors Goch, Anglesey. 300+ counted at MOD Sennybridge, Brecon in 2000.

Large numbers pass through Bardsey, Caernarfon. Max. passage counts from there were:
1995   c1,000 Apr. 10th, 800 on the 24th and 30th, counts of over 300 on 5 other days in April
1996   800 Apr. 27th, over 100 on 12 days. 2,000 Aug. 20th and 100 on Sept. 20th
1997   450 Apr. 7th, 80 on the 26th, over 100 on 13 days in April. 350 Aug. 8th and 500 on the 17th
1998   very poor spring and autumn passage. Max. 130 Apr. 20th
1999   500 May 8th, 450 on the 9th and 200 on Aug. 10th
Max. counts in Pembroke were 800 at Ramsey on Apr. 27th and 300 at Skokholm on Aug. 8th, both 1999.
Individuals of the northern race *acredula* trapped on Bardsey, May 10th & 22nd 1998.

GOLDCREST *Regulus regulus*   DRYW EURBEN
(C) An abundant resident, passage migrant and winter visitor. Increasing.

Birds in Wales quotes 171,000 ha of conifer forest in Wales and that no species have benefited from this post war change in land use more than the Goldcrest.
Using data from the National Atlas produces a Welsh population estimate of 68,000 pairs, which corresponds to estimates from CBC data of 60,000. Results from BBS, 1994-1999 suggests a population change of 125%, producing a revised population estimate of 85,000 pairs.

Exceptional numbers at Bardsey, Caernarfon during spring migration 1998. Influx started Feb. 25th and went on until mid March, climaxing Mar. 29th when 2150 were counted in the plantation, 1,000 were still there the next day but only 25 remained on the 31st. Autumn peak is usually more pronounced but that year there was only a max. of 70 on Sept. 17th. 150 recorded Ramsey, Pembroke Sept. 15th.
In 2000, 200 were recorded at Bardsey in March and 50 at Ramsey on Oct. 15th.

FIRECREST *Regulus ignicapillus*   DRYW PENFFLAMGOCH
(E) Formerly a scarce breeding resident. An uncommon passage migrant.

Up to the early 1960s, Firecrests were rare vagrants to Wales with only 5 records. In the 1960s and 1970s small numbers were recorded with regularity, particularly on Bardsey and Skokholm and mainly on autumn passage. There was a dramatic increase in records from 1974 onwards and breeding took place in Gwent later in that decade. *Birds in Wales* states that this population increased, with 21 singing males in Wentwood in 1989 and smaller numbers at other Gwent sites, Lake Vyrnwy (Montgomery), Mynydd Du (Brecon) and Nercwys (Flint). All these populations collapsed at the beginning of the 1990s, possibly due to cold weather.

Since then breeding was only confirmed at one site, Nercwys in Flint, where a pair was present throughout the 1998 breeding season and a pair with young was seen the following year. There have been 5 other late spring / summer records: at a former site in Montgomery in 1992, one singing in Radnor Forest June 1994, one in Wentwood, Gwent in July 1996 and in May 2000 and one at Whitland, Carmarthen in spring 1998.

As a mainly passage and wintering migrant, Firecrests were recorded from most counties. The total number of individuals recorded in each month during the period is shown in the chart opposite (not including the breeding records), while the total for each county is shown on the map.

Most records were of ones or twos but 6 possibly 8 were at Kenfig, Glamorgan Dec. 4th 1998, with 3 remaining into 1999.

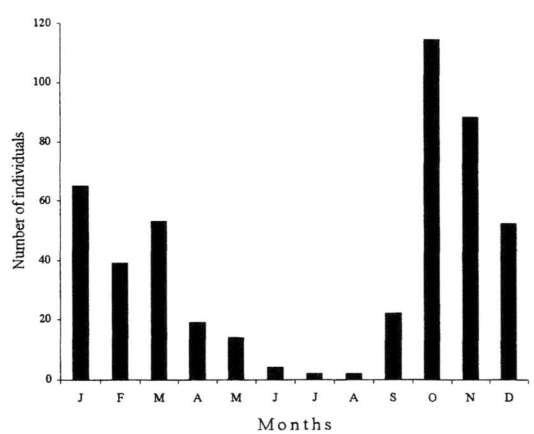

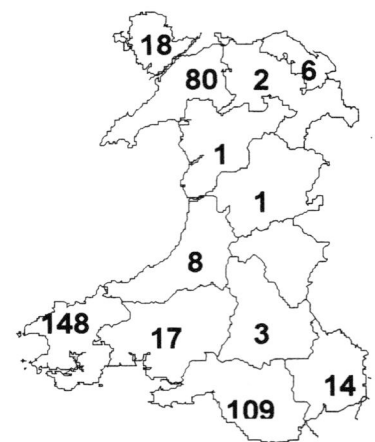

SPOTTED FLYCATCHER *Muscicapa striata* GWYBEDOG MANNOG
(E) Once a common summer visitor, now appears to be declining; passage migrant, May and August / September.

Spotted Flycatchers are one of the latest summer visitors to arrive in Wales, usually mid April – 1st week of May. Although well distributed in Wales the population is thought to be declining. Using data from the National Atlas produces a Welsh population estimate of 10,500 pairs. Results from BBS, 1994-1999 suggests a population change of 89%, producing a revised population estimate of 9,400 pairs.

By 2000, described as declining in Caernarfon, Anglesey and Meirionnydd. Increasingly scarce in Meirionnydd and Montgomery, uncommon in Gower. The breeding population in Brecon was estimated at 350 pairs in 1990, since then there has been a decline, especially in the north and west of the county, with a revised estimate of 250-300 pairs by 1999.

Only systematic counts of breeding birds were:

1995	25 along the River Usk, adjacent to the canal, Brecon – Pencelli
1997	Survey of 1-km squares in Radnor found presence in 28 squares, distributed in 13 10-km squares, sparse distribution with only 28 – 30 pairs
1999	6 prs. in 1 km^2 in Montgomery.

Large counts on passage included: in 1999, 20 on Skomer, Pembroke May 15th, 40 Bardsey, Caernarfon on May 17th and 9 on Sept. 20th. In 2000, 72 at Bardsey May 13th. Extremely late individual was at Llanishen/Lisvane Res., Glamorgan Dec. 29th 2000.

RED-BREASTED FLYCATCHER *Ficedula parva*   GWYBEDOG BRONGOCH
(B) Scarce but regular autumn passage migrant, particularly to Bardsey where it can be considered something of an island specialty.

Birds in Wales quotes 126 individuals up to and including 1991, since then there has been a further 23 individuals. All records were:
Pembroke: at Skokholm Aug. 31st, Sept. 2nd and Nov. 3rd 1993, at Skomer Oct. 19th – 20th 1993, Oct. 3rd & 10th 1999 and Oct. 16th 2000, at Ramsey Oct. 31st – Nov. 1st 1994, at Porthceli Sept. 24th 1993, at Strumble Head Oct. 2nd 1993 and at Nine Wells Oct. 21st 1996.
Caernarfon: at Bardsey June 1st – 2nd & Sept. 29th 1992, May 18th, June 5th, Oct. 22nd & Nov. 2nd 1994, Oct. 22nd 1995, Sept. 21st, Sept. 22nd & Oct. 23rd 1996.
Anglesey: at the Skerries June 1st 1998 and at Soldier's Point Oct. 12th 1999.

A breakdown of all Welsh records by county is shown on the map (sites with multiple records are given, e.g. of the 43 recorded in Pembroke, 20 were on Skokholm, 10 on Skomer, 6 on other islands and 7 on the mainland, a total of 86 in Caernarfon, of which 83 were on Bardsey.

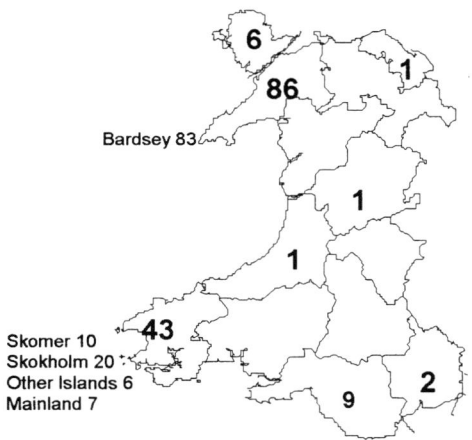

There was one winter record at Powys Castle, Montgomery on Feb. 3rd 1983 and there have only been a total of 14 spring records, one in April, 5 in May and 8 in June. A breakdown of autumn records by weeks is shown in the chart.

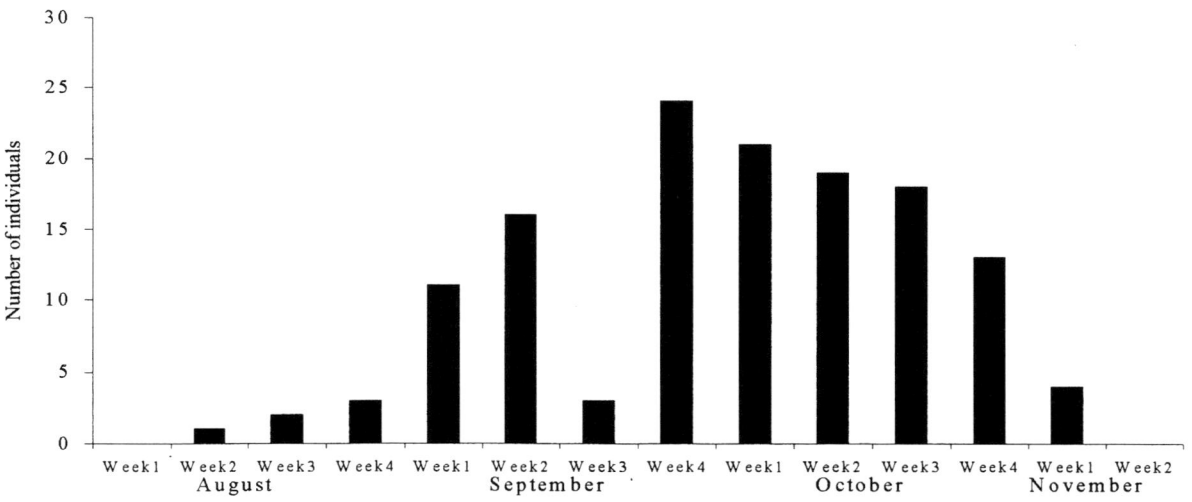

PIED FLYCATCHER *Ficedula hypoleuca*   GWYBEDOG BRITH
(E) A numerous summer visitor in most mainland counties, but a rare breeding bird on Anglesey. Now declining.

	1992	1993	1994	1995	1996	1997	1998	1999	2000
RSPB Dinas / Gwenffrwd Carm.	310 prs	286 prs 1688 fl.	183 prs 1092 fl.	271 prs 1653fl.	252 prs 926fl.	225 prs	218 prs 911 fl.	160 prs 489 pu.	
Lake Vyrnwy Mont.	600+ pu.		114 prs 420 pu.	114 prs 262 fl.	104 prs 381fl.	99 prs 426fl.		66 prs 164 fl.	
Cwm Clydach Glam.		114 prs 482 pu.	115 prs 293 pu.	105 prs 387 pu.			114 prs	101 prs 4.2 y/box	68
Gelli Aur Carm.		29 prs, 91 pu.	22 prs 51 fl.	21 prs 54 fl.	21 prs 75 fl.	20 pr 61 fl.	10 prs 19 fl.		9 prs. 15 fl.
Abbey Woods Carm.			18 prs 99 fl.	19 prs 64 fl.	17 prs 6 fl.	19 prs 47 fl.	19 prs 49 fl.		3 prs. 22 fl.
Ynyshir Cere.		91 nest boxes		70 nest boxes				69 prs 136 fl.	

fl. used to represent the number fledged, pu. represents the number ringed as pulli.
1995 figure at Dinas / Gwenffrwd represented a 48 % increase compared to 1994.

Pied Flycatchers are characteristic of the open sessile oakwoods of upland Wales, arriving from mid-April onwards. Numbers vary from one year to the next and breeding success is dependent on the weather and its effect on the supply of caterpillars. Using data from the National Atlas produces an estimate for the total Welsh population of 21,000 pairs. Results from BBS, 1994-1999 suggests a population change of 87%, with a revised population estimate of 18,000 pairs.

The table above summarises breeding data at selected sites: RSPB reserves: Dinas / Gwenffrwd, Carmarthen, Ynyshir, Ceredigion, Lake Vyrnwy, Montgomery and Cwm Clydach, Glamorgan; and Gelli Aur & Abbey Woods both Carmarthen. The poor productivity from north Brecon's woodlands in 1995 linked to the cold weather in late spring. A survey of 60 nest boxes in the Llangollen area, Denbigh 1997, found 28 were occupied, mean clutch size of 6.9, 5.02 young / nest fledged – an increase of 29% on 1996 figures.

Breeding performance based on records from nest box schemes, tabulated below. The decline in breeding performance shown is significant (r_s= -.086, n=7, P<0.05). Note that young fledged includes young which had reached ringing age. No data available from 2000.

1998 figures were generally poor due to the cold wet spring, indeed the productivity at Cwm Clydach (3.3 young per brood) was the worst since the survey began in 1986. At Ynyshir 25 pairs deserted at the egg stage, 8 at the juvenile stage and 8 pairs were predated by weasels. In contrast then, the Llangollen survey recorded an 38% occupancy rate but 5.8 young fledged / box – reflecting that this area missed the worst of the June downpours.

In 1999 the Llangollen survey of 60 boxes, recorded a 39% occupancy rate but only 3.5 young fledged / box (a 47% occupation rate in 2000). At nearby World's End, 11 out of 23 boxes were occupied with a productivity of 4.9 young fledged / box.

By 2000, this species was thought to be declining in Caernarfon & Meirionnydd. Indeed of 106 available nest boxes at the Mawddach RSPB reserve, Meirionnydd only 44 were occupied in 2000.

	1993	1994	1995	1996	1997	1998	1999
Pairs checked	420	452	511	373	166	316	361
Young fledged	2261	1950	2356	1313	674	1115	1135
Young fledged/pair	5.38	4.31	4.61	3.52	4.06	3.53	3.15

BEARDED TIT *Panurus biarmicus*   TITW BARFOG
(B) An erratic breeding species in one or two locations in very small numbers from the mid-1960's to the late 1980's; otherwise a rare autumn and winter visitor to reedbed areas.

Only 30 records since 1991, all records were:
Gwent: Uskmouth Nov. 28th 1992 & 2 there Mar. 14th 1993, 4 including 2 males Nov. 27th – 30th 1993 and a single at Magor on Mar. 30th 1996.
Glamorgan: individuals at Kenfig: 2 Nov. 13th – 19th 1992, a male Dec. 5th – 7th 1992, presumed the same Apr. 14th 1993. A male and female Oct. 24th 1993, 2 males were trapped on the 25th and all three birds were present until the 29th. At Oxwich: 5 birds, of which at least 2 were males, Oct. 18th 1992, birds heard there until Dec. 23rd.
Carmarthen: 3 at Witchett Pool Oct. 12th 1992.
Pembroke: 2 in bracken on Skokholm Oct. 24th 1993.
Ceredigion: at Ynyshir females, Apr. 15th – Sept. 17th 1993, Apr. 14th – Oct. 4th 1994, Apr. 11th – July 10th 1995 and Mar. 3rd 1996. One heard Teifi Marshes Jan. 30th 1995 and a female seen there July 6th 1999.

LONG-TAILED TIT *Aegithalos caudatus*   TITW GYNFFON-HIR
(D) A relatively common resident, widely distributed except on parts of Anglesey and Llyn. Numbers fluctuate widely from year to year in response to the severity of winter weather. Increasing.

Using data from the National Atlas produces a Welsh population estimate of 24,000 pairs. Results from BBS, 1994-1999 suggests a population change of 136%, producing a revised population estimate of 33,000 pairs.

MARSH TIT *Parus palustris*   TITW'R WERN
(E) A resident species, strongly sedentary, widespread but patchily distributed in the eastern half of Wales and in Pembroke. Scarce or absent in other western and northern areas. Thought to be declining across its range.

Birds in Wales states that this species is more plentiful on the eastern side of the mountain spine but is reasonably numerous across the south of Wales. There is very little concrete data from the period under review but it now appears that this species is declining across much of Wales. By 2000, it was described as common / fairly common in Gower, Ceredigion & Montgomery but scarce in Meirionnydd and only one record from Anglesey. Declining in Radnor and in Brecon where the population was estimated at 200 pairs in 1990 but has declined since (25 pairs in 4 10-km^2 in southeast), a revised estimate for 1999 is 100-150 pairs. Thought to be virtually extinct in Eastern Glamorgan by 1999.

Survey work in Carmarthen in 1993 discovered 40 prs. in 10 km square SN52, 16 pairs in the mid-Tywi valley in 1999. No evidence of any change in this county.

Using data from the National Atlas produces a Welsh population estimate of 9,400 pairs.

WILLOW TIT *Parus montanus*   TITW'R HELYG
(E) A thinly distributed resident, most frequent in south-east and east Wales and absent from most of north-east Wales. Thought to be declining across its range.

Using data from the National Atlas produces a Welsh population estimate of 4,000 pairs. Results from BBS, 1994-1999 suggests a population change of 58%, producing a revised population estimate of 2,300 pairs.

By 2000, it was described as uncommon & localized in Gower, Ceredigion and Meirionnydd. A handful of breeding records from 1995 onwards on Anglesey suggests that the population may be increasing there. Elsewhere a decline noted in Eastern Glamorgan, Pembroke, Montgomery and Radnor, where it has disappeared from some areas. In Brecon the population was estimated in 1990 at 250 pairs, since when it has declined. Most pairs are in conifers with areas of damp willow or in riverside alders; a revised estimate for 1999 is 125-175 pairs. There is no evidence of any change in Carmarthen, where 25 pairs were found in 10-km square SN52 in 1993.

COAL TIT *Parus ater* — TITW PENDDU
(C) A widespread and fairly common resident. Most numerous in conifer woodlands. Has increased in the second half of this century with the spread of conifer plantations.

Using data from the National Atlas produces a Welsh population estimate of 63,000 pairs.

Influx in autumn 1997, recorded at Bardsey, Caernarfon and Ramsey, Pembroke. On Bardsey the influx started Sept. 25th onwards lasting into November, with 15 on Sept. 25th, 20 on Oct. 7th and at least 35 were ringed 170 were recorded on Ramsey, some showing the characteristics of the continental race *ater*.

BLUE TIT *Parus caeruleus* — TITW TOMOS LAS
(C) An abundant resident species, most numerous in deciduous woodland but not restricted to this preferred habitat, especially in winter when it is more numerous. Increasing.

Using data from the National Atlas produces a Welsh population estimate of 340,000 pairs. Results from BBS, 1994-1999 suggests a population change of 111%, producing a revised population estimate of 377,000 pairs. One large count of 130 in oats at Pennorth, Brecon Aug. 30th 1996.

GREAT TIT *Parus major* — TITW MAWR
(D) A widespread and numerous species throughout Wales up to approximately 500m. Increasing.

Using data from the National Atlas produces a Welsh population estimate of 163,000 pairs, which corresponds to estimates from CBC data of 140-160,000. Results from BBS, 1994-1999 suggests a population change of 121%, producing a revised population estimate of 200,000 pairs.

NUTHATCH *Sitta europaea* — DELOR Y CNAU
(D) A widely distributed and fairly numerous resident of mature deciduous and mixed woodland throughout Wales. Increasing.

Birds in Wales quotes a westward spread of the Nuthatch in Wales (which was simultaneous with a northward expansion in England) can be detected from the early part of the 20th century, gathering momentum in the 1940s and 1950s. This increase is interesting in light of a loss of 50% of deciduous woodland in Wales over the same time scale.

Using data from the National Atlas produces a Welsh population estimate of 30,000 pairs. Estimates from CBC data however, put the Welsh population at only 22,000 pairs. Results from BBS, 1994-1999 suggests a population change of 126%, producing a revised population estimate of 38,000 pairs.

TREECREEPER *Certhia familiaris* — DRINGWR BACH
(C) Resident. Widely distributed in woodlands throughout Wales. Increasing.

Using data from the National Atlas produces a Welsh population estimate of 23,600 pairs, this corresponds to estimates from CBC data. Results from BBS, 1994-1999 suggests a population change of 175%, producing a revised population estimate of 41,300 pairs.

A single recorded on Skomer, Pembroke March 22nd 1998, and only the 12th island record.

PENDULINE TIT *Remiz pendulinus* — TITW PENDIL
(B) A vagrant.

The 2nd Welsh record was at Llyn Rhos Ddu, Anglesey Oct. 21st 1992. A male and at least one other bird was found at Kenfig, Glamorgan Nov. 11th 1996, staying there until Mar. 9th 1997. The only previous record was of a male at Bardsey, Caernarfon May $9^{th} - 13^{th}$ 1981.

GOLDEN ORIOLE *Oriolus oriolus* — EURYN
(B) A scarce summer visitor, recorded in all counties and particularly in Pembroke. Most of the records fall in May and June as is to be expected with a species overshooting its continental breeding areas.

A rare summer visitor, principally to Pembroke and Caernarfon. The table below summarises all records since *Birds in Wales*.

	1992	1993	1994	1995	1996	1997	1998	1999	2000
Gwent					1	1			
Glam.	1							2	
Pemb.	3	2	3		2	2			2
Cere.		2	2			1		1	1
Brecon	1								
Caern.	3	3-4	5			2		2	2
Angle.		1	2						
Flint						1			

Birds in Wales quotes 5 records for Carmarthen, all of these are now thought to be unreliable. *Welsh Birds* published a record of 2 individuals in Radnor in 1996, these however were never confirmed and should therefore not stand as the first record for that county. A breakdown of all Welsh records by month (arrival dates) is shown in the chart and by county as shown on the map.

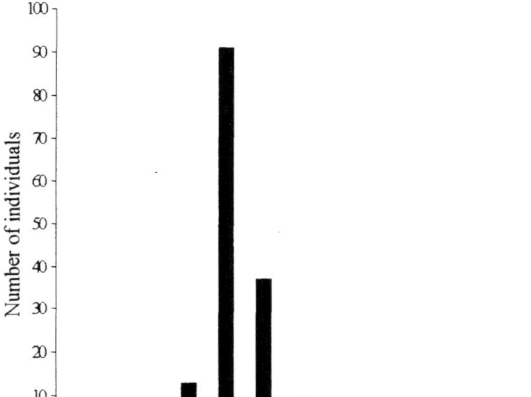

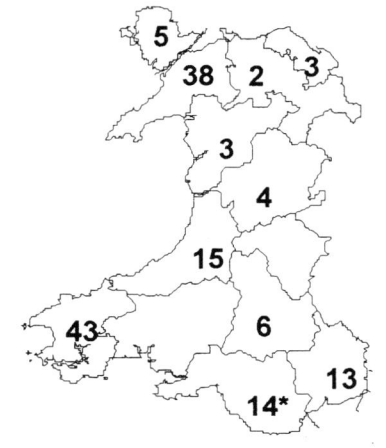

* 4 breeding records in 1800s

ISABELLINE SHRIKE *Lanius isabellinus*   CIGYDD GWDW
(B) A vagrant.

The second Welsh record was of a 1st winter at Nine Wells, St. David's, Pembroke Oct. 27th 1995. This was followed by an immature at Bardsey, Caernarfon Oct. 25th – 26th 1996 and a female at Cemlyn, Anglesey July 2nd – Aug. 16th 1998. The only previous Welsh occurrence of this eastern shrike was of a first winter bird at Holyhead, Anglesey Oct. 25th 1985.

RED-BACKED SHRIKE *Lanius collurio*   CIGYDD CEFNGOCH
(B) A scarce passage migrant in spring and autumn chiefly to coastal areas, formerly a widespread breeding species in small numbers.

Since 1991 there have only been 24 individuals recorded in Wales. All records were: Carmarthen: a male at Cynghordy July 9th 1996; Pembroke: at Skomer Oct. 18th 1993, May 24th, 27th & 31st 1994 (different birds), May 23rd 1995, Aug. 17th 1996 and June 5th 1997, Skokholm June 6th 1996, Freshwater West Sept. 27th 1992, St. Govan's May 30th 1997 and at the Gann Sept. 29th 1997; Ceredigion: a female at the Teifi Marshes Sept. 14th 1992, males at Cors Caron June 30th 1993 and near Tregaron June 7th – 8th 1999; Radnor: at Begwyns Aug. 20th 1995; Montgomery: at Newtown Nov. 4th 1992; Caernarfon: at Bardsey Sept. 21st 1992, Sept. 14th 1994, Sept. 6th 1996, a male June 3rd 1997, May 28th 1999, a female Aug. 11th and a male Aug. 26th 2000; Anglesey: a male at the Inland Sea June 23rd 1993, at Penysarn Sept. 23rd 1993 and a juv. Alaw Est. Aug. 24th 2000; Flint: a male at Moel Fammau June 6th 1993 and at Prestatyn Golf Course June 6th 1997.

Skomer is the top site for this species over the last 9 years, with 7 individuals, a mere 5 from Bardsey. Of interest is why so few on Skokholm, only one record, yet it is so close to Skomer.

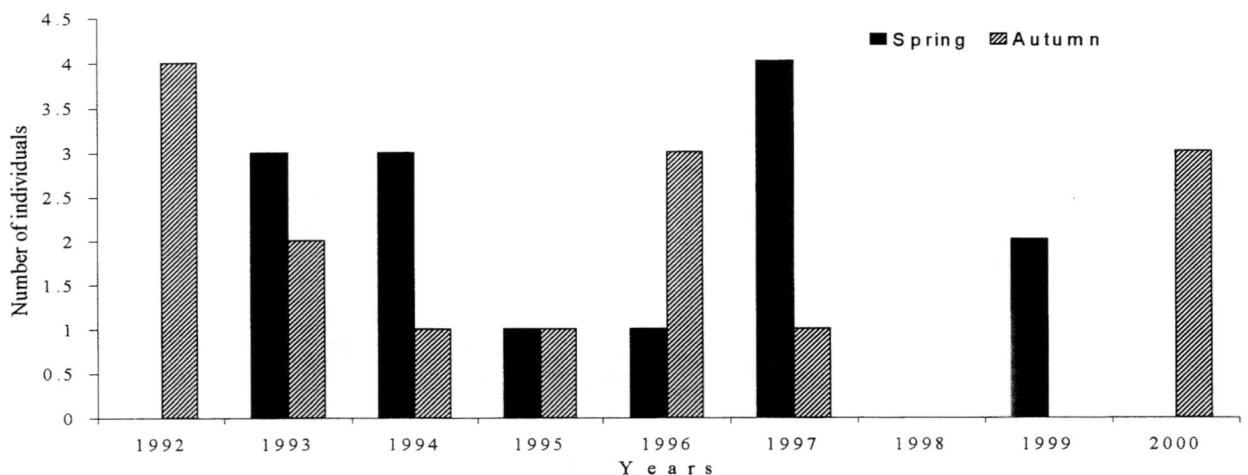

Analysis of Red-backed Shrikes 1992-2000

LESSER GREY SHRIKE *Lanius minor*   CIGYDD GLAS
(B) A vagrant.

Two individuals since 1991, both in Pembroke. One near St. David's June 24th was thought to be the same bird that was on Skomer July 2nd – 4th 1993. One at Lower Carnhedryn, St. David's Sept. 22nd 1998 was the other record.

Of the previous 7 records, 2 were in Caernarfon, at Pen y groes in June 1967 and the long staying individual at Abersoch, October - November 1986. Pembroke has had one other record, at Skomer in September 1974. The other individuals were at South Stack, Anglesey May 1961, at Shotton, Flint September 1961, at Ferryside, Carmarthen in October 1975 and at Fan Pool, Montgomery in May 1982.

GREAT GREY SHRIKE *Lanius excubitor*   CIGYDD MAWR
(C) A scarce but annual winter visitor, occasionally recorded on passage.

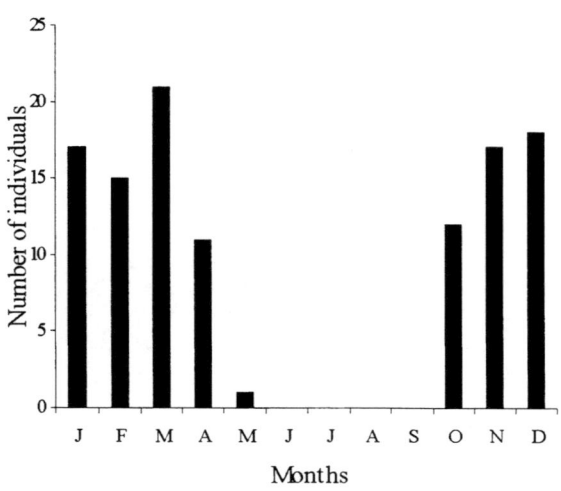
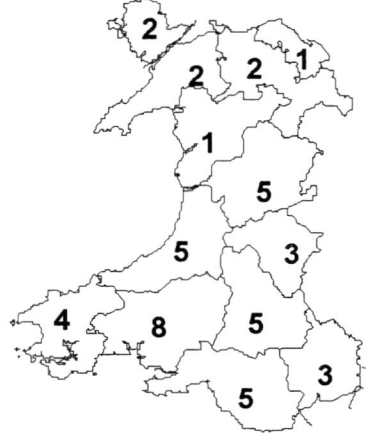

Birds in Wales states that the first arrivals from their Scandinavian breeding grounds reach Wales around the middle of October, with individuals seen until early May. This is born out by analysis of records from the period 1992-2000, as plotted in the chart above. A small number of wintering individuals and an influx of presumably passage birds in March / April. No records from the summer months except for one at Kilvery Hill, Glamorgan on May 13th 1996. The total number of individuals recorded in each county is shown on the map.

There were only 4 sites that recorded multiple records of wintering individuals: Mynydd Bach / Magor Forest in Glamorgan, Crychan in Brecon / Carmarthen, Usk Res. in Brecon / Carmarthen and Clocaenog Forest in Denbigh. A summary table is given below.

	92/93	93/94	94/95	95/96	96/97	97/98	98/99	99/00
Mynydd Bach							1	1
Crychan	1	2	1		1		1	
Usk	1	1			1		2	
Clocaenog	1	1						

WOODCHAT SHRIKE *Lanius senator*   CIGYDD PENGOCH
(B) A scarce visitor.

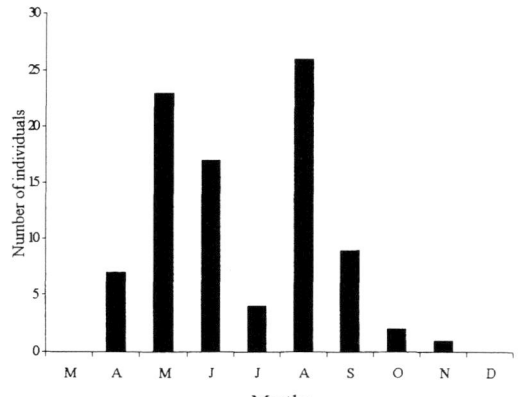
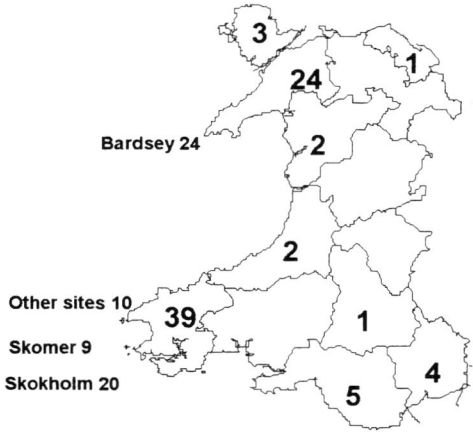

Birds in Wales quotes 67 individuals recorded up to and including 1991, since then there have been a further 14. The chart above and map summaries all the Welsh records.
All records since 1991 were:
Gwent: at Magor May 14th and at Caldicot Moor Aug. 8th 1993.
Pembroke: at Skomer May 24th – 28th 1994 and Apr. 29th 2000, at Skokholm May 31st 1997, at Dowrog Common June 6th 1992, at St. David's Head May 14th – 15th 1994 and an adult male at Strumble Head May 15th 1998.

Ceredigion: a male at Llangranog Apr. 30th 1994.
Brecon: an adult at Llanwrtyd Wells Aug. 25th – 27th 1996.
Caernarfon: at Bardsey June 7th 1992, Aug. 9th 1997, May 11th 1998 and a male May 1st – 4th 2000.

JAY *Garrulus glandarius*   YSGRECH Y COED
(C) Widespread resident, common in most deciduous woodland areas, least so in western coastal fringes; scarce in west Llyn and absent from large parts of Anglesey. Autumn / winter influxes from the east occur irregularly.

Using data from the National Atlas produces a Welsh population estimate of 21,600 pairs.
A marked irruptive movement noted on the Pembroke coast in October 1996 where the species is not normally seen, with groups of 3-7 noted at several places in the coastal plain, max. 34 at Porth Clais on Oct. 21st, following a failure of the acorn crop in that county.

Birds in Wales 1992-2000

MAGPIE *Pica pica*   *PIODEN*
(E) An abundant and widespread resident; numbers still increasing.
Using data from the National Atlas produces a Welsh population estimate of 85,000 pairs. Results from BBS, 1994-1999 suggests a population change of 116%, producing a revised population estimate of 100,000 pairs.

NUTCRACKER *Nucifraga caryocatactes*   *MALWR CNAU*
(B) An irruptive vagrant.
Additional record to the eight individuals published in *Birds in Wales*, of one at Beddgelert, Caernarfon Aug. 27th 1968. The others were from Mostyn, Flint Oct. 1753, Swansea in the early 1800's, Newport, Gwent Oct. 1954, Llangattock, Brecon autumn 1957 and in 1968, 2 at Henllan, Carmarthen August, at Whiteford, Glamorgan also in August and at Llanhilleth, Gwent in November.

CHOUGH *Pyrrhocorax pyrrhocorax*   *BRAN GOESGOCH*
(C) A resident in small numbers, mainly on cliff coasts of the western seaboard.
The Welsh population appears to have increased since 1963 (see table below). To what extent this represents a true increase rather than reflecting increased observer effort and improved survey methods is unknown.

Year	1963	1982	1992	1998
Number of pairs	110	139-144	151-178	173-185

More evident are the shifts in distribution within Wales, with the recolonisation of Anglesey and increases on the coast mirrored by declines in the population of inland Wales (Ceredigion, Montgomery and southern Meirionnydd).
The map opposite shows the maximum number of occupied sites of Chough for each county in 1992, data from the last decadal survey (the next survey is due in 2002).

Through the 1990s increased survey efforts have revealed the population to be relatively stable but with the continuation of the patterns of distribution mentioned above. The range has also increased to the east along the south coast beyond Gower.
The map opposite shows the maximum number of occupied sites of Chough for each county in 1998.

Elsewhere in the UK, declines in the population have increased the significance of the Welsh population in a UK context, so that Wales now has approximately three-quarters of the UK population.

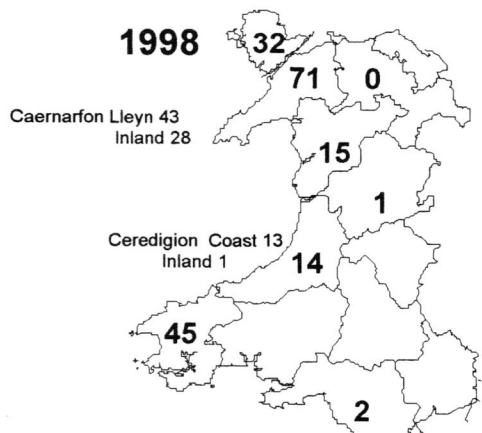

In 1996 – 1997 co-ordinated counts were conducted along suitable coasts, the results are tabulated below:

	Feb. 3rd / 4th 1996	Sept. 7th / 8th 1996	Feb. 1st 1997	September 1997
Gower	4	4	3	3
Pembroke	101-106	190	82	160
Ceredigion	29	58	36	95
Montgomery	7	0	1	2
Meirionnydd	36	58	18	24
Caernarfon	78	178	95	180
Anglesey	53	62	48	52

Colour ringing studies investigating the population dynamics and movements of Welsh Choughs have been carried out in north and mid Wales since 1990 (Cross and Stratford in prep.) and in south Wales since the mid 1990s (Haycock pers. Comm.). These have shown a high mobility amongst young birds, particularly between the Pembroke & Ceredigion populations moving to winter at Craig yr Aderyn, Meirionnydd and birds from the Lleyn & Anglesey wintering at Llanberis, Caernarfon. The start of a new population in Glamorgan is though to have originated with young birds moving from Pembroke.

Regular roost sites and associated feeding sites have been discovered and monitored in mid and north Wales since the mid 1990s. Numbers in roosts fluctuate both seasonally and between years, with peak concentrations in northern Snowdonia in the autumn months (July – November) when some large concentrations have been found e.g. 67 near Llanberis in October 1996. Numbers then drop off rapidly with few birds present in the winter months. Numbers at coastal roosts have shown a more even attendance throughout the year and there are suggestions that in the winter months birds are in smaller groups and that roosts are more mobile. The largest wintering concentrations were at Craig yr Aderyn, where counts were conducted on a monthly basis:

	J	F	M	A	M	J	J	A	S	O	N	D
1992	39	46	36	31			40	47	47	62	43	53
1993	50	47	37	32	21	29	52	55	54	50	47	44
1994	39	34	28	30			37	42		36		40
1995												
1996	49	32		13	14		37		45	20	19	9
1997	11	14	14	13	13	18	33	35	38	42	28	24

JACKDAW *Corvus monedula* JAC Y DO
(C) An abundant and widespread resident. Increasing.

Using data from the National Atlas produces a Welsh population estimate of 50,000 pairs. Results from BBS, 1994-1999 suggests a population change of 119%, producing a revised population estimate of 59,000 pairs. Records of large roosts / flocks, over 1000, were:

- 1994 5,000 Ystrad Llwypia, Jan. 1st with 3,000 there Oct. 15th, 3,400 Danygraig, Porthcawl Jan. 3rd and over 4,000 flying over Skewen at both ends of the year, all Glamorgan
- 1995 1,500 Llanfihangel, Radnor in November
- 1996 3-5,000 at Crymlyn Bog at both ends of the year, 2,200 at Gowerton Oct. 18th both Gower and 1,200 in January at the Newborough Warren corvid roost, Anglesey
- 1997 2,500 at Tredegar Park, 1,000 at Goytre in January, 1,000+ at Brynmawr on Dec. 27th, all Gwent; 4,500 at Jersey Marine on Jan. 23rd, 1,800 at Ystrad Llwynpia on Feb. 8th, 1845 at Nantyffyllon on Sept. 19th, 2,500 at Aberdare on Dec. 2nd and 3,200 at Gowerton on the 31st, all Glamorgan
- 1999 2,000+ roosting in Penrhyn Park, Caernarfon in autumn/winter

ROOK *Corvus frugilegus* YDFRAN
(C) A plentiful and widespread resident, breeding mainly below 305m.
The 1996 BTO survey estimated the Welsh population at 53,140 pairs (95% confidence limits of 35,900 – 73,600 pairs) [Marchant & Gregory *Bird Study* 46:258-273]. Densities varied widely but were predictably higher in lowland districts. The present population compares to estimates of 38,916 pairs in 1975 and 98,260 in 1944-46.

1992	93	94	95	96	97	99	2000
987	1046 in 35	986 in 35	1088 in 40	1157 in 42	1108 in 43	1036 in 55	651 in 29

The only area with an annual census was Gower. Data tabulated above. The first figure the number of nests, the second the number of Rookeries. Numbers slow a slight increase. A total of 453 nests were counted in the whole of Glamorgan in 2000.

In Ceredigion, A. Chater censused 600 km² north of grid line 70, see *Welsh Birds* 1 (3): 3-28. Data: 2466 nests in 59 rookeries in 1992, 2550 in 57 in 1993 and 2228 in 61 in 1994. The largest Rookery was at Aberllolwyn and numbered 200 nests in 1992. The 1993 figure was the highest recorded in the 19 years of this study.

In Carmarthen, 37 colonies totaling 808 nests were surveyed in 1994, 54 colonies and 1368 nests in the following year.

In Pembroke 770 nests were counted within a 10 mile radius of Haverfordwest in 1994, a decline of 20% when compared to the average for the previous four years, however 817 were counted in the following year – some sort of recovery. In 1996 1585 nests in 48 colonies were counted south of the Milford Haven waterway, east to Penally, an area of c135 km², approx. 9% of the county's land mass. This equates to 11.7 pairs/km², a density less than half of that produced by a random sample of 14 km squares counted for the Rook Survey.

At Llangollen, Denbigh, 101 nests in 3 colonies were surveyed in 1997. Five of which in one tree had the very low mean clutch size of 2.75 and a mean fledged b/s of 0.75 – the lowest levels since 1986 and probably reflecting the dry and cold early spring. Rookeries in other counties were counted in 1999:317 nests in 10 rookeries were counted in Gwent 317, while in Brecon 1123 in 32 rookeries to the south & east of the Epynt (cf 1286 in 1980).

CARRION CROW *Corvus corone*   *BRAN DYDDYN*

(C) An abundant resident and partial migrant throughout Wales. Often regarded as the principal avian pest species in rural areas. The Hooded Crow is an irregular wanderer to Wales, mainly in winter and most frequent to coastal areas.

The Welsh population of this extremely abundant species is estimated at 126,000 pairs, using data from the National Atlas.

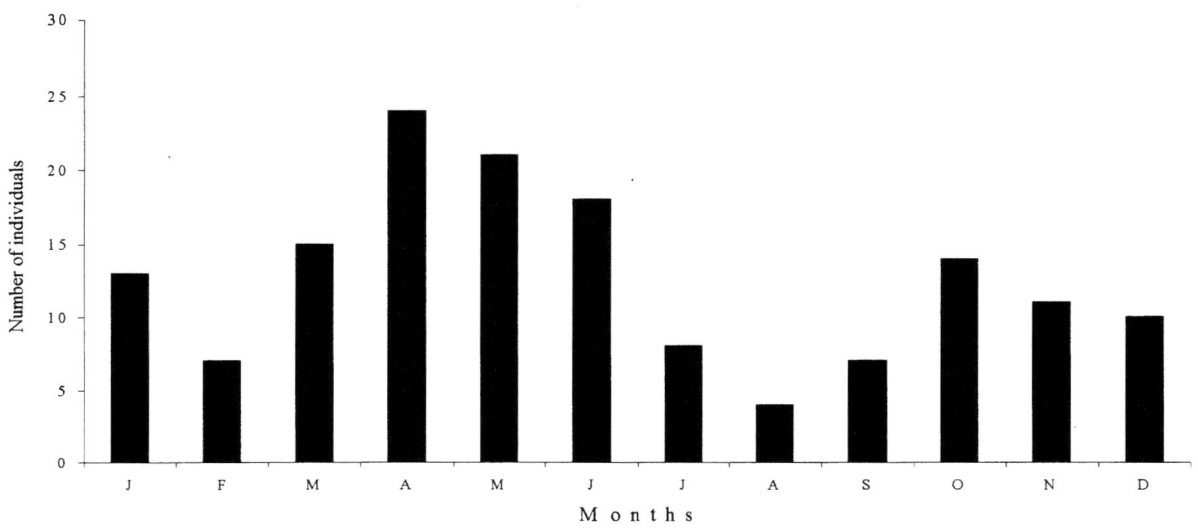

Monthly distribution of Hooded Crows 1992-2000

Hooded Crows were recorded regularly in 7 counties with a pair breeding on Anglesey in 1996. The table below summarises these records. A pair nested near South Stack, Anglesey in 1996 fledging at least 2 young (*Welsh Birds* 2(4): 228) and one bred with a Carrion at Clarach, Ceredigion in 1994.

Hybrids were reported from Sealyham, Pembroke Mar. 29th 1999; in Caernarfon at Rhyd Du Sept. 1993, Bangor Nov. 1997 and Bardsey April 1998; on Anglesey at Aberffraw July 1993, Llyn Traffwll 1993 – 96 & July 4th 1999, Penrhos April 1998, Llyn Alaw July 13th & Aug. 7th 1999 and a pair at South Stack in June 1997, a single there Nov. 18th 1999; in Denbigh at Llangollen 1997 – 8.

	1992	1993	1994	1995	1996	1997	1998	1999	2000
Pembroke	2	1	3	6	5	5	3+	2	3+
Ceredigion	1	2							1
Meirionnydd	1	1							
Caernarfon	8	2	4+	10	3+	2	5	5	3
Anglesey	1	3	6	5	1		2	4-5	3
Denbigh									1
Flint	1						1		

RAVEN *Corvus corax*   CIGFRAN

(C) Resident: distributed throughout the whole of Wales. Numbers are very high in many of the uplands and coastal-cliff areas.

The range of the Raven in Britain is still much reduced from what it was 200 years ago (Holloway 1996). In Wales and the Marches however, the species appears to be undergoing a considerable increase and the population here must be back to something near their pre-persecution heyday. Generally associated with rugged coasts and mountainous terrain, the Raven, in the absence of persecution, is equally at home in lowland farmland, parkland, and open forest and it is in these habitats that the increases have largely occurred.

Being an omnivore and an opportunist, large numbers tend to congregate at regular food sources such as rubbish tips and abattoirs. During August and September 1997 hide watches at a landfill site at Tylwch near Llanidloes, Montgomery showed that of the reasonably constant c.100 Ravens present at any one time, approximately 1 in 10 were colour-ringed. Over the two month period 113 different individuals were recorded suggesting that the site was being used by over 1,000 Ravens. Identified birds included 9 birds of the year ringed in Shropshire and one from Gwynedd (Cross pers obs). Nearby in Rhayader, Radnor, the Red Kite feeding station at Gigrin Farm, run as a tourist attraction, has also become important for the local Raven population and regularly attracts between 50-100 (Cross pers obs) some of which also utilise the Llanidloes refuse site. For unknown reasons other regular kite feeding stations at Tregaron, Talsarn and Ponterwyd, Ceredigion attract far smaller numbers of Ravens and certainly not in proportion to the number of kites present.

Parts of Wales, always one of the species strongholds in Britain and Ireland, hold some of the highest recorded breeding densities anywhere in its extensive global range, reflecting the abundance of suitable nest-sites and the high availability of carrion. Only in parts of Idaho (Kochert et al 1976 in Nogales, 1994), on El Hierro Island, Canaries (M. Nogales, 1994) and Wolgast, Germany (Sellin 1987 in Ratcliffe 1997) are there published studies showing a greater breeding density over a large area - 72.6, 32-35.6 and 18.7 pairs /100km^2 respectively. A figure of 72.6 pairs /100km^2, from the Snake River Raptor Reserve in Idaho is truly astonishing but was derived from a long linear study area in otherwise featureless plains so is not very comparable.

Recent work by the Shropshire Raven Study Group has shown a density of up to 14 territorial pairs/100km^2, in the southern hills of this border county. Population densities in Wales as a whole must be fairly similar to those in southern Shropshire. A study of the raven population of 399km^2 of the Brecon Beacons (Dixon 199) found a total of 45 territories used in one of the three years 1993 - 1995 representing a territory density of 11.3 pairs/100km^2. An annual occupancy rate of 92% at these territories gave an annual breeding density of 10.4 pairs/100km^2. Taking a mean figure of 10 pairs/100km^2 over the whole of Wales gives a population estimate of some 2,078 breeding pairs. Assuming a similar number of non-breeding birds, this puts the total population in Wales in excess of 10,000 individuals. This figure is twice that given in *Birds in Wales*. The discrepancy is partly because the number in lowland areas of Wales has increased in recent years and partly because the previous estimates were over-cautious.

Roberts and Jones (*Welsh Birds* 2(3): 121-130), have documented big increase on a local scale on a North Wales grouse moor. Between 1988 and 1997 the number of pairs nesting around the periphery increased from 1 to 10 with birds starting to occupy neighboring lowlands. In 2000 an additional two pairs were present (Welsh Birds) suggesting that the increase is ongoing. Signs of this recovery are nowhere more apparent than along the Welsh border counties and West Midlands which may involve birds reared in Wales. The Shropshire Raven Study Group proved a breeding population in excess of 115 pairs within the county in 1999 - the estimate for the Shropshire Breeding Atlas, 1992 based on fieldwork undertaken in 1985 - 1990 was only 30 - 35 pairs. Some of this increase is due to better coverage but much is due to a real increase in the population.

A long running colour-ringing scheme in Wales during the years 1986 - 1999, involving over 1,500 individually colour-ringed chicks, failed to reveal much about the movements of Ravens (Cross pers obs.).

In the USA several radio-tagging studies have shown that individual Ravens may undertake significant daily movements (>50 kms) between communal roost sites and daytime feeding areas (Engel and Young, 1992, Heinrich et al 1994). Little equivalent data exists for Britain and Ireland but colour-ringed Ravens have been noted making return movements of 47 km between a rubbish tip at Onibury in Shropshire and the site at Tylwch (Cross own obs.). Recruitment of colour-ringed birds within the study area was very low suggesting either high mortality or low natal philopatry. As few records were obtained of birds travelling great distances it can only be assumed that it was due to high juvenile mortality. The recovery rate from BTO rings was well below the national average and it is assumed that many birds were being killed illegally and any rings present were not reported (this was known to be the case in several instances). The reporting rate of colour-ringed birds by birdwatchers or the general public was similarly extremely low.

Ravens have long been regarded as intelligent and playful, and recent observations in Wales confirm this. Several reporters have recorded Ravens sliding down snow banks on their backs and then hopping or flying back to the top in order to repeat the performance (pers obs, Williams & Rees (1998), Moffet 1984). Adrienne Stratford (pers comm.) has recently recorded similar behaviour on steep sand dunes at Aberffraw on Anglesey.

Monthly counts of the Newborough Warren Raven roost, Anglesey by Bangor University Bird Group since 1995 reveal a seasonal pattern with numbers peaking December – February followed by a steady decrease in early spring. There usually follows a slight increase in May (with up to 900 birds) then another decline to a low point of 200-300 in August and early September. Finally numbers of birds build up once more throughout the autumn period.

It appears that the majority of the roost – using Ravens are immature birds. Breeding adult birds remain on territory all year round. Where so many immatures come from remains a bit of a mystery – one or two ringed birds have been identified from south Caernarfon and mid-Wales but the origin of the vast majority is not known. Certainly there seem to be more than a local north Wales population on it's own could produce. This raises the likelihood that there is long distance immigration of Ravens to Anglesey in autumn and winter, perhaps from the southern uplands of Scotland, the Isle of Man or possibly Ireland.

The highest densities of roosting birds are found in mature Corsican and Monterey Pines on or close to the rocky ridge of higher ground which dissects the 769 hectare Forestry Enterprise plantation in a NE-SW direction.

The analysis of the content of pellets of undigested food produced by roosting birds showed that individuals roosting together had usually fed upon the same food items almost certainly from the same sites, e.g. many birds using trees in the east of the roost had fed on arable land judging by the high concentration of grain, whereas those roosting in the west produced bone rich pellets from different food sites. This was further tested using sheep carcasses laced with coloured plastic beads at several sites within a 20 km radius of the Newborough roost and subsequent monitoring of pellets, which revealed concentrations of beads of the same colour in localised parts of the roost. Furthermore the temporal sequence of appearance of coloured beads beneath trees and their increasing distance from the initial dropping sites over a period of a few days suggested that the discovery of a carcass by one Raven is then communicated to others, which join in the scavenging exercise and likewise produce the beads to prove it.

STARLING *Sturnus vulgaris* *DRUDWEN*
(C) An abundant resident; passage migrant and winter visitor in large numbers.

Using data from the National Atlas produces a Welsh population estimate of 80,300 pairs. Records of large roosts, over 5,000 strong during the period under review were:

1992	6,000 at Kenfig, Glamorgan Dec. 12th, 5,000 at Llangorse, Brecon Nov. 29th with 15,000 there Dec. 13th
1994	20,000 at Kenfig on Jan. 25th, 10,000 at Llangorse & 10,000 at Mynydd Epynt in December, both Brecon and 50,000 at Llanfihangel, Radnor in January
1995	100-150,000 Teifi Marshes, Pemb/Cere January – March, September – December, 50,000 Epynt in January, 6-7,000 Llangorse Oct. 20th & Dec. 23rd, 20,000 Llanwrthwl Nov. 1st – 9th, 50,000 Abereithon Turbary, Newbridge on Wye for 2 weeks in January, 20,000 Llanfihangel in November and 20,000 at Malltraeth, Anglesey Dec. 17th
1996	5,000 Sandfields estate, Gower Feb. 9th, 6,000 Oxwich, Gower Oct. 27th, 100,000 Teifi Marshes in January, 5,000 Ffrwdgrech, Brecon January – March, 16,000 Llangorse Nov. 30th & 12,000 in December
1997	5,000 Crymlyn Bog, Gower February, 12,500 Kenfig in December, 65,000 Teifi Marshes Oct. 17th, 15,000 Ffrwdgrech in January, 5,000 at Llangorse in October, 50,000 Conwy RSPB, Caernarfon October – December, 5,000 Malltraeth in January & 25,000 in October, 50,000 Rhyl Harbour, Flint Jan. 6th & 7-10,000 on Dec. 26th
1998	7,500 at Kenfig in November, 20,000 Ffrwdgrech on Feb. 8th & 33,000 on Dec. 28th, 250,000 at RSPB Conwy Nov. 7th, 250,000 at Malltraeth Nov. 28th and 500,000 flying west over Red Wharf Bay, Anglesey, just after dawn on Jan. 24th
1999	10,000 Kenfig November & December, 6,500 Llangorse Jan. 10th with 20,000 on Oct. 28th, 20,000 at Ffrwdgrech Jan. 31st, 10,000 Conwy RSPB January – February & November – December, 40,000 Ystymllyn Dec. 18th, 15,000 Rhyd Wen Nov. 9th, 15-20,000 Llanfaelog Dec. 31st, 50,000 Kinmel Bay Nov. 20th and 5-6,000 Rhyl Harbour in January
2000	15,000 Kenfig Nov. 19th, 5,000 Cynghordy, Carmarthen Feb. 10th, 5,000 Llangorse Nov. 3rd, 16,000 Ffrwdgrech Dec. 29th, 20,000 Conwy RSPB Dec. 21st and 50,000 Llanfaelog Nov. 27th

ROSE-COLOURED STARLING *Sturnus roseus*   *DRUDWEN WRIDOG*
(B) A rare visitor, most often recorded from Pembroke and Caernarfon.
This species appears to be occurring more often, with 20 records in the last 9 years compared to only 29 individuals previously recorded in Wales. Breakdown of records since 1991:

Year	1992	1993	1994	1995	1996	1997	1998	1999	2000
Adults			2			1	1	3	2
Juveniles	1	1	1	1	1	1	1	1	3

All records were:
Glamorgan: an adult at Lansbury Park, Caerphilly June 19th 1998 and Aberdare late August 2000, juveniles at Sker/Kenfig Sept. 20th – Oct. 4th 1992, at Newton, Swansea Nov. 13th – 30th 1998 and at Nash Point Sept. 23rd 2000.
Pembroke: adults at St. David's May 16th – June 22nd 1999 and Skokholm June 8th – 10th 2000, juveniles at Strumble Head Sept. 18th – 29th 1995, Aug. 29th – Sept. 5th 1999 & Sept. 10th – 12th 2000 and at Pembroke Dock Jan. 13th – 14th 1996.
Ceredigion: adults at Ynyslas June 12th 1994 and at New Quay on the same date but in 1999 !Radnor: a juvenile at Llansantffraid Sept. 22nd - 27th 1994.
Caernarfon: at Bardsey an adult July 27th – 28th 1999 and juveniles Sept. 21st – 29th 1993 & Oct. 14th 2000.
Anglesey: adults at Moelfre July 26th – Aug. 21st 1994 and at Rhosneigr Aug. 20th – 28th 1997.
Denbigh: a juvenile at Glan Conwy Oct. 22nd 1997.

Birds in Wales quotes 7 records of shot specimens from the 19th century. A breakdown of all Welsh 20th century individuals by county is shown on the map:

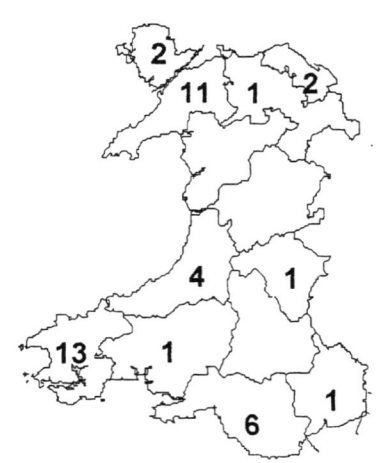

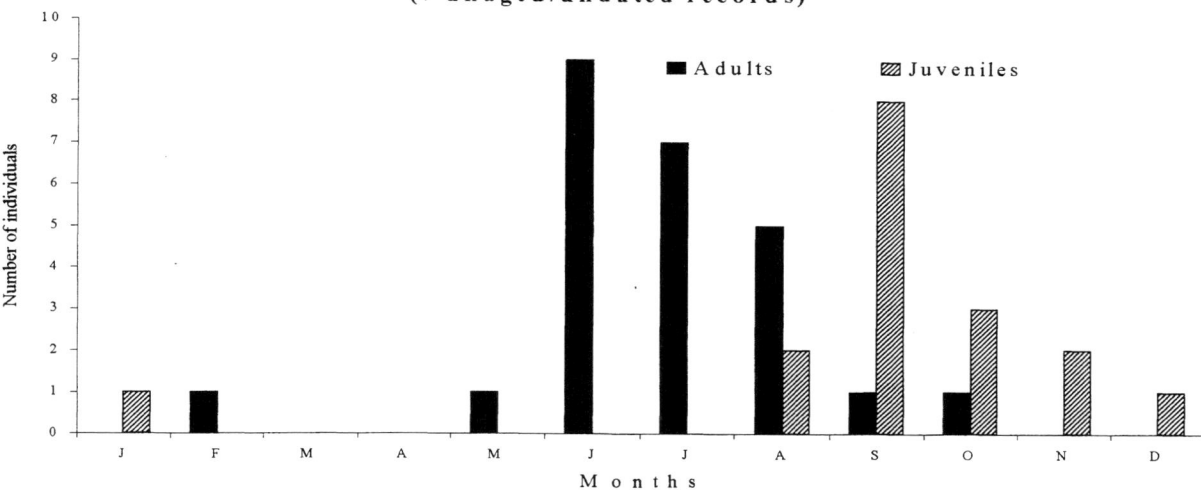

Monthly distribution of Welsh records of Rose-coloured Starlings (7 unaged/undated records)

HOUSE SPARROW *Passer domesticus*   ADERYN Y TO
(E) Resident, breeding throughout Wales in all counties. Not ubiquitous, however, and absent from many upland settlements. Significant declines noted in most areas during 1970 –1990 but this may have been reversed by the end of the century.

The National Atlas shows that this species breeds in all bar a handful of 10-km squares in Wales. Since the 1970s this species has been declining in Wales and by using data from the National Atlas produces a Welsh population estimate of 300,000 pairs. Results from BBS, 1994-1999 suggest a 162% change producing a revised population estimate of 484,000 pairs.

TREE SPARROW *Passer montanus*   GOLFAN Y MYNYDD
(E) Breeding resident in small numbers with a patchy distribution. Absent from most western areas. Declining in most areas.

Birds in Wales states that this species is in decline, since publication the decline has continued. The size of the decline is hard to determine due to the mobility of pairs in utilising undiscovered nest sites. In some areas this has been reversed by the use of artificial nest boxes (e.g. Tywi valley and Newtown).

Using data from the National Atlas produces a Welsh population estimate of 4,600 pairs.

A summary of breeding records per county is shown below and the maximum number of pairs in each county on the map opposite. No data was available from Denbigh.

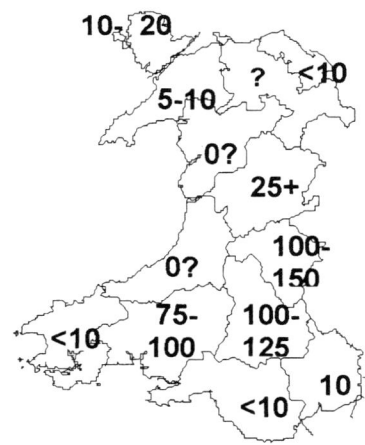

Gwent: 8 broods fledged from 5 nest boxes in 1993 but no breeding at Llandegfedd. Described as scarce but plentiful in the NW of the county in 1995. In 1997 pairs bred at 2 sites and were present at another 4, 6-7 pairs bred in 1999, bred at 3 sites in 2000.
Glamorgan: 6 pairs bred in the east of the county in 1998 and pairs present at 2 sites in the west in 1992.
Carmarthen: a small population in the Tywi valley, a post breeding flock of 40 birds present in 1992, 8 pairs bred at 4 sites in 1998 increasing to 38 pairs in 1999, some of which were in nest boxes. 10 pairs fledged 30 young in 2000. The estimated county population is 75 – 100 pairs.
Pembroke: irregular breeder, 2 pairs in 1997.

Brecon: 9 pairs at 5 sites in 1992, present at 10 sites in 1993, county population estimated at 90 – 95 prs. in 1995, 40 – 45 prs. in 1996, 70 prs. in the SE & 15 prs. in the N in 1997, total of 90 prs. in 1998 and 80 in 1999. The county population was estimated at 250-300 prs. in 1990, now thought to be 100- 125 prs. all in the SE with only one site to the north of the Epynt.

Radnor: survey in 1997 found presence in 15 1-km squares, a total of 20 pairs. Declining due to a lack of suitable nest sites.

Montgomery: 15 prs. in nest boxes near Newtown in 1996, 4-5 prs. in 1997, 23 prs. in 1998 (21 of which were in nest boxes), 25 prs. in 1999 (22 of which were in next boxes) and in 2000, 15 prs. in nest boxes near Newtown, 1-2 prs. at Dolydd Hafren and 9 prs. at Bishop's Castle, again in nest boxes.

Anglesey: only one pair found in a survey of 73 1-km squares in 1997 (cf. 6 in 1986).

Flint: 2 prs. at one site in 1999.

Flocks over 50 recorded at:
1992 130 at Garth January – February, 100+ near Brecon Dec. 20th both Brecon, 200 at Llanasa, Flint in January.
1993 50 in stubble at Glasbury, Brecon in September
1994 60 at Bouaghwood, in January, 70 at Garth Nov. 4th, 72 at Llyn Helyg, Flint Feb. 6th
1995 80 at Garth Jan. 4th with 110 there Dec. 21st, 50 at Llyswen, Brecon November – December
1996 120 at Garth in February, 50 at Pipton all January with 74 there in December, both Brecon. 200 in undersown barley stubble Llyn Helyg in early November
1997 50 at Llandegfedd Res., Gwent at both ends of the year, 70 flew east over Warren Farm, Point of Air, Flint Oct. 18th
1998 150 at Pipton Dec. 13th in weedy roots
1999 50 Ffairfach, Carmarthen Dec. 16th, 150 at Pipton until Jan. 21st at least, up to 100 at Dolydd Hafren, Montgomery from August onwards
2000 40-60 at Garth during January & February. In Radnor, a total of 95 individuals in 3 flocks at Glasbury in December and 50 near Old Radnor on Sept. 28th. At Dolydd Hafren, Montgomery 80 on Feb. 14th and 100 on Sept. 27th

SPANISH SPARROW *Passer hispaniolensis* GOLFAN SBAEN
(A) A vagrant.
A male at Martin's Haven, Pembroke May 18th 1993 was the first and only record for Wales.

RED-EYED VIREO *Vireo olivaceus* TELOR LLYGATGOCH
(B) A vagrant.
The fourth Welsh record was of an individual at Porth Clais, Pembroke Oct. 18th – 19th 1995. On Sept. 19th 1998 one was found dead under the lighthouse at Bardsey, Caernarfon. The three previous records were from Skokholm, Pembroke Oct. 14th 1967, Aberdaron Sept. 25th – 26th 1975 & Bardsey Oct. 15th 1985, both Caernarfon.

CHAFFINCH *Fringilla coelebs* JI-BINC
(E) An abundant resident, numerous passage migrant and winter visitor. Possibly declining.

Birds in Wales states that the intimate pattern of small fields with many coppices, larger areas of woodland and abundant hedgerow trees, which occupies much of the Welsh countryside produces an ideal countryside for Chaffinches. Using data from the National Atlas produces a Welsh population estimate of 540,000 pairs. Results from BBS, 1994-1999 suggests a population change of 86%, producing a revised population estimate of 464,000 pairs.

A large movement was recorded over Lavernock, Glamorgan in October 1992, 1747 in two hours on the 7th and 2300 in 3 hours on the 11th. 649 counted passing south over Bardsey, Caernarfon Nov. 2nd 2000.

The annual autumn passage recorded at Strumble has declined dramatically over the nine years, with a maxima of only 200 counted passing west on Oct. 23rd and 265 going NW on Nov. 6th 1998.

Large flocks, over 300 were recorded at:
1992 500 in linseed stubble, Marloes & Newgale, Pembroke in December, 300 at Garth, Brecon Jan. 1st
1993 900 at Marloes in January, 400 at Talgarth, Brecon in January and 600 at Garth in November
1998 370 at Braich y Pwll, Caernarfon Oct. 23rd
1999 400 at Monmouth, Gwent Dec. 29th, 500 at Talybont roost Oct. 29th, 2,000+ at Glan y-Mor Elias, Caernarfon Nov. 7th
2000 400+ at Monmouth, Gwent in January & February, 307 at Merthyr Mawr, Glamorgan Jan. 2nd, 500 at Margam Park, Glamorgan Oct. 1st, 300 at Broad Oak, Carmarthen Feb. 1st, 300 on Skomer, Pembroke Nov. 21st and 400+ at Whitebridge, Caernarfon Sept. 18th

BRAMBLING *Fringilla montifringilla*   *PINC Y MYNYDD*
(C) Winter visitor in variable numbers; seldom numerous and usually locally distributed.
Large flocks, over 100 were recorded:

1992	150 at Pentre Mawr Park Dec. 5th and 200 roosting at Gwysaney on the 24th, both Clwyd
1993	100 Rhyader, Radnor January – Mar. 23rd, 100 Greenfield Valley, Flint Jan. 8th, 300 Cilcain, Flint Mar. 20th, in Gwent 200 at Caldicot Castle Nov. 22nd, 350 Wentwood Forest Dec. 1st
1994	800 at Gigrin Farm, Radnor at both ends of the year
1995	150 at Llanafan, Brecon during the early months. Large influx in late October and early November, which appeared to be confined to Ceredigion, Brecon, Radnor and Montgomery. The largest flocks were noted in Ceredigion, with 200 at Hafod Ash, Cwm Ystwyth on Nov. 15th, 800 – 1000 at Pwll Peiran on the 22nd increasing to 1500 by Dec. 27th and a roost of 800+ Llyn Fanod in December, many flocks of up to 50 noted in the Ystwyth and Rheidol valleys. There were also 300 – 400 on beech mast in the Elan Valley, Radnor in late October – early November, 300 at Lake Vyrnwy, Montgomery in November and 220 feeding in a single rowan tree at Llanwrtyd, Brecon Oct. 29th
1996	following the major influx in late October 1995, large flocks were widespread in central and north Wales. 800 Dinefwr Park, Carmarthen Jan. 7th, up to 500 Hafod Woods, Cwm Ystwyth, Mar. 19th and similar numbers at Llysdinam Newbridge, Brecon / Rador until early April. 800 at Gregynog Hall, Montgomery on Mar. 16th. 100 at Ogwen Bank, Caernarfon Feb. 28th. By contrast only 2 birds were seen in Gower and 3 in Pembroke
1997	fewer and generally smaller flocks, mostly in Gwent. Largest flock 100 Llanfair Discoed on Jan. 1st
1998	scattering of records and mostly of small numbers in both winter periods. Max. 200 at Pen y fan industrial estate, Gwent Jan. 19th – Mar. 12th, with some birds remaining until the 27th, 100 Cwm Ystwyth on Nov. 14th
1999	100+ at Loggerheads, Denbigh Feb. 28th

Late records were: a singing male on Ruabon Moor, Denbigh June 21st 1993, in 1997 a male at Llyn Helyg, Flint Apr. 14th and a singing male at Llanbadarn Fynydd, Radnor on Apr. 23rd, at Wentwood, Gwent Apr. 27th and at Glog, Pembroke May 9th – 10th.

SERIN *Serinus serinus*   *LLINOS FRECH*
(B) A rare passage migrant.
Birds in Wales quote 14 records, 9 in Pembroke, 2 in Carmarthen and singles in Glamorgan, Ceredigion and Flint. Since then there have been four further records, three from Pembroke and one Caernarfon. In Pembroke at Skomer Nov. 12th 1993, a male on Ramsey May 22nd 1994 and at Porth Clais Nov. 4th 1998. Bardsey entertained a male May 23rd 1997.

GREENFINCH *Carduelis chloris*   *LLINOS WERDD*
(E) Common resident, increasing; partial post-breeding migrant.
Generally a common species, occurring throughout lowland areas in all counties of Wales. Using data from the National Atlas produces a Welsh population estimate of 35,500 pairs. Results from BBS, 1994-1999 suggests a population change of 123%, producing a revised population estimate of 44,000 pairs. Estimates from CBC data however suggest an even larger Welsh population of 63,000 pairs.

Thought to be declining as a breeding bird in north and west Brecon 1996 but appears to be increasing in east Montgomery, probably as a result of *leylandii* and feeding in gardens.

Large flocks over 100 were:
120 at Sker Farm, Glamorgan Jan. 24th & Nov. 19th 1992, 150 at Talgarth, Brecon Nov. 14th 1992 with 400 there November 1993, 140 Talybont Res. Brecon Feb. 21st 1993 and 100 at Nantyglo, Gwent February & March 1993.

In 1999: 100-110 at Treforest, Glamorgan in the early months and 175 there at the end of the year, 150 at Hendre Lake Sept. 11th, 200 Sker Farm Sept. 28th, 250 at Broad Oak, Carmarthen in August, 100 at Carnhedyn, Pembroke Sept. 18th and 200 at Ffrwdgrech roosts, Brecon Dec. 5th.

In 2000: 210 at Treforest Industrial Estate Jan. 13th, 110 on the Blackpill Foreshore in January & February, 160 at Lavernock Sept. 8th, all Glamorgan; 200 at Ffrwdgresh, Brecon on Jan. 18th with 100 there on Dec. 29th and 300 at the William Condry Reserve, Montgomery on Nov. 12th.

GOLDFINCH *Carduelis carduelis*   NICO

(E) Resident, summer visitor and passage migrant; possibly winter visitor in small numbers. Population is increasing.

A common species breeding throughout the country. Using data from the National Atlas produces a Welsh population estimate of 25,000 pairs. Results from BBS, 1994-1999 suggests a population change of 157%, producing a revised population estimate of 39,000 pairs. This corresponds to a population estimate of 30,000 from CBC data.

Passage noted at Lavernock, Glamorgan during September 1992, when 700 were counted on the 9th, 1,000 on the 11th in just 3 hours. During September – November 1997, a total of c1,600 birds involved, mainly on Oct. 19th when c1,000 passed NE. Passage also in September 1998, max. 500 on the 22nd. A good spring passage on Bardsey 1998, max. 67 on May 9th and in autumn max. 50 Sept. 29th. Large flocks, over 100 were recorded:

1992	180 Llangorse Lake, Brecon Sept. 13th, two flocks of 130 each at Pembrey, Carmarthen Jan. 17th and 350 Pendine, Carmarthen Sept. 17th
1993	100+ at Gobion, Gwent Feb. 7th, 150 at Llandegfedd, Gwent Sept. 13th, 150 Waunarlwydd Glamorgan Sept. 30th and 100 Llangammarch, Brecon Oct. 10th
1995	120 on linseed stubble, Sandy Haven, Pembroke Feb. 4th
1996	200 on linseed stubble, Martletwy, Pembroke Jan. 15th
1997	330 Woebley, Gower Oct. 4th
1999	200 at Llandevaud, Gwent September/October, 100 Pembrey, Sept. 26th and 200 on linseed at Rhodiad y Brenin in January, both Carmarthen, 100 Brechfa on Aug. 12th, and at Llangorse on Sept. 6th, and at Llanwrtyd Wells on the 20th, all Brecon, 108 at Penmon Point, Anglesey Sept. 25th and 410 attracted to the Bardsey light on Oct. 24th (the largest count for the island)
2000	170 at Uskmouth, Gwent Sept. 1st, 130 at Landy Way, Cardiff Sept. 28th, 100 at Efail Isaf Sept. 22nd, 160 at Ginst Point, Carmarthen Oct. 1st, 200+ at Castle Martin, Pembroke Oct. 7th – 8th, 100 at Tredomen, Brecon Oct. 21st, 186 at Ynyshir, Ceredigion Aug. 28th, 150 at Ystumuen, Ceredigion Sept. 3rd, 100 at Trawscoed, Meirionnydd Sept. 15th, 119 on Bardsey Sept. 15th and 200 at Malltraeth RSPB, Anglesey Sept. 10th

SISKIN *Carduelis spinus*   PILA GWYRDD

(C) An increasing breeding species; numerous winter visitor and passage migrant.

Birds in Wales states that the establishment of extensive areas of conifer forests in Wales in the middle of the 20th century gradually opened up opportunities for colonisation by this species, which it has been quick to exploit. By 1991 this species was thought to be breeding in every county except Flint. Using data from the National Atlas produces a population estimate of 27,000 pairs in Wales.

Few breeding surveys; 25 on MOD Epynt and 12 in Crychan in 1992, both Brecon, present at 12 sites in Gower 1993 (1st breeding noted there in 1987), appears to be spreading SW through Ceredigion, a pair bred at Llyn Cefni, Anglesey in 1993 (a scarce breeder there) and recorded in 9 plantations in the Llangollen / Ruabon area of Denbigh in April – July 1997.

Passage noted in October 1993, when 250 counted on Skomer, 293 Strumble Head both on the 11th and 180 on Skokholm on the 14th, all Pembroke. Three flocks totaling several hundreds reported at Llyn Cefni, Anglesey on the 15th.

There was also a marked autumn movement in Gwent 1997 during September and early October, involving at least 1620 birds and 1000 an hour passed east over Dunraven, Glamorgan on Oct. 22nd. Large flocks, over 100 recorded:

1992	100 at Cathedine, Brecon Jan. 26th
1993	200 along the River Monnow above Monmouth, with similar numbers in Wentwood, both Gwent in November and December. 100 at Dinefwr Ponds Oct. 10th and at Rhosmaen Oct. 12th, both Carmarthen.
1994	250 at Llansadwrn Mar. 17th and 150 at Ffairfach Oct. 30th, both Carmarthen
1997	influx into the north west, mainly Denbigh, with 3,000 at Clocaenog Jan. 18th, 500+ Llandegla in February & March and 300 at Cynwyd in March. 350 + at Talybont, Brecon Mar. 18th was the only other sizeable flock reported
1999	100 at Llangorse, Brecon Dec. 5th and 150 in alders at Inner Marsh Farm, Flint Jan. 12th
2000	150 at Beacon Hill, Gwent Oct. 6th, 100 at Screthog, Brecon Oct. 22nd and 200 Cors Caron Sept. 22nd

TWITE *Carduelis flavirostrus*　　　　　　　　　　　　　　　　*LLINOS Y MYNYDD*
(E) Scarce breeding species, presumably resident, declining in range, now apparently extinct in Meirionnydd, a former stronghold; scarce winter visitor and passage migrant in very small numbers.

Birds in Wales quote spring migrants in Burry Port, Carmarthen, these records are now thought to be unreliable. It also quotes that a small population of this species breeds regularly in Caernarfon, Meirionnydd and Denbigh. During the period of this review the only confirmed breeding was: of a pair at Nant Ffrancon Apr. $20^{th} - 22^{nd}$, a pair + juv. at Llugwy Valley June 27th, 1 pr. Ffynnon Llugwy Reservoir June 27th, all Caernarfon in 1995.

As part of the 1999 UK survey, the RSPB found only 2 pairs. This species was absent from the Migneint, Meirionnydd, an area that has recently held a small breeding population. There were a scattering of reports during the breeding season from Flint & Ceredigion but none from the previously known sites of Rhinogau, Hiraethog, Ruabon Moors and Pumlumon. The main breeding stronghold continues to be the Carneddau, Glyderau and adjacent slopes in Caernarfon.

The large autumn flocks on the Caernarfon / Flint coast in late September of up to 200 birds (some of which were young birds still being fed by adults) suggests a Welsh breeding population of 20-40 pairs.

Counts at Nant Ffrancon, Caernarfon include 60 birds Sept. 7th 1995, in 1999 there were 40 on Aug. 15th, 70 on Sept. 17th and 200 on the 26th and in autumn 2000 there were 75 there on Aug. 28th.

Other summer records include 3 at Llandewi Brefi, Ceredigion June 1st 1996 and in Caernarfon a single Aber Valley June 11th 1998, a female at Llyn Eigian June 26th 2000 and 2 at Dinas Gromlech, Llanberis Pass on June 14th 2000.

Approximate monthly distribution of non-breeding birds during 1992-2000 is shown in the following chart and the total number of individuals recorded in each county is shown on the map.

Two records of passage in Pembroke, singles at Skokholm May 3rd and at Ramsey June 9th.

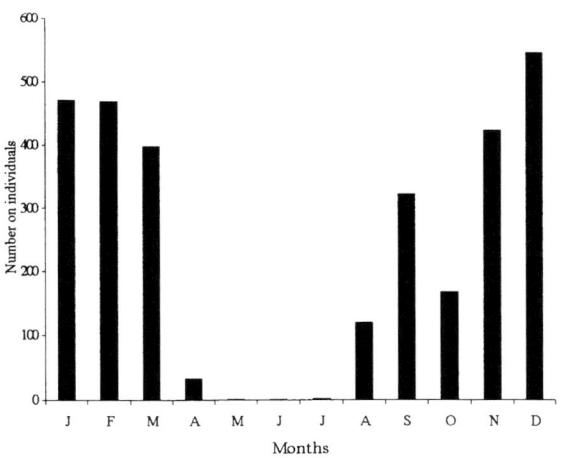
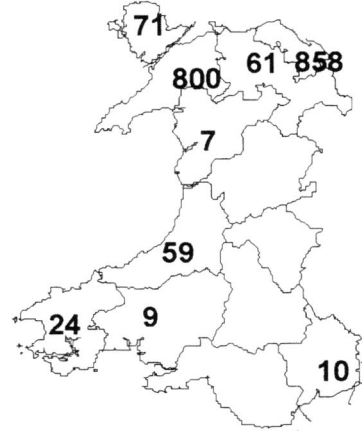

The only regular wintering sites were on the north Wales coast at Llanfairfechan / Morfa Madryn / Conwy in Caernarfon and Pensarn /Gronant / Point of Air, Flint / Denbigh, summary of records tabulated below:

Llanfairfechan / Morfa Madryn

Year	Jan.	Feb	Mar			Sept.	Oct	Nov	Dec
1993			15						
1995								70	
1997			24				4		
1998			20			50			
1999	50	50	38	38				30	30
2000	30	33							

Pensarn / Gronant / Point of Air

Year	Jan.	Feb	Mar	Sept.	Oct	Nov	Dec
1996	80	95	77		21	85	65
1997	42	76	60		35	62	83
1998	110	50	55			58	75
1999	70	87	70		7	40	20
2000		8	8				60

The only other regular wintering site for this species was on the Dyfi estuary / Ynyslas in Ceredigion, none in 1992 or 2000:

Year	Jan.	Feb	Mar	Apr	Sept.	Oct	Nov	Dec
1993				1-2				
1994	6							
1995	8	8	8				4	21
1996	8	8					4	6
1997	5							
1999							2	1

The only other large flock was of 20 at Foryd Bay, Caernarfon on Nov. 11th 1999.

LINNET *Carduelis cannabina*   LLINOS

(C) Breeding resident, summer visitor and passage migrant.

A widespread and common bird throughout Wales. Using data from the National Atlas produces a Welsh population estimate of 53,000 pairs. Estimates from CBC data however put the Welsh population as 42,000. Results from BBS, 1994-1999 suggests a population change of 117%, producing a revised population estimate of 62,000 pairs.

During the period under review there were few detailed surveys. A systematic survey of 39 km^2 of commons in central Brecon found 66 territories / singing males 1995. 85 – 90 pairs found in a systematic survey of 10-km squares SO03 & 13 in Brecon 1997. 22 were found on 120 ha of common at Llanafan, Brecon in 1997. A rather clumped distribution was discovered in 18 1-km squares in Radnor, 130 were counted in only 6 10-km squares 1997.

30 pairs were counted along 8 km of cliff top, Caswell to Little Tor, Gower 1998, 54 pairs in 442 ha of bog & 39 ha of common in the north of Brecon, a density of 4-6 pairs / km^2, a 41% increase on the sites counted in 1996 & 97.

In 1999 a GOS/RSPB survey of unenclosed cliff on the Gower, Mumbles to Rhosilli, found 112 prs.

36 prs. Range / South Stack, Anglesey in 1999.

Large flocks, over 500 were counted at:
1992   500 Rhymney estuary & 700 Aberthaw, both Glamorgan Sept. 22nd
1993   700 in linseed stubble, Newgale, Pembroke Dec. 26th
1994   SE bias of flocks, relating to arable farming bias, mainly in linseed, oil seed rape, weedy rootfields and cereal stubble. 4,000 + Lavernock Sept. 21st, 600 Ogmore on the Sept. 25th both Glamorgan and 500 Paviland Mawr Farm, Brecon 24th
1995   600 at Talgarth Sept. 10th and 700 Garthbrengy in January, both Brecon
1996   1,000 Wernffrwd, Gower Feb. 7th
1997   500 at Glasbury, Brecon Aug. 3rd
1998   500 St. Ann's Head, Pembroke Sept. 26th
2000   561 at Bardsey, Caernarfon Sept. 19th

LESSER REDPOLL *Carduelis cabaret*   LLINOS BENGOCH LEIAF

(E) Partial migrant, widely distributed as a breeding bird but probably declining. Passage migrant and winter visitor in varying numbers.

Birds in Wales states that this species is widely distributed, occurring as a breeding species in all counties, most numerous in those along the central mountain spine and least common in the southwest and Llyn. Using data from the National Atlas produces a Welsh population estimate of 15,000 pairs. Results from BBS, 1994-1999 suggests a population change of 82%, producing a revised population estimate of 13,000 pairs.

During 1992-2000 there were few detailed surveys. 26 males were counted during a systematic survey at Craig-y-Llyn, Gower 1996. 66 prs. were counted in Brecon in 2000, 25 of which were on the MOD Sennybrige range.

The Radnor survey of 1997, found a total of 26 – 27 pairs in 21 1-km squares, in 8 10-km squares but 43% were in SN97.

Large counts were: in 1999, 100 Parc Padarn, Caernarfon Oct. 4th & Nov. 25th, 80 at Ynyshir, Ceredigion in September and 70 at Blaencanaid, Glamorgan Oct. 3rd.

94 were at Wernddu, Glamorgan Apr. 16th 2000.

COMMON REDPOLL *Carduelis flammea*   *LLINOS BENGOCH*
(E) An unusual winter migrant.
BOURC split this from the above species as of January 2001. There are very few documented records of this new species in Wales. Records from pre-1992, include individuals at Skokholm, Pembroke Oct. 4th – 11th 1985, Franksbridge Nov. 20th 1987 – Jan. 2nd 1988, 3 at Newbridge-on-Wye Dec. 2nd 1993 and 5 in the Elan Valley Feb. 3rd 1988, all Radnor.

Since 1991 there have been only 7 records: 5 at Cors Erddreiniog, Anglesey Dec. 29th 1995, a flock of 60 at Marford Quarry, Denbigh Mar. 10th 1996, the majority if not all, of a flock of 29 Redpolls at Bardsey on May 6th 1999 were of this species and a single at Glasbury, Brecon Mar. 26th – 30th 1999. A small movement through Bardsey in spring 2000, 2 on Apr. 30th, up to 15 between May 8th - 14th and a single July 29th – 30th.

ARCTIC REDPOLL *Carduelis hornemanni*   *LLINOS BENGOCH ARCTIG*
(B) A vagrant.
Birds in Wales quotes one record at Bardsey May 3rd – 4th 1987. Following the major influx into eastern England during the 1995 autumn, there were three records in Wales 1996. The first at Goodwick, Pembroke Feb. 6th – Mar. 17th, 2 at Marford, Denbigh Mar. 6th – 11th and one on the Great Orme, Caernarfon Apr. 11th.

TWO-BARRED CROSSBILL *Loxia leucoptera*   *CROESBIG WENADEN*
(B) A vagrant.
The second Welsh record was of a male at Llanfihangel Glyn Myfyr, Denbigh Mar. 3rd – 26th 1991. The only other Welsh record was of a dead male found at Llandrindod Wells, Radnor in November 1912.

CROSSBILL *Loxia curvirostra*   *GYLFIN GROES*
(C) An increasing resident species although numbers fluctuate; occasional irruptive immigrant.
Birds in Wales states that the Crossbill is now a well-established resident in Wales, having become one of the principal species to benefit from the maturing of the post-war conifer plantations. It was virtually unknown as a breeding species at the time of the first National Atlas and a relatively scarce bird. The New Atlas recorded its presence in 82 10-km squares, breeding in 45. During the period under review there were many breeding records, summarised in the table below.

1992	no reports from Glamorgan. Bred at Strady Woods, Carmarthen, Llangnidr, Brecon and at four sites in Gwent
1993	six nests known Wentwood, 2 young reared Abercregan, Glamorgan; bred at 4 sites in Brecon. Present at 8 other sites, in Gwent / Glamorgan/ Brecon / Carmarthen
1994	in Glamorgan there were several at Wern Du and the Warren, present in two areas of Maesteg, Rheol forest & Crynant; in Carmarthen seen at 11 sites, several with juveniles; in Brecon – a total of 10 pairs at 5 sites; in Radnor at Elan Valley and Aberycwmhir
1995	bred at Usk Res., Carmarthen.
1996	possible breeding in 4 counties; at 6 sites in Gower, 5 in Brecon, one in Ceredigion and a few pairs in the Elan Valley
1997	several family parties seen in April / May, Gwent; at 4 sites in Glamorgan, 4 sites in Brecon and 3 sites in Denbigh. 100 adults & juvs. at Rheola forest in June, Glamorgan
1998	bred at two sites and present at 3 others in Glamorgan, 8 pairs at Pembrey, Carmarthen, one in Ceredigion and a family party seen in Denbigh. Several family parties were seen in Newborough Forest, Anglesey
1999	in Glamorgan: a pair + juv. at Ystrad/Llwynpia, 1+2j at Forest Fawr, 18 ad/juvs. at Cwm Garn Fechan, 5 singing males plus 5 females at Crychant, 4 singing males at Mynydd Margam, singing males at Penlena; juveniles seen at Pembrey, Carmarthen; 2prs. Crychan, Brecon plus one at Battle Hill, 2prs. Llyn Elsi, Caernarfon.
2000	a pair with 2 juv.s at Dingestow Court, Gwent

Large flocks, over 50 reported at:

1992	up to 50 in two areas of Wentwood, Gwent October – December
1993	80 – 100 Wentwood in August & 150 in October – December, 200 in three flocks Ebbw forest, Gwent in Oct. 26th and 60 Myherin, Ceredigion Oct. 26th
1994	72 Tywi forest, Ceredigion Mar. 1st , 80 – 100 Llyn Brianne, Brecon Apr. 17th – 18th and 200 at Wentwood January – March
1995	largest flock recorded was of 20 at Lake Vyrnwy, Montgomery
1996	largest flock was 40 Cwm Ysywyth, Ceredigion in December

1997   small invasion into south Wales from mid May onwards: 60% of records were from Gwent, where a max. 250 at Wentwood on June 24th; 50 Cynnant, Cynghordy, Carmarthen June 20th and 50 Pembrey Nov. 3rd; in Pembroke, 1-20 reported from 18 sites, max. 22 at Coed Glyn Aeron, 30 at Cwm Rheinol and 50 at Stackpole
1998   50 Pentraeth Forest, Anglesey May 31st
1999   54 Rheola forest- Abergarwed, Glamorgan Jan. 21st, 94 at Foel Fynddu, Glamorgan Feb. 9th
2000   70 in the Newborough Forest, Anglesey Jan. 24th

COMMON ROSEFINCH *Carpodacus erythrinus*   *LLINOS GOCH*
(B) A scarce migrant.

Birds in Wales quote 45 individuals being recorded in Wales up to 1991 but this included an unsubstantiated record from Cemlyn, Anglesey on Aug. 31st 1991.

Since then there has been a further 38 accepted individuals. The most favoured site was Bardsey, Caernarfon with 18 individuals compared to 10 from Skokholm, 3 from Skomer and one on Ramsey, all Pembroke. On the mainland, Strumble Head, Pembroke has paid host to 4 individuals, single records from Pwllheli, Llanfairfechan and Conwy RSPB, all Caernarfon. All records were:

Pembroke: at Skokholm 2 June 16th – 18th 1992, singles Sept. 13th – 16th 1993, June 6th – 9th, Oct. 6th – 8th and 11th – 12th 1994, Sept. 13th & 16th 1995, Sept. 16th – 17th 1997, Sept. 11th 1999 and Oct. 7th – 8th 2000, at Skomer May 20th 1992, June 5th – 7th 1998 and Oct. 19th 2000, at Ramsey Aug. 25th 1993 and at Strumble Head Oct. 10th 1994, a singing immature male May 20th & 29th 1997, a different immature Nov. 30th 1997, male May 30th 1999. One individual over-wintered, December 1991 – January 1992 at a garden in Newport.

Caernarfon: at Bardsey Sept. 6th – 7th, 9th – 10th, 12th & 21st 1992 (different birds), June 2nd & Sept. 17th 1993, June 2nd - 5th & 23rd – 26th 1994, 4 birds May 24th – June 12th 1995, male May 31st, adult & 1st summer male June 6th, different bird June 16th 1996, Sept. 11th – 12th 1999 and a male June 8th 2000, a singing male at Pwllheli June 8th – 13th 1995, at Llanfairfechan June 8th & at Conwy RSPB Sept. 25th 1996.

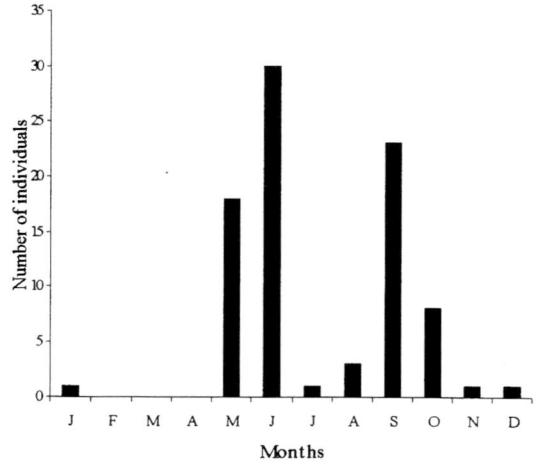
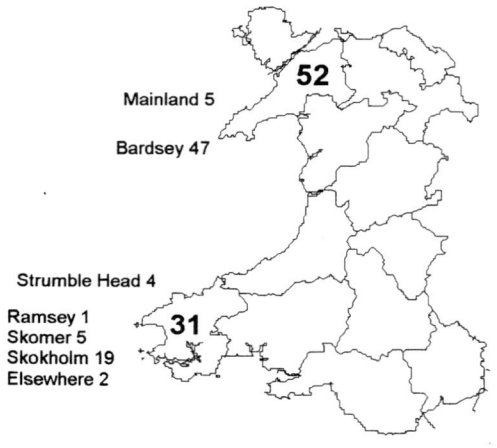

BULLFINCH *Pyrrhula pyrrhula*   *COCH Y BERLLAN*
(E) A fairly numerous resident in the lowlands, patchier in its distribution in the far west and scarce in the uplands, other than in some areas of recent afforestation. Now thought to be declining.

Although numerous in lowland areas, with the data from the National Atlas, producing a Welsh population estimate of 19,000 pairs, results from BBS, 1994-1999 suggest a 50% population change, producing a revised population estimate of 9,500 pairs.

1997 was an exceptionally good breeding season for this species in Pembroke and Montgomery. 34 pairs located in mixed farming in 10-km squares SO 03 & 13 in Brecon 1997. 45 prs. located in the south of Brecon in 2000.

HAWFINCH *Coccothraustes coccothraustes* GYLFINBRAFF
(C) Resident in small numbers in lowland areas, principally in east Wales. Formerly scarcer. Some evidence of small-scale immigration in winter.

Hawfinches are one of the least known and most under-recorded birds in Wales. Using data from the National Atlas produces a total Welsh population estimate of 300 pairs. During the period under review breeding was recorded in 7 counties, mainly in the traditional stronghold of Gwent but more recently in Meirionnydd and Caernarfon where small colonies and large roosts have been discovered. The breeding information from the period is summarised below.

1993	pair at Elan village, Radnor in May
1994	bred at one site and present at three others in Gwent
1995	in Gwent 3 pairs bred lower Wye, fledging broods of 2, 3 and 4 and a pair bred in the south of the county. One at Penybanc, Llandeilo, Carmarthen from May 30th to early June was feeding a nest of Greenfinches
1996	six nests found in a Chepstow, Gwent wood but 4 failed. Two adults and 2 juveniles at Cynghordy, Carmarthen Aug. 13th. Pair seen regularly in Shade Oak Wood Guilsfield, Montgomery possibly bred
1997	recorded in 10 sites in Gwent and 3 nests found near Chepstow. Two adults and a juvenile at Castell Coch, Glamorgan July 7th
1999	a male singing at Wentwood Apr. 11th, 4 nests found near Chepstow, a female with nest material seen at Whitchurch, Cardiff, a pair at Penybanc, Carmarthen and at Pontynyswen, 25-30 territories in Meirionnydd
2000	3 nests found with eggs near Chepstow, 2 were later predated, reported at 5 other sites during the breeding season. 16+ territories located in Meirionnydd in the Dolgellau district, suggesting a estimated population of 25-30 pairs. One small loose breeding colony was located of at least 6 pairs in another wood in the Mawddach valley. A flock of 10 individuals including 2 juvs. being fed in Caernarfon.

The total number of individuals during the period 1992-2000 in each county is shown on the map opposite. Most records were of ones and twos but larger flocks were recorded in many places. These are summarised by county below.

Gwent: 15 at a garden at Wyesham Monmouth Mar. 19th 1995 had been there a least a week, 9-10 near Tintern on Mar. 16th and 10 at Lords Grove Monmouth May 5th 1996.

Glamorgan: 6 at Tongwynlais Nov. 27th and 5 at Cyncoed in February – March 1992.

Radnor: 7 near Newbridge-on-Wye during the early months of 1995.

Meirionnydd: at least 2 roosts were identified in the Dolgellau area, with 3 at Llanelltyd Church Mawddach Dec. 23rd 1996, two there at both ends of 1997, up to 7 on Mar. 23rd, 8 roosting there on Nov. 14th, 12 in the churchyard Nov. 17th 1998, 4 at Abergwynant June 29th 1998 and 20 at Dolgellau on June 27th 1998, 40 on Dec. 29th 1999 and 40 at a second roost in late summer. The highest roost counts to date were made in August 2000, when both sites were occupied, holding a total of 80 individuals (63 & 17 respectively) with good numbers of juveniles being seen. Movement does occur between the two roost sites and both are not occupied simultaneously throughout the year.

Caernarfon: roosts of 20-23 at Llanbedr-y-Cennin on Mar. 3rd, up to 7 at Caerhun March – May, 3 at Penrhyn Castle Nov. 9th and 2 at Bardsey Oct. 19th.

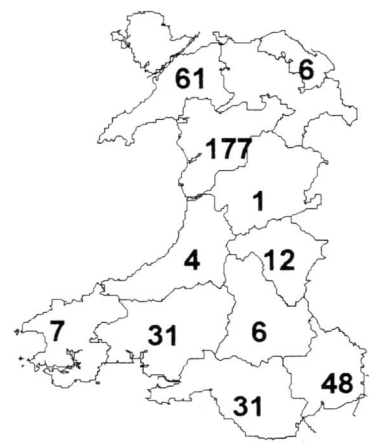

COMMON YELLOWTHROAT *Geothlypis trichas* *GYDDF – FELYN*
(A) A vagrant.
The first Welsh and only 5th European record of this American warbler was of a female at Bardsey, Caernarfon Sept. 27th 1996.

YELLOW-RUMPED WARBLER *Dendroica coronata* *TELOR TIN-FELIN*
(A) A vagrant.
The first and only Welsh record was of a first winter male on Ramsey, Pembroke Oct. 31st – Nov. 4th 1994.

LAPLAND BUNTING *Calcarius lapponicus* *BRAS Y GOGLEDD*
(C) An uncommon but regular passage migrant in small numbers, especially in autumn, and an occasional winter resident.
Birds in Wales quotes the first Welsh record in 1936 and from 1960 onwards that this species was recorded annually more regularly from coastal localities. During the period under review individuals were recorded from 8 counties, the totals for which are shown on the map. The total number of individuals recorded in each month in Wales, during 1992-2000 is shown in the chart opposite. The data does not include the following records from Pembroke in 1993 and 1995, both from periods of significant autumn passage. In 1993 there were records from Sept. 17th to Nov. 19th, with 3 Dale Airfield, up to 5 on Skomer, 11 on Skokholm and 34 on Ramsey. In 1995 at least 10 individuals visited Strumble Head, Pembroke between Oct. 14th and Nov. 3rd, with up to 4 a day.
Passage was also recorded at other sites during this period. On Anglesey, 8 on the Range South Stack Sept. 18th, 11 the next day, 21 on Oct. 1st and 4 on the 6th. In Flint, 2 at Point of Air on Sept. 30th and singles there on Oct. 15th, Nov. 4th & Dec. 14th, one at Gronant Oct. 9th, 19 on the 17th and 4 on the 31st.

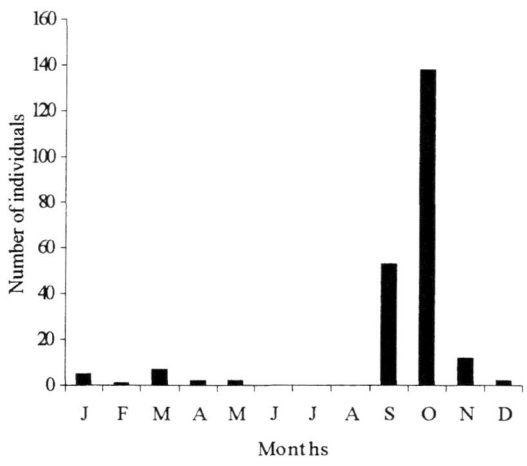
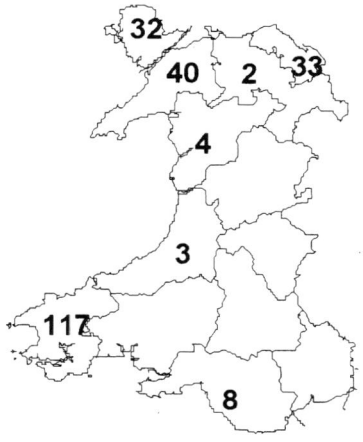

SNOW BUNTING *Plectrophenax nivalis*   BRAS YR EIRA
(C) Regular autumn passage migrant and winter visitor in small numbers.

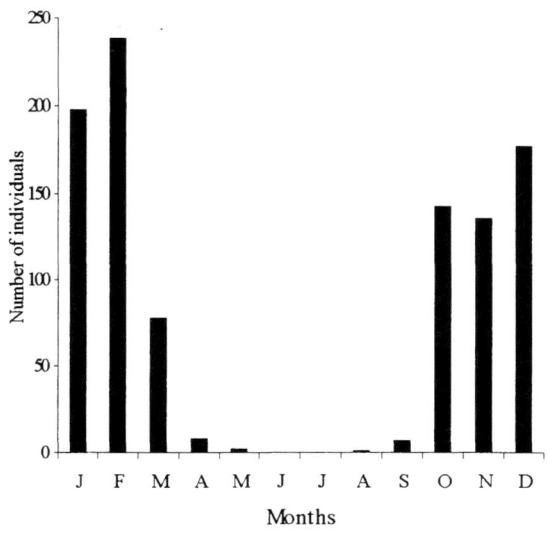
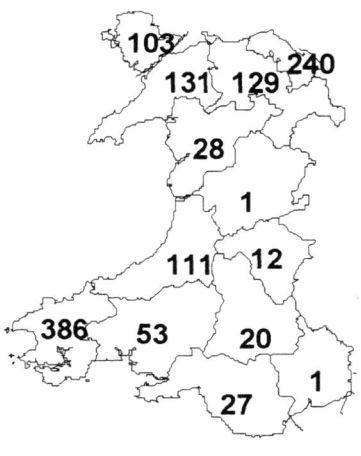

Small numbers have been known for many years to regularly winter on parts of the north Welsh coast. There is also a marked autumn passage on headland, peaking in October and November. The total number of individuals recorded in each month during the period is shown in the chart above, while the totals for each county is shown on the map.

There was an exceptional influx in autumn 1996. In Carmarthen there were monthly totals of 4 in October, 13 in November and 14 in December in Carmarthen. In Pembroke a total of 160 noted on 10 dates between Oct. 24th and Dec. 8th at Strumble Head and individuals were recorded elsewhere, with totals of 2 in September, 18 in October, 9 in November and one in December, max. 8 at St. Govan's Head on Oct. 14th. In Ceredigion a total of 22 in singles or small groups scattered all along the coast in 7 localities from Mwnt to the Dyfi estuary, Oct. 3rd – Dec. 16th. Max. 5 at Ynyshir on Oct. 3rd. 5 recorded in Brecon and 10 in Radnor, max. 9 in Radnor Forest mid December. 7 in Caernarfon while on Anglesey 13 were at Soldiers Point on Nov. 1st and one at Red Wharf Bay on the 29th. In Denbigh 7 at Kinmel Bay, Denbigh on Oct. 29th and 14 at Abergele on Nov. 26th. In Flint, 8 Gronant Oct. 17th, increasing to 25 on Nov. 8th, 32 on 9th, 46 on 17th, and 53 on 18th, thereafter numbers dropped rapidly. 3 at Point of Air Oct. 24th, 16 on Nov. 13th but only a single remaining the next day. The only other large passage was of 38 at Strumble Head on Oct. 16th 1999.

Wintering flocks were recorded at several sites. The main wintering area is the north coast of Denbigh and Flint. On Dec. 21st a count along the Denbigh / Flint coast on Dec. 21st found 18 at Abergele, 20 at Pensarn, 5 at Gronant and 2 at Point of Air.
A summary for each of the main sites is given below:

At Point of Air, Flint: up to 7 in during 1992/3 and 6 during 1993/4 (with individuals then moving to Gronant).
At Gronant, Flint: 9 during 1994/5, up to 5 in 1995/6 and 37 in 1999/2000.
At Abergele, Denbigh: 9 during 1994/5 and 15 during 1997/8
At Pensarn, Denbigh: up to 26 during 1998/9, 20 during 1999/2000 and 5 during 2000/01.
At Towyn, Denbigh: 6 during 1991/2.
At Kinmel Bay, Denbigh: 7 during 1992/3.

Elsewhere there were winter records from:
Two records from Carmarthen: 11 at Ginst Point, Carmarthen Jan. 14th 1993 and 13 at Cefn Sidan, Carmarthen Dec. 12th 1996 – Jan. 18th 1997
Two records from Ceredigion: 3 at Borth / Ynyslas Jan. 2nd – Feb. 19th 1994 and 60 on Cardigan island, Ceredigion Feb. 1st 1997
In Pembroke the only winter record was of 23 at Ramsey, Pembroke Jan. 6th 2000.
On Anglesey up to 4 Red Wharf Bay, Anglesey Jan. 7th – Feb. 13th, 3 at Cemlyn Feb. 20th – Mar. 5th, all 1994.

Birds in Wales states that as well as a coastal bird, Snow Buntings use the high hills in winter. During the period under review there were 37 records, summarised by county as follows.

Gwent	one at Mynydd Henllys Oct. 30th – Nov. 1st 1997
Glamorgan	one Mynydd Llangeinan Dec. 31st 1994
Carmarthen	a pair in the upper Tywi valley Apr. 21st 1992. One Tair Carn Uchaf Feb. 17th 1997, a male at Llyn y Fan Fach Nov. 23rd 1997, a single at Garreg Llwyd Sept. 21st 1999 and 7 at Mynydd Du Oct. 17th 1999
Ceredigion	one on the summit of Pumlumon Nov. 23rd 1993 & a pair there Feb. 26th 1995 and a single at Nant y Maen Dec. 7th 1999
Brecon	singles at Craig Cwm Oergwm on Feb. 24th and near Pen y Fan Apr. 4th 1992, 2 in the Brecon Beacons Oct. 25th, one on a Brecon birdtable Nov. 24th – 25th, and 2 on Mynydd Trawsnant on Dec. 31st 1996, 6 at Carn Du Dec. 16th 1998 were probably the same as the 6 at Cefn Cwm Llwch Mar. 10th 1999. Single Mynydd Epynt Jan. 15th 2000
Radnor	one at Pont ar Elan Nov. 12th & 9 in Radnor forest mid. Dec. 1996 and singles Elan Valley Dec. 28th 1998 & Oct. 30th 1999.
Montgomery	singles at Glaslyn Oct. 24th 1997 and at Llangurig on Mar. 26th 2000.
Meirionnydd	one over Cadair Idris Nov. 18th 1995, 8 on Cadair Idris on Oct. 31st 1999 and 6 there on Dec. 17th 2000
Caernarfon	singles at Foel Grach and at Foel Goch in November 1996, at Foel Fras Jan. 25th and at Carnedd Ugain Dec. 20th 1997, 30 – 40 at Carnedd Llewelyn Mar. 8th 1997, 6 at Moel Siabod on Jan. 16th 1999, one Clogwyn station, Snowdon Jan. 10th 1999, 2 Foel Goch Nov. 17th 1999 and 11 at the summit of Foel Grach on Dec. 9th 2000.
Anglesey	singles at Mynydd Bodafon Mar. 4th 1995 and Oct. 11th & 21st 2000.

YELLOWHAMMER *Emberiza citrinella* — *MELYN YR EITHIN*
(E) Declining as a resident breeding species over large parts of Wales.

Although this is Wales' commonest bunting, being well distributed throughout Wales, its population has declined over the past few decades. Data from the National Atlas puts the Welsh population at 81,600 in 1988-89. Results from BBS for the years 1994-1999 suggest a 69% population change. Assuming a similar decline in the years 1989-1994 then the total Welsh population at the end of 1999 is in the order of 38,850 pairs.
This approximate halving of the population in the last decade appears to be about right in light of comparative data from Brecon. The county's population was estimated at 450-500pairs in 1990 but now largely confined to mixed farming areas in the SE, with a few on ffridd and common. Declining in such habitats in north of the county. In 2000 the county's population estimate was revised to 250 pairs.

In most counties there was very little information collected about this species breeding population. What these was is summarised by county below:
Glamorgan: decreased in western areas but still quite widespread. A survey of Gower cliff tops found 32 pairs in 1999.
Pembroke: although overall distribution has not changed since 1988, the population has declined particularly in non-arable areas.
Ceredigion: formerly very common, now very much reduced. By 2000 described as a scarce resident but locally common in arable areas. 19 males in a 5km^2 of arable farmland at Mwnt in 1998.
Brecon: recent counts were: 47 prs. on 39 km^2 of common in central Brecon in 1995, 110 prs. on farmland and 34 on commons in SE and 24 on ffridd & commons in the north and west in 1998, total of 160 prs. in SE. 6 prs. located by RSPB survey of 15,000 ha of Mynydd Du (Carmarthen /

Brecon) in 1996. 10 prs. on the northern commons and 6 in the Chwefru valley in 2000.

Radnor: 1997 survey found 106 prs. in 69 1-km squares.

Montgomery: widespread only on the thin arable eastern strip along the English border.

Meirionnydd: declining on coastal belt. Still relatively widely distributed on suitable ffridd and unimproved habitats, nowhere numerous.

Anglesey: the 1997 RSPB survey of 72 1-km squares found this species in only 3 (cf. 18 in 1986) a 83% decline.

Clwyd: formerly common, now declining.

Very few large flocks, over 50 were recorded in Wales during the period under review. Those that were are tabulated below.

1992	70 in wheat stubble at Llanhamlach, Brecon Dec. 31st
1993	120 on unharvested wheat at Slwch Mar. 31st
1996	50 Felinfach, Brecon in Feb. 17th, 80 Ffostyll Oct. 22nd
1997	80 at Pandy, Gwent Dec. 30th, 60 in Caernarfon Oct. 18th, 50+ at Battle Hill Oct. 18th and 80 at Ffostyll Oct. 19th, both Brecon
1998	50 Llanfilo / Tredusten Oct. 19th, 100 in unharvested wheat Llanigon in early part of the year, both Brecon
1999	50 at Llanasa, Flint Feb. 18th
2000	50 at Tregare, Gwent Mar. 5th, 100+ at Ewenny Mon, Glamorgan and 50 at Afon Cegin, Caernarfon on Jan. 1st

ORTOLAN BUNTING *Emberiza hortulana*   *BRAS Y GERDDI*

(B) A scarce and irregular passage migrant to Wales.

Correction to Birds in Wales, 107 individuals recorded up to and including 1991. Since then there have been a further 31 individuals recorded in Wales. The top site for this species is still Skokholm, with a grand total of 52 individuals recorded (although there have only been 13 there since 1980) compared with 12 on Skomer and five on Ramsey and a mere 20 on Bardsey. The top mainland site is Lavernock Point in Glamorgan.

All records were:

Glamorgan: at Rhoose Point June 17th 1992, at Rhosilli Sept. 18th 1992, Lisvane Res. Sept. 17th 1996 and at Lavernock 7 on Sept. 12th – 13th 1992, 2 on Aug. 29th – 31st 1998 and a single Sept. 10th 2000.

Pembroke: at Skokholm 2 on Sept. 16th – 30th 1992, 3 on Sept. 2nd 1994, a male on May 11th 1995, single 1st winter birds Sept. 20th and a different bird on the 30th 1998, at Skomer Sept. 11th – 12th 1992, at Ramsey Sept. 21st 1997 and at Strumble Head Oct. 12th & 16th 1992 (presumably the same individual).

Radnor / Brecon: single at Glasbury Sept. 4th 1993.

Caernarfon: at Bardsey May 2nd 1994, Oct. 17th – 18th with a different bird 25th – 29th 1996, at Morfa Dinlle Sept. 17th 1996 and at Conwy RSPB male May 9th – 11th 1998.

Flint: at Shotwick Sept. 8th 1992.

Breakdown of all Welsh individuals, by month in the chart and by county on the map (sites with multiple records are given, e.g. of the 76 individuals recorded in Pembroke, 52 were on Skokholm, 12 on Skomer, 5 on Ramsey and 7 at other sites):

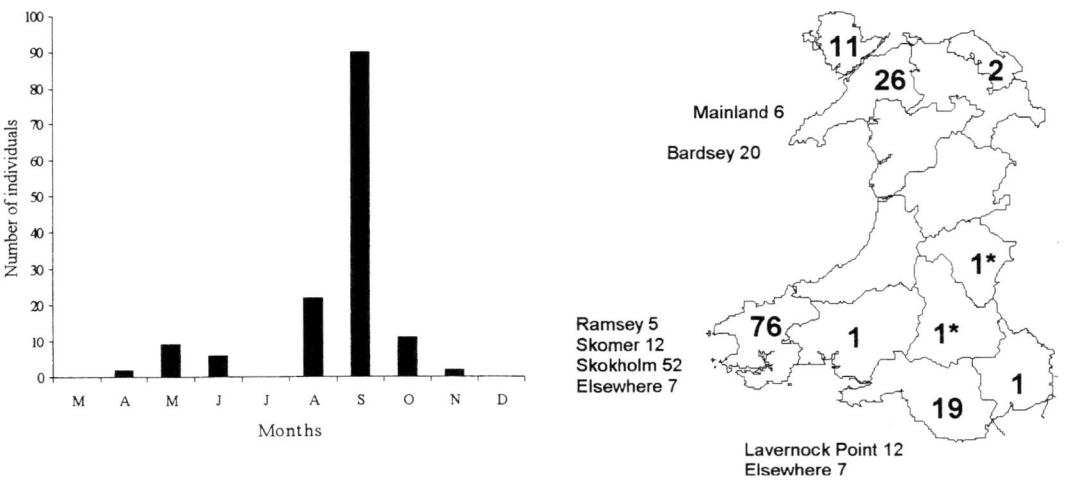

RUSTIC BUNTING *Emberiza rustica* *BRAS GWLEDIG*
(B) A vagrant.
Birds in Wales quote 4 records, two on Skokholm (June 1953 & 1975) one on Skomer (May 1987) both Pembroke and one on Bardsey, Caernarfon (March 1981). Since then there have been a further 5 records. The first, an adult at Arthog Bog, Meirionnydd Apr. $5^{th} - 10^{th}$ 1991, singles on Skokholm, Pembroke May 20th 1993 & May 23rd 1994. A first winter male was at Cors Caron, Ceredigion on Nov. 11th 1997. In 2000 a male was in an Abergele garden Apr. $7^{th} - 9^{th}$, finally crashing into a window and dying on the last date.

LITTLE BUNTING *Emberiza pusilla* *BRAS LLEIAF*
(B) A rare visitor to Wales.
Of the 11 quoted in *Birds in Wales*, Bardsey, Caernarfon notched up 7, others at Llanddeusant, Anglesey in 1957, Oxwich, Glamorgan in 1957, Skerries, Anglesey in 1961 and Skokholm, Pembroke in 1967. Since then a further 4 records all of singles: on Bardsey, Caernarfon Oct. 14th 1994 and a 1st winter on Oct. 13th 1999, at Skomer, Pembroke May 7th 1999 and at Ffairfach, Carmarthen Dec. 17th 1998.

REED BUNTING *Emberiza schoeniclus* *BRAS Y CYRS*
(C) A fairly common resident breeding species, mainly in waterside habitats; evidence of a small autumn emigration.
A fairly common bird over the whole of Wales. Using data from the National Atlas produces a Welsh population estimate of 16,000 pairs. Few systematic breeding counts were undertaken. A summary of available data is given below.

1994	In Carmarthen: 170 territories in fenland in the SE from Penallt to Hendy, with the main concentrations at Ffrwd fen (18), Pinged (16) and Llangennech (20) and 70 territories along the River Tywi, Llandovery to Carmarthen.
1995	In Brecon: 20 territories around Llangorse, 14 on MOD Epynt and 14 on 39 km^2 of common around the centre of the county. RSPB /CCW survey of 150 km^2 of the Elenydd found only 12 pairs
1996	31 territories found in the RSPB survey of 15,000 ha of Mynydd Du, Brecon / Carmarthen and 5 on the neighbouring 4,000 ha of the Cnewr estate, Brecon.
1997	14 prs. RSPB Ynyshir, Ceredigion, in Brecon, 17 prs. in 220 ha of bogs at Llanwrtyd, 6 in 100ha of similar habitat at Gors Llwyn and 10 at Llangorse lake, where it is decreasing. Radnor survey found this species in 56 1-km squares, with a total of 70 – 77 pairs
1998	c100 singing males in Brecon, in the north of the county 48 males found in 442 ha of bog and 4 prs. in 139 ha of common, an overall density of 9 pairs / km^2, an increase of 64% at Bronffynon
1999	12 prs. Mynydd Garn Clochddu, Gwent, 6 prs. at Kenfig, Glamorgan, 16 males at Penclacwydd, Carmarthen. In Ceredigion: 16 Ynyshir, 71 Cors Caron, 24 Cors Fochno, 27 Teifi Marshes. A total of 105 in Brecon, 4 at Conwy RSPB, Caernarfon and on Anglesey: 12 at Malltraeth and 17 at Cors Goch.

Largest concentration reported was 120 at Clyne Common, Glamorgan on Jan. 24th 1995.

CORN BUNTING *Milaria calandra* *BRAS YR YD*
(E) Formerly breeding commonly on coastal and nearby farmland all round Wales but scarce inland and absent except as a vagrant from Radnor, Brecon and Gwent (2 breeding records). Currently confined to a few pairs at one site in northeast Wales with the likelihood that this species will become extinct as a breeder in Wales.
A small breeding population continues to hold on in the Shotwick/Sealand area of Flint. Recorded at 3 sites in that area in 1993. 12 pairs located in 1998 but only 4 pairs in 1999 and only one male in 2000. The only sizeable winter flocks in this area were of 26 at Shotwick on Jan. 26th 1996 and 17 at Cop Hole Shotton Apr. 3rd 1995.

Elsewhere records were few and far between, all of singles:
Gwent	at Peterstone-super-Ely on July 11th 1998
Glamorgan	at Sker farm Apr. $10^{th} - 22^{nd}$ 1994
Carmarthen	Cynnant Cynghordy Dec. 22nd 1994
Pembroke	on Skokholm May 31st 1992, a singing male at Llanycefn May 22nd 1992 and on Ramsey Aug. 26th 1993
Caernarfon	on Bardsey May 4th 1994

BLACK-HEADED BUNTING *Emberiza melanocephala.*   *BRAS PENDDU*
(B) A rare visitor.
Birds in Wales quote 11 records of which 4 have been at Bardsey and 3 on Skokholm. Since then there have been a further 7: at Llanarmon-yn-Ial, Denbigh June 9th – 10th 1992, at Cemlyn, Anglesey June 12th 1993, at Aberdaron, Caernarfon June 2nd – 5th 1995, a male at Lower Town Fishguard, Pembroke May 18th – 20th 1995 and at Skokholm May 29th – June 5th before moving to Skomer June 6th – 12th 1996. A male at Bardsey, Caernarfon June 6th 2000 and another there on the 30th.

Eleven out of the 18 records have been in May, 6 in June with the remaining record in October.

INDIGO BUNTING *Passerina cyanea*   *BRAS DULAS*
(A) A vagrant.
A first winter male at Ramsey, Pembroke Oct. 17th – 26th 1996 was the first accepted record of this species in Britain.

BOBOLINK *Dolichonyx oryzivorus*   *BOBOLINC*
(A) A vagrant.
The first and only Welsh record was of a first winter individual at Skokholm, Pembroke Oct. 13th – 14th 1999.

REFERENCES

The main points of reference for this book come from *Birds in* Wales and the Welsh Bird Reports, published by the Welsh Ornithological Society formerly as bird reports and latterly so as part of the Society's journal, *Welsh Birds*.

Lovegrove, R., Williams, G. & Williams, I. 1994. Birds in Wales. Poyser.

Shrubb, M. 1993. Welsh Bird Report 1992.
Shrubb, M. 1994. Welsh Bird Report 1993.
Shrubb, M. 1995. Welsh Bird Report 1994. *Welsh Birds* 1:2, 1-84.
Shrubb, M. 1996. Welsh Bird Report 1995. *Welsh Birds* 1:4, 1-80.
Shrubb, M. 1997. Welsh Bird Report 1996. *Welsh Birds* 1:6, 1-80.
Shrubb, M. 1998. Welsh Bird Report 1997. *Welsh Birds* 2:2, 1-84.
Shrubb, M. 1999. Welsh Bird Report 1998. *Welsh Birds* 2:4, 142-228.
Shrubb, M. & Green, J. 2000. Welsh Bird Report 1999. *Welsh Birds* 2:6, 293-380.
Green, J. et al 2001. Welsh Bird Report 2000. *Welsh Birds* 3:2, 65-140.

Additional material and data was gathered from the various County Bird Reports:

Gwent Bird Report.
Glamorgan Bird Report.
Gower Bird Report.
Carmarthenshire Birds.
Pembrokeshire Bird Report.
Ceredigion Bird Report.
Brecon Bird Report.
Cambrian Bird Report.
Bardsey Bird Observatory Report.

For certain species and sections of text data was extracted from the following publications.

Baines, D. 1993. The Black Grouse Report: First approaches towards the restoration of Black Grouse numbers in Britain. Report to English Nature and Scottish Natural Heritage.

Baines, D. 1994. Seasonal differences in habitat selection by Black Grouse *Tetrao tetrix*. *Ibis* 138: 177-180.

Baines, D. 1996. Seasonal variation in lek attendance and lekking behaviour by male Black Grouse *Tetrao tetrix* in the northern Pennines, England. *Ibis* 136: 39-43.

Baines, M.E. & Earl, S.J. 1998. Breeding seabird surveys of south-west Wales colonies 1996-98. CCW Sea Empress Contract Report 323 ST41. A report commissioned on behalf of the Sea Empress Environmental Evaluation Committee (SEEEC).

Birkhead, T.R. 1997. Skomer guillemot studies, 1997. University of Sheffield Report to the Countryside Council for Wales.

Bristow, P. 2000. Cardiff Bay. *Welsh Birding* 2.2 17-19.

BTO/WWT/RSPB/JNCC, 1999. The Wetland Bird Survey 1997-98 Wildfowl and Wader Counts.

BTO/WWT/RSPB/JNCC, 2000. The Wetland Bird Survey 1998-99 Wildfowl and Wader Counts.

Cayford, J.T. & Hope-Jones, P. 1989. Black Grouse in Wales. RSPB Conservation Review 1989. Sandy: RSPB.

Colombe, S.V., Reid, J.B. & Webb, A.. 1996. Seabird studies off south-west Wales and south-east Ireland following the *Sea Empress* incident at Milford Haven, February 1996. Joint Nature Conservation Committee Report No. 225.

Cowley, M., Thomas, C., J. & Warren, M. 2000. Assessing butterflies' status and decline. *British Wildlife* 11:242-249.

Cross A. V. & Davis P. E. Monitoring of Ravens and land use in Central Wales. Unpublished report to the Nature Conservancy Council. 1986

Dare P. J. (1986). Raven Corvus corax populations in 2 upland regions of north Wales. Bird Study, 33, 179-189.

Deans, P., Sankey, J., Smith, L., Tucker, J., Whittles, C. & Wright, C. (eds.) 1992 An Atlas of the Breeding Birds of Shropshire. Shropshire Ornithological Society.

Dixon, A. (199) Breeding biology of a Raven *Corvus corax* population in the Brecon Beacons

Engel K. A. & Young L. S. (1992) Movements and habitat use by Common Ravens from roost sites in southwestern Idaho. J. Wildlife Manage. 56(3):596-602.

Gibbons, D.W., Reid, J.B. & Chapman, R.A. 1993. The New Atlas of Breeding Birds in Britain and Ireland 1988-1991. Poyser.

Gillian, G. , Gibbons, D.W. & Evans, J. 1998. Bird Monitoring Methods. RSPB.

Grove, S.J., Hope-Jones, P., Malkinson, A.R.& Thomas, D. H. 1986. Number oand distribution of Black Grouse in Wales. RSPB Unpublished Report.

Hamer, K.C. & Turner, V. 1997. Productivity and nest attendance patterns of kittiwakes on Skomer following the *Sea Empress* oil spill. University of Durham, Dept. Of Biological Sciences, Report to SEEEC, Dyfed Wildlife Trust and the Countryside Council for Wales.

Heinrich, B., Kaye, D., Knight, E. & Schaumburg, K . Dispersal and Association among Common Ravens. (1994). The Condor, 96: 545-551

Holloway S. J. (1996) Historical Atlas

Hughes, B., Stewart, B., Brown, M. & Hearn, R. 1997. The effect of the *Sea Empress* oil spill on wintering Common Scoter *Melanitta nigra* in Carmarthen Bay, Pembrokeshire. Wildfowl and Wetlands Trust Report to SEEEC.

Hurford, C & Lansdown, P. 1995. Birds of Glamorgan.

Law, R.J., Kelly, C.A., Graham, K.L. & Woodward, R.J. 1997. Hydrocarbons and PAH in fish and shellfish from south west Wales following the Sea Empress oil spill. Pp. 205-211 In: *Proc. 1997 International Oil Spill conference, Fort Lauderdale, Florida, April 7-10 1997.* American Petroleum Institute Publ. No. 4651, Washington.

Moffet A. T. (1984). Ravens sliding in the snow. British Birds **77**:321-322

Musgrove, A.J., Pollit, M.S., Hall, C., Heard, R.D., Holloway, S.J., Marshall, P.E., Robinson, J.A. & Cranswick, P.A. 2001. The Wetland Bird Survey 1999-2000. Wildfowl and Wader Counts. BTO/WWT/RSPB/JNCC. Slimbridge.

Nogales, M. 1994. High Density and distribution patterns of a Raven Corvus corax population on an oceanic island (El Hierro, Canary Islands). J. Avian Biol. 25 80-84.

Lovegrove, R., Hume, R.A. & McLean, I. 1980. The status of Breeding Wildfowl in Wales. *Nature in Wales* 17:4-10.

Parr, S. J., Haycock, R. J. & Smith, M. E. 1997. The impact of the *Sea Empress* on birds of the Pembrokeshire coast and islands. International Oil spill conference. Improving environmental protection. Progress, challenges, responsibilities. American Petroleum Institute publication, no. 4651, Washington DC, US.

Picozzi, N. & Hepburn, L.V.1986. A study of Black Grouse in north-east Scotland. In Lovel, T.W.I. (ed.) Second International Grouse Symposium: 462-480. York: World Pheasant Association.

Poole, J. 1997. Razorbill breeding success on Skomer Island, 1997.Report to the Countryside Council for Wales, from the Wildlife Trust, West Wales.

Poole, J. & Smith, S. 1996. Seabird Monitoring on Skomer Island in 1996. A Report from the Dyfed Wildlife Trust to the JNCC.

Ratcliffe, D. A. (1997) The Raven, London: T & AD Poyser

Roberts, J. L. & Jones, M. S. Increase of a population of Ravens (Corvus corax) in N. E. Wales - its dynamics and possible causation. (1999) Welsh Birds 2:3 121-130

Roberts, J. L. & Jones M. S. (1999) Increase of a population of Ravens *Corvus corax* in NE Wales - Its dynamics and possible causation. Welsh Birds Vol 2. No. 3. 121-130

Robinson, R.A., Wilson, J.D. & Crick, H.Q.P. 2001. The importance of arable habitat for farmland birds in grassland landscapes. *J. Appl. Ecol.* 38:1059-1069.

SEEEC, 1998. The Environmental Impact of the Sea Empress Oil Spill. Final report of the Sea Empress Environmental Evaluation Committee, London: The Stationary office.

Sellers, R.M. & Hughes, B. 1996. Status and breeding success of cormorants *Phalacrocorax carbo* in Wales in 1996: The effect of the *Sea Empress* Oil Spill. Wildfowl & Wetlands Trust Report to SEEEC.

Sharrock, J.T.R., 1976. The Atlas of Breeding Birds in Britain and Ireland. Poyser.

Shropshire Raven Study Group The Raven in Shropshire (in prep).

Stewart, B. 1995. Survey of Common Scoter *Melanitta nigra* in Carmarthen Bay 1994/5. Report to The Countryside Council for Wales, Aberystwyth.

Stewart, B. Bullock, I, Haycock, R J. & Hughes, B. 1997. Land-based and air-based monitoring of common scoters in Carmarthen Bay, 1996-1997. Wildfowl and Wetlands Trust Report to SEEEC.

Thompson K. R., Brindley E. & Heubeck, M. 1997. Seabird numbers and breeding success in Britain and Ireland, 1996. Peterborough, Joint Nature Conservation Committee. (UK Nature Conservation, No. 21).

Walsh, P.M., Halley, D.J., Harris, M.P., del Nevo, A., Sim, I.M.W. & Tasker, M.L. 1995. *Seabird Monitoring Handbook for Britain & Ireland*. JNCC/RSPB/ITE/Seabird Group, Peterborough.

Williams, I., King, A., Cowan, T. & Hughes, B. 1997. Black Grouse in Wales, spring 1997. RSPB Unpublished Report.

Williams, Iolo & Rees, David (1998) Ravens playing in snow Welsh Birds Vol 2 No.1 57.

SPECIES INDEX

Accentor, Alpine 192
Auk, Little 170
Avocet 113
Bee-eater 178
Bittern 64
　Little 65
Blackbird 198
Blackcap 209
Bluethroat 193
Bobolink 243
Brambling 231
Bullfinch 236
Bunting, Black-headed 243
　Corn 242
　Indigo 243
　Lapland 238
　Little 242
　Ortolan 241
　Reed 242
　Rustic 242
　Snow 239
Buzzard 99
　Honey 94
　Rough-legged 100
Chaffinch 230
Chiffchaff 212
Chough 222
Coot 111
Cormorant 61
Corncrake 110
Crake, Spotted 110
Crane 111
Crossbill 235
　Two-barred 235
Crow, Carrion 225
Cuckoo 172
　Yellow-billed 172
Curlew 133
　Stone 114
Dipper 192
Diver, Black-throated 50
　Great Northern 50
　Red-throated 49
　White-billed 50
Dotterel 116
Dove, Collared 172
　Stock 171
　Turtle 172
Dowitcher, Long-billed 129
Duck, Ferruginous 84
　Long-tailed 87
　Mandarin 76
　Ring-necked 84
　Ruddy 93
　Tufted 84
Dunlin 126

Dunnock 192
Eagle, White-tailed 96
Egret, Cattle 66
　Great White 66
　Little 67
Eider 86
Falcon, Red-footed 101
Fieldfare 198
Firecrest 213
Flycatcher, Pied 216
　Red-breasted 215
　Spotted 214
Frigatebird 63
Fulmar 55
Gadwall 78
Gannet 61
Gargeney 80
Godwit, Bar-tailed 131
　Black-tailed 131
Goldcrest 213
Goldeneye 90
Goldfinch 232
Goosander 92
Goose, Barnacle 73
　Bean 71
　Brent 74
　Canada 73
　Egyptian 75
　Greylag 73
　Lesser White-fronted 72
　Pink-footed 71
　Red-breasted 74
　White-fronted 72
Goshawk 99
Grebe, Black-necked 54
　Great Crested 51
　Little 51
　Pied-billed 51
　Red-necked 52
　Slavonian 53
Greenfinch 231
Greenshank 136
Grouse, Black 104
　Red 104
Guillemot 167
　Black 169
Gull, Black-headed 150
　Bonaparte's 150
　Common 152
　Franklin's 148
　Glaucous 157
　Great Black-backed 158
　Herring 154
　Iceland 157
　Ivory 160
　Lesser Black-backed 153

Gull, Little 148
　Mediterranean 146
　Ring-billed 151
　Ross's 159
　Sabine's 149
　Yellow-legged 156
Harrier, Hen 97
　Marsh 97
　Montagu's 98
Hawfinch 237
Heron, Grey 68
　Night 65
　Purple 68
　Squacco 66
Hobby 102
Hoopoe 179
Ibis, Glossy 69
Jackdaw 223
Jay 221
Kestrel 101
　Lesser 100
Killdeer 116
Kingfisher 178
Kite, Black 94
　Red 95
Kittiwake 159
Knot 121
Lapwing 119
Lark, Shore 184
　Short-toed 183
Linnet 234
Magpie 222
Mallard 79
Martin, House 186
　Sand 185
Merganser, Hooded 91
　Red-breasted 92
Merlin 101
Moorhen 110
Nightingale 192
Nightjar 176
Nutcracker 222
Nuthatch 218
Oriole, Golden 218
Osprey 100
Ouzel, Ring 198
Owl, Barn 173
　Little 173
　Long-eared 174
　Short-eared 175
　Tawny 174
Oystercatcher 112
Partridge, Grey 107
　Red-legged 107
Peregrine 103
Petrel, Leach's 60

Birds in Wales 1992-2000

Petrel, Soft-plumaged 56
   Storm 60
   Wilson's 59
Phalarope, Grey 141
   Red-necked 141
   Wilson's 141
Pheasant 108
   Golden 109
   Lady Amherst's 109
Pintail 80
Pipit, Meadow 188
   Red-throated 188
   Richard's 187
   Rock 188
   Tawny 188
   Tree 188
   Water 189
Plover, Golden 117
   Grey 117
   Kentish 116
   Little Ringed 114
   Pacific Golden 119
   Ringed 115
Pochard 83
   Red-crested 82
Puffin 171
Quail 108
Rail, Water 109
Raven 226
Razorbill 168
Redpoll, Arctic 235
   Common 235
   Lesser 234
Redshank 135
   Spotted 134
Redstart 194
   Black 193
Redwing 200
Robin 192
Rook 224
Rosefinch, Common 236
Ruff 128
Sanderling 122
Sandpiper, Baird's 124
   Broad-billed 127
   Buff-breasted 127
   Common 138
   Curlew 125
   Green 137
   Marsh 136
   Pectoral 124
   Purple 126
   Sharp-tailed 125
   Terek 138
   White-rumped 124
   Wood 138
Scaup 85
Scoter, Common 88
   Surf 89
   Velvet 90

Serin 231
Shag 62
Shearwater, Cory's 56
   Great 57
   Little 59
   Manx 58
   Mediterranean 58
   Sooty 57
Shelduck 75
   Ruddy 75
Shoveler 82
Shrike, Great Grey 220
   Isabelline 219
   Lesser-Grey 220
   Red-backed 219
   Woodchat 221
Siskin 232
Skua, Arctic 143
   Great 145
   Long-tailed 144
   Pomarine 142
Skylark 183
Smew 91
Snipe 129
   Jack 128
Sparrow, House 229
   Spanish 230
   Tree 229
Sparrowhawk 99
Spoonbill 69
Starling 227
   Rose-coloured 228
Stilt, Black-winged 113
Stint, Little 122
   Temminck's 123
Stonechat 195
Stork, White 68
Swallow 185
   Red-rumped 186
Swan, Bewick 70
   Mute 69
   Whooper 71
Swift 177
   Alpine 177
Teal 78
   Blue-winged 81
   Green-winged 79
Tern, Arctic 163
   Black 165
   Caspian 161
   Common 162
   Forster's 164
   Gull-billed 160
   Little 164
   Roseate 162
   Sandwich 161
   Whiskered 165
   White-winged Black 166
Thrush, Blue Rock 197
   Eye-browed 198

Thrush, Mistle 200
   Song 200
Tit, Bearded 216
   Blue 218
   Coal 218
   Great 218
   Long-tailed 217
   Marsh 217
   Penduline 218
   Willow 217
Treecreeper 218
Turnstone 139
Twite 233
Vireo, Red-eyed 230
Wagtail, Citrine 190
   Grey 191
   Pied 191
   Yellow 189
Warbler, Aquatic 202
   Barred 207
   Booted 204
   Cetti's 200
   Dartford 206
   Dusky 212
   Garden 208
   Grasshopper 201
   Great Reed 204
   Greenish 210
   Hume's Leaf 212
   Icterine 204
   Lanceolated 201
   Marsh 203
   Melodious 205
   Olivaceous 204
   Pallas' 210
   Reed 203
   Ruppell's 207
   Sardininan 207
   Savi's 201
   Sedge 203
   Subalpine 206
   Western Bonelli's 212
   Willow 213
   Wood 212
   Yellow-browed 211
   Yellow-rumped 238
Waxwing 191
Wheatear 196
   Black-eared 197
   Desert 197
   Isabelline 197
   Pied 197
Whimbrel 132
Whinchat 195
Whitethroat 208
   Lesser 208
Wigeon 76
   American 77
Woodcock 130
Woodlark 183

		Birds in Wales 1992-2000
Woodpecker, Great Spotted 181	Woodpigeon 172	Yellowhammer 240
Green 181	Wren 192	Yellowlegs, Lesser 137
Lesser Spotted 182	Wryneck 180	Yellowthroat 238